RESPONSIBLE CONDUCT OF

RESEARCH

RESPONSIBLE CONDUCT OF

RESEARCH

Adil E. Shamoo and David B. Resnik

SECOND EDITION

OXFORD
UNIVERSITY PRESS

2009

OXFORD
UNIVERSITY PRESS

Oxford University Press, Inc., publishes works that further
Oxford University's objective of excellence
in research, scholarship, and education.

Oxford New York
Auckland Cape Town Dar es Salaam Hong Kong Karachi
Kuala Lumpur Madrid Melbourne Mexico City Nairobi
New Delhi Shanghai Taipei Toronto

With offices in
Argentina Austria Brazil Chile Czech Republic France Greece
Guatemala Hungary Italy Japan Poland Portugal Singapore
South Korea Switzerland Thailand Turkey Ukraine Vietnam

Copyright © 2009 by Oxford University Press, Inc.

Published by Oxford University Press, Inc.
198 Madison Avenue, New York, New York 10016

www.oup.com

Oxford is a registered trademark of Oxford University Press

Library of Congress Cataloging-in-Publication Data
Shamoo, Adil E.
Responsible conduct of research / Adil E. Shamoo and David B. Resnik.—2nd ed.
p. ; cm.
Includes bibliographical references and index.
ISBN 978-0-19-536824-6
1. Medicine—Research—Moral and ethical aspects. 2. Medical ethics.
3. Bioethics. 4. Human experimentation in medicine—Moral and ethical aspects.
I. Resnik, David B. II. Title.
[DNLM: 1. Biomedical Research—ethics. 2. Ethics, Research. 3. Bioethics.
4. Ethics, Medical. 5. Human Experimentation—ethics. W 20.5 S527r 2009]
R852.S47 2009
174.2—dc22   2008022079

9 8 7 6 5 4 3 2 1

Printed in the United States of America
on acid-free paper

Preface to the Second Edition

When the first edition of this textbook went to press in 2002, the field of responsible conduct of research (RCR) was in its infancy. Since then, there has been a great deal of change at many different levels—governmental, institutional, and individual. The Office of Research Integrity (ORI), part of the U.S. government, has funded empirical research, conferences, and course development on RCR. Another part of the federal government, the National Science Foundation (NSF), has adopted new RCR education and training requirements. At the institutional level, many universities have developed RCR policies and implemented RCR education and training policies, and scientific journals have also developed rules and policies. At the individual level, researchers have published numerous books and articles on RCR and developed RCR courses, class materials, and training modules. However, it is not yet clear whether all of this activity has had a significant impact on scientific attitudes and behavior, and there are still many serious challenges in the field of RCR.

The media headlines featuring research misconduct in American universities continue to focus public attention on the dramatic ethical problems that can arise in research. In some instances, investigators have been accused and occasionally found guilty of falsifying, fabricating, or plagiarizing data. Since the first edition, there have been major fraud cases involving stem cells, nanotechnology, and clinical trials. There is widespread concern that public confidence in the scientific research establishment has been undermined. In the current atmosphere of accountability, the once-exempt research enterprise continues to be under increased scrutiny by the media, Congress, and the public. Issues related to conflict of interest and how it adversely affects research have been at the forefront of concerns among the congressional leaders, the public, and the media. Most clinical trials are now conducted outside the United States, often in developing countries. Many other issues also have gained in importance and attention, such as genetics and stem cells, research with vulnerable groups, and international research. In response to these pressures, there have been many calls for reforms in dealing with conflict of interest. This concern reached its heights when the U.S. Congress demanded and received the strongest conflict of interest rules governing the employees of the National Institutes of Health (NIH).

For some years, all trainees (students, fellows, and others) on training grants from the NIH have been required to have some exposure to issues related to research ethics. In the fall of 2000, the ORI announced its long-awaited mandate for training and education in RCR. The agency recommended that all personnel involved in research funded by U.S. Public Health Service grants or contracts, including principal investigators, postdoctoral students, graduate students, technicians, and staff, receive some education in RCR. The ORI also outlined nine core areas for RCR education—all of which are covered in this book—and recommended that institutions also provide ongoing education in RCR. Although the ORI mandate was suspended when the George W. Bush administration took control of the executive branch of government in January 2001, many universities have implemented RCR training programs. The NSF now requires that all recipients of NSF funding have an RCR education plan for all their students on those grants. The NIH adopted ORI's recommended training requirements for its intramural researchers shortly after ORI announced them. Some universities, such as Duke University, have adopted RCR training requirements for doctoral students.

The second edition of our textbook, *Responsible Conduct of Research*, presents a comprehensive introduction to the ethical issues at stake in the conduct of research. A total of 16 chapters, compared with 12 chapters in the first edition, range in scope from the broad issues relating to social responsibility, research funding, and freedom of inquiry to more narrow topics such as the ethical aspects of entering data into lab notebooks, designing experiments, citing published works, and deciding authorship matters. The chapters also contain questions for discussion and more case studies. Due to the increased importance of certain areas, we have added four new chapters, on collaboration between academia and private industry, protecting vulnerable subjects in research, genetics and stem cell research, and international research. Because this book deals with ethical issues in the use of humans and animals in research, its primary focus is on biomedical research. However, the chapters on such topics as intellectual property, authorship, peer review, and conflict of interest make this text suitable for most research-focused disciplines. We have also included some case studies that deal with issues in physics, statistics, and the social sciences.

We apologize for any errors or oversights in this second edition. Please feel free to send your comments and suggestions to Adil E. Shamoo, PhD, Department of Biochemistry and Molecular Biology, University of Maryland School of Medicine, 108 North Greene Street, Baltimore, Maryland 21201–1503; e-mail: ashamoo@umaryland.edu.

Acknowledgments

Dr. Shamoo is grateful for all of the guest lecturers in his course since 1991, among them, Jousef Giffels, Jack Schwartz, Leslie Katzel, and Meryl Eddy. Dr. Shamoo also thanks the students in his Responsible Conduct of Research classes since 1991 for their input and discussions. For useful discussions and insight about ethics in research, Dr. Resnik is especially grateful to Dr. Loretta Kopelman, Dr. Kenneth De Ville, Dr. Thomas Feldbush, Dr. John Bradfield, Dr. Jeremy Sugarman, Frank Grassner, John Doll, and students and co-instructors in his research ethics classes taught in 1999 and 2001 at the Brody School of Medicine. Finally, we thank the many anonymous reviewers of the prospectus during the publisher's review process for their considerable and valued suggestions for improving the textbook in content and style. Moreover, we thank Patrick L. Taylor, Children's Hospital Boston, Harvard Medical School, for his thorough review of the second edition. Research for the second edition of *Responsible Conduct of Research* was supported, in part, by the Intramural Program of the National Institute of Environmental Health Sciences, National Institutes of Health. It does not represent the views of the National Institute of Environmental Health Sciences or the National Institutes of Health.

Contents

RESPONSIBLE CONDUCT OF

RESEARCH

1

Scientific Research and Ethics

There is a growing recognition among scientists, government officials, research institutions, and the public that ethical conduct is essential to scientific research. Ethical conduct is important to foster collaboration, cooperation, and trust among scientists, to advance the goals of research, to and fulfill scientists' social responsibilities, and to avoid or minimize damaging scandals resulting from unethical or illegal behavior. This chapter discusses the importance of ethics in research and the nature of scientific professionalism.

Ethical problems, issues, and dilemmas occur for most people on a daily basis. Whenever we ask the question, "What should I do?" there is a good chance that an ethical issue or concern lurks in the background. In everyday life, such questions frequently arise as we make choices among different interests and commitments, such as career, family, community, church, society, prestige, and money. Professional researchers—scientists, engineers, and scholars—also frequently face ethical problems, issues, and dilemmas. Consider the following cases.

Case 1

You are graduate student in pharmacology at a large university working under the direction of a senior researcher. You notice, on reading a paper on the pharmacology of a new serotonin reuptake inhibitor, which your senior director published in a top journal in your field, that there are problems with the diagrams in the paper. You cannot seem to reconcile the diagrams with the published data. You approach her with this problem, and she shrugs it off, saying that you do not understand the research well enough to make a judgment about it. What should you do?

Case 2

You are a graduate student in sociology working on a dissertation on attitudes toward the environment, comparing rural, suburban, and urban populations. You have collected data from focus groups, phone interviews, and surveys distributed by mail and e-mail. You have compiled the surveys, and now you have begun to edit and clean up your data. It appears that 20% of the subjects who took the survey misunderstood a couple of the questions, because their answers to these questions are inconsistent with other questions and their written comments. You also discover that for some of the other questions, 15% of the subjects gave no answer, and that for other questions, 12% of the answers were difficult to read. All of these issues may affect the statistical significance of your work and the analysis and interpretation of the data. How should you deal with these problems with your data?

Case 3

You are a postdoctoral fellow in epidemiology at a university collaborating with a senior researcher. You have just read an important paper in your field and contact the author about it. The paper is one of several the author has published from a large, publicly funded database. You ask the author if you can have access to the database to confirm your own work. The author says he will share data with you only if you agree to a formal collaboration with him and name him as a coauthor in publications that use the database. What should you do?

Case 4

You are the director of the Department of Veterinary Medicine at a veterinary school associated with a university. You are also the chair of your institution's animal care and use committee, which oversees animal research. The group People for the Ethical Treatment of Animals (PETA) has staged some protests recently in your community. A local reporter calls you on the phone and wants to do an interview with you about animal research and animal rights. How should you handle this situation?

Case 5

You are professor of pediatrics and an expert on child abuse. You have been asked to serve as an expert witness for the defense in a case. The prosecutors have hired two experts who are prepared to testify that child abuse probably occurred. You know these experts from professional meetings and from their published work, and you have a great deal of respect for them. After examining the X-rays, photographs, and other data, you are not convinced that child abuse has occurred, and you could offer testimony that supports the defense in this case. As a long-standing advocate of child welfare and rights, you would have a hard time testifying for the defense and contradicting the testimony offered by your colleagues. You want to protect your reputation, in any case. The defense will find another expert if you refuse their request. What should you do?

These cases typify some of the complex ethical, social, and legal dilemmas that can arise in the conduct of research. To make well-informed, reasonable, and responsible choices in cases such as these, as well as to guide conduct in similar cases, researchers need some understanding of the moral and social dimensions of research. The purpose of this book is to provide researchers with some of the education that is required for the responsible conduct of research.

In recent years, universities, research organizations, professional societies, funding agencies, politicians, the media, and most scientists have come to realize that ensuring ethical conduct is an essential part of basic, applied, and clinical research. Little more than a few decades ago, many scientists would not have accepted this statement. According to a view that has held sway among scientists, humanists, and the general public for centuries, science is objective (Bronowski 1956, Snow 1964) and ethics are subjective, so scientists need not deal with ethical issues and concerns when conducting research. Ethical and social questions, according to this view, occur in the applications of science, but not in the conduct of science. Humanists, politicians, and the public can grapple with the ethical (or moral) aspects of research; the main task of the scientist is to do research for its own sake (Rescher 1965). (Note that, although some scholars distinguish between "ethics" and "morality," in this book we use these terms interchangeably.)

While it is important for scientists to strive for objectivity, this does not mean that ethical questions, problems, and concerns have no place in research conduct. The idea that ethics is an important part of research is by no means new. Charles Babbage (1830 [1970]) wrote about the lack of ethics, honesty, and integrity in British science. During the Nuremberg trials after World War II, researchers became painfully aware of the need for ethical reflection on research when the world learned about the horrors of Nazi research on human beings. Such physicists as Albert Einstein and Robert Oppenheimer came to terms with their moral responsibilities in the war effort, both as supporters of research on atomic weapons during the war and as advocates for the peaceful use of atomic energy after the war (Zuckerman 1966).

In the 1960s, ecologist Rachel Carson published *Silent Spring* (1961), which alerted the public to the dangers of pesticides and helped to launch the environmentalist movement. Clinical researcher Henry Beecher (1966) published an influential essay in the *New England Journal of Medicine* that exposed unethical or questionable research involving human subjects, including the now infamous Willowbrook experiments, where children were exposed to hepatitis virus. In 1972, the public learned about the Tuskegee syphilis study, and congressional investigations exposed many ethical problems, including violations of informed consent and failure to provide subjects with adequate medical care (Jones 1981). (Chapter 12 discusses this case in depth.)

Today, scientists need to pay special attention to research ethics in their own work and in teaching students about how to conduct research for many reasons (Sigma Xi 1986, Shamoo 1989, Shamoo and Dunigan 2000). First, modern research is not a solitary activity. Scientists collaborate with students and colleagues in their own institutions as well as with other scientists in other institutions (Merton 1973, Ziman 1984). Although Isaac Newton was able to develop his laws of motion with little help from other physicists, today's physicists work together in research teams. It is not uncommon for more than 100 researchers from across the world to collaborate on a single experiment in high-energy physics. Collaboration is also important in the evaluation of research proposals and projects that occurs in peer review. Peer review functions as a gate-keeping mechanism to ensure that research proposals or publications meet methodological and professional standards. Because scientists must collaborate in conducting research, they need to adhere to standards of conduct that foster effective collaboration and promote the goals of science (Hull 1988). Thus, ethical concepts and principles, such as honesty, integrity, trust, accountability, respect, confidentiality, and fairness, play a key role in shaping research conduct within science (Shamoo and Annau 1987, American Association for the Advancement of Science–American Bar Association 1988, LaFollette 1992, National Academy of Sciences 1992, 1994, Resnik 1998c, Macrina 2005, Steneck 2006).

Second, research always takes place within a social context. Economic and political interests as well as social, cultural, and religious values influence scientific goals, resources, and practices (Ziman 1984, Boyd 1988, Jasanoff 1990, 1995, Longino 1990, Resnik 1998c, 2007b, Guston 2000, Washburn 2006, Shulman 2007). Most universities and research organizations would not exist without an ample supply of public and private funds. Money plays a crucial role in deciding whether a particular problem will be studied, who will study it, how it will be studied, and even whether the results will be published. For example, many illnesses are not sufficiently studied because of a lack of funding. In the private sector, the prospect of profit is the main reason for investing money in research on treatments for diseases. Pharmaceutical companies tend to invest heavily in studying diseases that afflict many people, such as heart disease or hypertension, while virtually ignoring illnesses that affect fewer people, such as personality disorders and malnutrition. Political interests and social values can also influence the research process. For example, research in HIV/AIDS languished during the 1980s but took off during the 1990s, due in large part to intense lobbying efforts by AIDS activists. Federal funding for AIDS research is now more than $1 billion annually, outpacing even cancer research. In contrast, in the 1980s, anti-abortion activists pushed for a ban on the use of federally funded research involving human embryos. In August 2001, President George W. Bush announced a policy that significantly restricted the use of federal government money for research on human embryonic stem cells. In July 2006, Congress passed legislation to expand the use of federal funds for human embryonic stem cell research, but President Bush vetoed the bill (Mooney 2005, Shulman 2007).

Because science takes place within society, with the help of considerable public support and funding, scientists have social responsibilities (National Academy of Sciences 1994, Shrader-Frechette 1994, Resnik 1998a,c). Scientists have an obligation to benefit society and avoid causing harm to people, communities, and the environment. Scientists must also be accountable to the public. Scientists can fulfill their social responsibilities in many different ways, such as conducting useful research, educating the public about science and its social implications, providing expert testimony and advice on scientific issues, or engaging in policy debates concerning issues related to the applications or implications of science and technology, such as global climate change, air and water pollution, food and drug safety, public health, disaster preparedness, and national security. Since people use the results of research to form social policy and to address practical problems in medicine, engineering, agriculture, industry, and public health, scientists must strive to earn the public's support and trust. If research results are erroneous or unreliable, then people may be killed or harmed, the environment may be degraded, money and resources may be misused or wasted, and misguided laws or policies may be enacted. Research results also play an important role in public education. School-age children, as well as adults, learn about science every day, and science can have a profound impact on how they view the world. Because science can influence how we relate to each other, to society, and to nature, it is important for research results to be reliable and relevant.

A third reason why scientists should pay special attention to ethics is that illicit or unwise research activities can generate negative publicity, legal liability, and poisonous political fallout. To avoid these damaging repercussions, scientists need to be concerned about ethics in the laboratory and in their interactions with society. Since the 1980s, science has experienced its share of scandals, which have had an adverse impact on the public's support for research and its trust in science. The following are but a few examples:

- Allegations of data fabrication and falsification against Massachusetts Institute of Technology (MIT) researcher Thereza Imanishi-Kari in an immunology paper coauthored by Nobel Laureate David Baltimore (Kevles 1998) and against University of Pittsburgh medical researcher Bernard Fisher related to the National Surgical Breast and Bowel Project (Buyse and Evans 2001)
- Confirmed cases of data fabrication or falsification involving Harvard postdoctoral fellow John Darsee (LaFollette 1992), University of Pittsburgh psychologist Stephen Breuning (Broad and Wade 1982 [1993]), Bell Labs physicist Jan Hendrik Schön (Resnik 2003b), South Korean embryonic stem cell researcher Woo Suk Hwang (Normile et al. 2006), and University of Vermont medical researcher Eric Poehlman (Kinitisch 2005)
- Suppression of data by private companies, such as Merck's failure to publish data, in a timely fashion, related to cardiovascular risks associated with its drug Vioxx (rofecoxib), and several companies' failure to publish data showing that children who take some types of selective serotonin uptake inhibitors have an increased risk of suicide (Resnik 2007b)
- Private companies' intimidation and harassment of scientists who sought to publish data that would adversely affect their products, such as Boots Pharmaceuticals' attempt to stop University of California pharmacologist Betty Dong from publishing a study of its hypothyroidism drug Synthroid, and Apotex's retaliation against Nancy Olivieri, who was attempting to tell scientists and patients about risks associated with its thalassemia drug deferiprone (Resnik 2007b)
- Tragic, avoidable deaths of research subjects in clinical trials, including Hoiyan Wan, Jesse Gelsinger, Ellen Roche, and Jolee Mohr (Steinbrook 2002, Edwards et al. 2004, Weiss 2007)
- Poorly managed risks to human research subjects in a variety of studies, such as the phase I trial of TGN1412, in which six research subjects developed a severe immune response to a monoclonal antibody and almost died (Wood and Darbyshire 2006, Shamoo and Woeckner 2007)

- Financial conflicts of interest in academic research at the National Institutes of Health (NIH) and on Food and Drug Administration (FDA) advisory panels (Resnik 2007b)
- Politically inflammatory research on the relationship between race and intelligence, including Herrnstein and Murray's controversial book *The Bell Curve* (1994)
- Publication of research that could be used by terrorists to make biological weapons or threaten public health and safety (National Research Council 2003, Resnik and Shamoo 2005, Wein and Lu 2005)
- Underreporting of adverse events in many clinical trials that have a direct impact on the integrity of data (Shamoo and Katzel 2007, 2008)

And the list goes on (Broad and Wade 1982 [1993], American Association for the Advancement of Science 1991, LaFollette 1992, 1994b,c, Advisory Committee on Human Radiation Experiments 1995, Resnik 1998a,c, Steneck 1999, Pascal 2000, Shamoo and Dunigan 2000).

Fourth, science's evolving relationship with private industry has raised many ethical issues in research, because the academic values of objectivity and openness frequently clash with the business interests in profit and competitiveness. Science–industry collaborations can create issues concerning conflicts of interest, biases, secrecy, insider trading, and social responsibility as well as intellectual property disputes (Bowie 1994, Spier 1995, Rule and Shamoo 1997, Resnik 1998c, 2007). These issues have become increasingly important as private investments in research and development (R&D) have skyrocketed in recent years: In the United States, total private funding of R&D rose from $50 billion annually in 1980 to $100 billion annually in 1995. As we begin the twenty-first century, private funding of R&D is more than $200 billion per year and now accounts for more than 60% of all R&D funding (Shamoo 1989, Jaffe 1996, National Science Foundation 1997, Malakoff 2000a, Resnik 2007).

For these reasons and many others, universities, funding agencies, research institutions, professional societies, and scientists are now very much aware of the importance of ethics in research. There have been a variety of responses to these ethical concerns, including workshops and conferences; books, journals articles, and edited volumes; debates about the definition of "research misconduct"; investigations of misconduct in research; revisions of journal policies; drafts and revisions of professional codes of ethics in research; and education in research ethics, including formal courses and seminars.

SCIENCE AS A PROFESSION

In discussions of ethics in research among scientists, science scholars, ethicists, and policy analysts, there has been a broad and emerging consensus that research ethics can be understood according to the professional model

(Shrader-Frechette 1994, Resnik 1998c). Each profession has its own ethical standards, which govern the practices in the profession. In medicine, physicians abide by rules such as "do no harm," "promote the patient's health," "maintain confidentiality," and "honor the patient's right to make decisions" (Beauchamp and Childress 2001). Science also has its own standards, such as "honestly report data and results," "give proper credit contributions," and "share data, methods, ideas, materials, and results." (Chapter 2 examines these standards in more detail.) Professions usually adopt codes of ethics to signal to members of the profession and the public at large the type of behavior that is expected in the profession (Bayles 1988). Ever since the time of Hippocrates (ca. 460–377 B.C.), physicians have sworn allegiance to a code of conduct. Although modern codes of conduct in health care differ from the Hippocratic Oath, they still play a key role in governing professional conduct. Since the 1980s, many different scientific disciplines and societies have adopted ethics codes, including the American Chemical Society (1994), the American Physical Society (2002), the American Society for Biochemistry and Molecular Biology (1998), the American Psychological Association (1992), the American Statistical Association (1999), among many others.

Professional ethical standards function (a) to promote the goals of the profession and (b) to prove to the public that members of the profession are trustworthy (Bayles 1988). For example, one of medicine's goals is to treat disease. A rule like "maintain the patient's confidentiality" is necessary to promote the goals of medicine, because patients may not tell physicians important information concerning their symptoms or medical history, if they cannot trust that physicians will keep this information in confidence. Without this information, medical diagnosis and treatment cannot be very effective. Confidentiality is also important for earning the public's trust in medicine. In science, some of the goals are (a) to develop systematic knowledge about the natural world, (b) to understand and explain natural phenomena, and (c) to predict and control natural phenomena. Ethical rules like "honestly report data and results" are necessary for developing knowledge, because knowledge cannot be built on false beliefs and deceptions. Honesty also helps to secure the public's trust in science.

Science shares several social characteristics with other recognized professions, such as medicine, law, engineering, and accounting. First, a profession is more than an occupation; it is a career or vocation (Davis 1995b). The first people to be recognized as professionals were physicians and ministers, who viewed themselves as being "called" to serve and devoted their lives to doing good works for society. Science is also a career or vocation. Second, because professionals provide valuable goods and services, they have social responsibilities and can be held publicly accountable (Davis 1995b). Physicians, for example, have professional duties to promote not only the health of their patients but also the public's health. As noted above, scientists also have social responsibilities and can be held publicly accountable. For example, scientists have an obligation to educate the public and to provide information and advice for policy debates.

Third, although professionals have social responsibilities, they are also granted a great deal of autonomy: Society allows professionals to be self-regulating (Bayles 1988). Professionals can make their own standards and rules, provided that they obey the law and fulfill their public responsibilities. Physicians, for example, set their own standard of care and determine what it takes to become a qualified member of the profession. Scientists are also self-regulating: Scientists make their own rules for designing experiments, drawing inferences from data, publishing results, and so on. Scientists determine what counts as "good scientific practice." Fourth, professions have codes of ethics to achieve self-regulation and promote social responsibility. As noted above, many scientific societies and professions have developed ethics codes. Finally, professionals are recognized as having expertise. Physicians, for example, have expertise when it comes to diagnosing, treating, and preventing diseases (Bayles 1988). Scientists also have expertise: They are recognized as experts within their domain of knowledge and in performing experiments, analyzing data, and other scientific activities.

Prior to the Scientific Revolution (ca. 1500–1700 A.D.), science was more of an avocation than a vocation. Scientists often worked in isolation and financed their own research. They did not publish very frequently—the printing press was not invented until the mid 1400s—and when they did, their works were not peer reviewed. There were no professional scientific societies or journals until the mid 1600s. Universities taught only a small number of scientific subjects, and many scientists could master several different subjects. For example, Newton made contributions in mechanics, astronomy, optics, and mathematics (Newton 1687 [1995]). Private businesses and governments saw little reason to invest in research. Science also did not have a great deal of social status or impact—the church and the state battled for social influence and political power (Ziman 1984, Burke 1995).

But all of that changed after the Scientific Revolution. At the beginning of the twenty-first century, there are thousands of scientific societies and professional journals. Peer review plays a key role in funding and publications decisions. As noted above, scientists now work in research groups, which may include laboratory assistants and data analysts as well as postdoctoral, graduate, and undergraduate students. Universities now offer study in hundreds of different scientific subjects, and it is virtually impossible to achieve scientific expertise without specialization. Governments and private corporations now invest billions of dollars each year in science. Science has become one the most influential social institutions in society (Ziman 1984). Most of the technologies and many of the ideas in our modern world are the direct or indirect result of scientific research. Scientists now publish millions of articles a year, and the information boom continues to increase. Scientists give expert testimony to congressional committees and government agencies, and they provide advice to presidents, governors, generals, and corporate executives. Children learn about science in school, and most professional careers require some type of scientific and technical knowledge.

Science is now also a sizable part of the world's economy: Total (private and public) R&D investments account for about 2.5% of the gross domestic product (GDP) in developed countries such as the United States, United Kingdom, and Germany (May 1998). Economic activity directly related to scientific research is estimated to be about 6% of the U.S. GDP (Resnik 2007). The indirect impacts of research are much larger than the direct impacts, since investments in R&D have led to economically significant innovations, such as computers, the Internet, airplanes, automobiles, nuclear energy, and radar. As the twenty-first century economy becomes more dependent on information, investments in R&D are likely to continue an upward trend that began in World War II (Dickson 1988). Literally millions of scientists are employed in universities, research institutions, private laboratories, or other organizations that conduct research (National Science Foundation 1997). It is estimated that there are more scientists alive today than all of the scientists who have lived during the past 2,500 years of human history (Dickson 1988).

For better or worse, the modern age is a scientific one. These increases in science's power carry added social responsibilities, and science should now be regarded as a profession, of which science has all the characteristics: It is a career, entails public responsibilities, is self-regulating, has ethical standards, and is endowed with expertise and intellectual authority. Ethical standards can play an important role in research by promoting the goals of science, such as pursuing objective knowledge and solving practical problems; by promoting cooperation and collaboration among scientists; and by helping to boost the public's trust in science. To understand professional conduct in science, we therefore need to have a better understanding of ethical theory and decision making.

Case Study: Fraud in South Korean Stem Cell Research

In 1998, James Thomson and colleagues perfected methods for growing human embryonic stem (HES) cells in culture (Thomson et al. 1998). Scientists from around the world soon became excited about the possibilities of some day using HES cells to treat diseases involving dead or poorly functioning tissue, such as paralysis due to nerve damage, type I diabetes, heart disease, Alzheimer's dementia, and many other conditions. However, there were several technical barriers to developing HES cell therapies, including coaxing the HES cells to differentiate into the appropriate cell types, getting the differentiated cells to function properly in the body, and overcoming the human immune system's reaction to transplanted cells or tissues made from those cells. Some scientists proposed that one way of dealing with this third problem would be to develop HES cells from the patient's own tissues (i.e., somatic cells), employing the nuclear transfer techniques used to create Dolly, the world's first cloned sheep. In this procedure, a nucleus from a somatic cell is transferred into a zygote that has had its nucleus removed. The cells created by the procedure should not be rejected by the patient's immune system, because genetically they would be almost identical to the patient's own cells. The procedure, also known as "therapeutic cloning," is ethically controversial (and illegal in some countries) because it involves creating embryos for the sake of research or therapy and destroying them (National Bioethics Advisory Commission 1999, Solter and Gearhart 1999, President's Council on Bioethics 2002).

In 2004 and 2005, Woo Suk Hwang, a professor at Seoul University in South Korea, published two papers in the journal *Science* reporting the derivation of HES cell lines by therapeutic cloning (Hwang et al. 2004, 2005). The papers claimed that the HES cell lines produced by this process were genetically identical to the donor somatic cells. Hwang had assistance from numerous colleagues at Seoul University and other collaborators. The two papers, especially the second one, had a tremendous impact on the stem cell research field, and Hwang attainted international recognition for his work and became a national hero. Hwang, a veterinarian by training, had published previous papers on cloning dogs and cows. In November 2005, one of Hwang's collaborators on the 2005 paper, University of Pittsburgh scientist Gerald Schatten, accused Hwang of misleading him about the source of the eggs used in the experiments. Hwang admitted that some of the eggs had been provided by women working in his laboratory, a practice that was legal in South Korea but regarded by many as unethical because using research subjects who are subordinates can be coercive. Hwang later admitted that egg donors were paid as much as $1,400 (Resnik et al. 2006).

In December 2005, the editors of *Science* received an anonymous tip that two of the photos of HES cells published in the 2005 paper were duplications. Later, one of Hwang's coauthors, Sung Roh, told the media that Hwang had fabricated 9 of the 11 cell lines presented in the paper. Hwang asked for the article to be withdrawn from *Science*, and a committee from Seoul University began investigating the 2005 paper and Hwang's other publications. The committee determined that none of HES cell lines were genetically identical to somatic donor cells. The committee also found that Hwang had used 273 eggs, not 185 as reported, and that the cell lines in the 2004 paper were also fabricated. And the committee found that Hwang had been involved in the egg procurement process—Hwang had helped donors fill out forms and had escorted them to clinics for egg removal—and that authorship on the 2004 and 2005 papers had been granted for minor contributions to the research. Hwang resigned his position at Seoul University at the end of December 2005. In May 2006, Hwang and five collaborators were indicted on charges of fraud, embezzlement ($3 million), and breach of bioethics laws (Whon and Normile 2006).

A committee at the University of Pittsburgh examined Schatten's collaboration with Hwang. The committee found that Schatten played no role in the data fabrication but that he had shirked his authorship responsibilities by failing to carefully review the data and the manuscript. The committee also found that Schatten had accepted unusually large consulting fees ($40,000) for his role in assisting Hwang's research group (Resnik et al. 2006).

Following the Hwang affair, the editors at *Science* decided to review their peer-review policies to try to prevent similar scandals from occurring in the future. Although peer review is not designed to catch fraud, the editors were embarrassed that an obvious problem (two duplicate images) had slipped through their peer-review process. The editors decided to start giving high-impact papers extra scrutiny and to pay greater attention to digital images (Kennedy 2006). Other journals have also revisited their peer-review and misconduct policies in response to the Hwang affair. The Committee on Publication Ethics (COPE) has also developed some guidelines and rules to help journals deal with misconduct (Committee on Publication Ethics 2007).

2

Ethical Decision Making

Ethics is an academic discipline that is concerned with answering age-old questions about duty, honor, integrity, virtue, justice, and the good life, Philosophers, theologians, and other scholars have developed many different theories and principles of ethics. This chapter describes some influential ethical theories and principles, articulates 12 principles of ethical conduct in science, and proposes a method for ethical decision making.

WHAT IS ETHICS?

Although there are many different definitions of "ethics," for our purposes we focus on four senses of the term: (a) ethics as standards of conduct that distinguish between right and wrong, good and bad, and so on; (b) ethics as an academic discipline that studies standards of conduct; (c) ethics as an approach to decision making; and (d) ethics as a state of character. This book focuses on senses a, b, and c, but we also briefly mention some important points about sense d.

Ethics as an academic discipline is a branch of philosophy (moral philosophy) that is concerned with answering age-old questions about duty, honor, integrity, virtue, justice, the good life, and so on. The questions asked by moral philosophy are normative (rather than descriptive) in that they have to do with how one ought to live or how society ought to be structured. Several disciplines in the social and behavioral sciences, such as psychology, sociology, anthropology, and political science, take a descriptive approach to ethical questions in that they attempt to describe and explain ethical beliefs, attitudes, and behaviors. Although facts pertaining to beliefs, attitudes, and behaviors can have some bearing on normative questions, they cannot, by themselves, solve ethical questions because ethical questions require normative answers.

For example, suppose that someone is trying to decide whether to cheat on her income taxes. Her question might be, "Should I cheat on my income taxes?" Suppose that a social scientist conducts a study that shows that 65% of people cheat on their income taxes. This scientific study still would not answer the person's question. She wants to know not how many people cheat, but whether she should cheat. The fact that most people cheat does not justify cheating. The person asking the ethical questions is requesting a normative justification for a particular course of action, but scientific studies do not provide this. Science delivers facts and explanations, not values and justifications.

The study of ethics can be subdivided into theoretical (or normative) ethics, which studies general theories, concepts, and principles of ethics;

meta-ethics, which studies the meaning and justification of ethical words, concepts, and principles; and applied (or practical) ethics, which studies ethical questions that arise in specific situations or areas of conduct, such as medicine or business (Frankena 1973). Research ethics is a branch of applied ethics that studies the ethical problems, dilemmas, and issues that arise in the conduct of research.

In this book, we do not explore meta-ethical issues in great depth, but we mention one issue that has some relevance for research ethics. One of the key questions of meta-ethics is whether ethical standards are universal (Frankena 1973, Pojman 1995). According to one school of thought known as objectivism, the same ethical (or moral) standards apply to all people at all times in all situations. A contrasting school of thought known as relativism holds that different ethical standards apply to different people in different situations: There are no universal moral rules or values. We mention this issue here because in some situations in research ethics one must take a stand on the relativism/objectivism dispute (Angell 1997a,b, Resnik 1998a, Emanuel et al. 2000). For example, different countries have different views about human rights, including the right to informed consent. Should a researcher in a developing nation conduct research that violates Western standards pertaining to informed consent? Although we believe that some standards of research ethics are universal, we recognize that difficult problems and issues can arise in applying those standards to particular situations (Hyder and Wali 2006).

Returning to our focus on the first sense of ethics, ethics as a standard of conduct, it is important to compare and contrast ethics and the law. Societies have had laws since ancient times. One of the first legal systems was the 282 rules of the Code of Hammurabi (1795–1750 B.C.), established nearly 4,000 years ago (Hammurabi 1795 B.C.). Modern legal systems are based, in large part, on laws developed in ancient Rome. Laws are like ethical standards in several ways. First, laws, like ethics, are standards of conduct: They tell people how they ought or ought not to behave. Second, ethical and legal standards share many concepts and terms, such as duty, responsibility, negligence, rights, benefits, and harms. Third, the methods of reasoning used in law and ethics are quite similar: Both disciplines give arguments and counterarguments, analyze concepts and principles, and discuss cases and examples.

However, ethics differ from laws in several important ways, as well. First, the scope of ethics is much different from the scope of law. There are many types of conduct that might be considered unethical but are not illegal. For instance, it may be perfectly legal to not give credit to someone who makes a major contribution to a research project, but this action would still be unethical because it would violate principles of fairness and honesty. Also, there are areas of law that have little to do with ethical concerns, such as laws pertaining to the regulation of various industries. We can think of ethics and law as two different circles that overlap in some areas. In general, since laws must be enforced by the coercive power of government, societies usually make laws pertaining to a behavior only when there is a social consensus concerning that behavior. The law usually sets a minimal standard of conduct, but ethics can go beyond that standard (Gert 2007).

Second, people can appeal to moral or ethical standards to evaluate or judge legal ones. People may decide that there needs to be a law against some type of unethical behavior, or they may decide that an existing law is unethical. If we consider a law to be unethical, then we may be morally obligated to change the law or perhaps even disobey it. For example, many people who considered South Africa's system of apartheid to be unethical fought to change the system. Some of them made a conscious decision to protest apartheid laws and engaged in kind of lawbreaking known as civil disobedience. Finally, ethical standards tend to be more informal and less arcane and convoluted than legal standards; ethical standards are not usually legalistic. Because ethics and the law are not the same, scientists must consider and weigh both legal and ethical obligations when making ethical decisions (more on this later).

To understand the relationship between law and ethics, imagine ethics and the law as two circles that intersect. Some behaviors fall into the circle of ethics but not the law, while others fall into the circle of law but not ethics. Many behaviors fall into both. If the intersection between the two circles does not cover much of the ethics circle, then people have a great deal of freedom to decide how to regulate their behavior, without fear of legal repercussion. As the circle of the law expands and covers more of the ethics circle, people lose some of freedom as more behavior falls within the legal domain. As the circle of the law shrinks, people gain freedom but lose personal security and social stability, because the state controls less behavior. In a totalitarian regime, people sacrifice freedom for security and stability; in anarchy, people have all the freedom they could want but almost no security or stability. Societies constantly struggle to find the right balance between freedom and security/stability.

For an example of the relationship between law and ethics, consider the regulation of clinical chemistry laboratories. Prior to 1988, many pathologists owned clinical chemistry laboratories and were able to refer their patients to their own laboratories, which created a conflict of interest (a topic we discuss later in this book). The conflict of interest led to various types of corruption, such as fraud and overbilling. Moreover, clinical chemistry laboratories all across the country had no quality standards, which led to unreliable test results and billions of dollars in waste. These two issues led Congress to enact the Clinical Laboratory Improvement Amendment in 1988. Pathologists lost some freedom, but society gained security and stability.

It is also important to distinguish between ethics and politics. Politics, like ethics, deals with standards for human conduct. However, political questions tend to focus on broad issues having to do with the structure of society and group dynamics, whereas ethical questions tend to focus on narrower issues pertaining to the conduct of individuals within society (Rawls 1971). Many of the controversial areas of human conduct have both ethical and political dimensions. For instance, abortion is an ethical issue for a woman trying to decide whether to have an abortion, but it is a political issue for legislators and judges who must decide whether laws against abortion would unjustly invade a woman's sphere of private choice. Thus, the distinction between ethics and politics is not absolute (Rawls 1971). Although this book focuses on the ethics of research, many of the issues it covers have political dimensions, as well, for example, public funding for human embryonic stem cell research.

The distinction between ethics and religion is also important for our purposes. Ethical theories and religious traditions have much in common in that they prescribe standards of human conduct and provide some account of the meaning and value of life. Many people use religious teachings, texts, and practices (e.g., prayer) for ethical guidance. We do not intend to devalue or belittle the importance of religion in inspiring and influencing ethical conduct. However, we stress that ethics is not the same as religion. First, people from different religious backgrounds can agree on some basic ethical principles and concepts. Christians, Jews, Muslims, Hindus, and Buddhists can all agree on the importance of honesty, integrity, justice, benevolence, respect for human life, and many other ethical values despite their theological disagreements. Second, the study of ethics, or moral philosophy, is a secular discipline that relies on human reasoning to analyze and interpret ethical concepts and principles. Although some ethicists adopt a theological approach to moral questions and issues, most use secular methods and theories and reason about arguments and conclusions. While our book focuses on research ethics, many of the issues it addresses have religious aspects, as well. For instance, various churches have developed opinions on specific issues arising from science and technology, such as cloning, assisted reproduction, DNA patenting, and genetic engineering.

ETHICAL THEORIES

For a better understanding of some of the different approaches to ethics, it is useful to provide a rough overview of some influential ethical theories. Philosophers, theologians, and others have defended a number of different ethical theories. An ethical theory is similar to a scientific theory in that it is a framework for systematizing and explaining evidence. In science, the evidence comes from observations we make through our senses or by means of instruments. We can test a scientific theory by making predictions from the theory and performing tests or experiments to determine whether those predictions are correct. For example, Albert Einstein's theory of general relativity predicted that objects with a strong gravitational force, such as the sun, should be able to bend light that passes nearby. In 1919, astronomers tested the theory during a solar eclipse by observing changes in the apparent positions of stars near the sun (Isaacson 2007). This test impressed many scientists, who did not expect light to bend under the influence of gravity. Since ancient times, many biologists and laypeople accepted the theory of spontaneous generation, which held that organisms can spontaneously arise from nonliving things, such as mud, dust, feces, and rotting food. In 1668, Italian physician Francesco Redi tested the theory by means of a controlled experiment. He placed meat in two jars, one open and the other closed. The meat in the open jar developed maggots, but the meat in the closed jar did not. He concluded that maggots do not spontaneously arise from meat but come from eggs laid by flies that land on meat (Miller and Levine 2005).

In ethics, evidence does not come from observations we make with our senses or by means of instruments, but from our judgments of right/wrong or good/bad in particular cases. We do not have a special "moral sense" that gives us information about the moral qualities of the world, in the way that we have other senses, such as vision or hearing, that give us information about the physical properties of the world. We make moral judgments not by observing a particular feature of a situation but from forming a holistic impression of the situation we are presented with. This holistic impression results from our sensory perception of the situation, our background beliefs about the situation, and our emotional reaction to the situation (Hauser 2006, Haidt 2007). For example, suppose I observe someone cutting some-one else with a knife, and I also believe that the person does not want to be cut, that the person doing the cutting is angry at the person he is cutting, and so on. I may experience revulsion, fear, anxiety, disgust, and other nega-tive emotions. I may have some empathy or compassion for the person being cut. I may conclude, from this whole impression, that what is happening is wrong. However, I might not reach this conclusion if I observe the same act of cutting but believe, instead, that the person wants to be cut, that the per-son doing the cutting is trying to help the person he is cutting by removing a tumor, and so on. Under these circumstances, I may experience no negative emotions, except perhaps some disgust at seeing a person being cut. The dif-ference between judging that an action is a vicious assault and judging that an action is a commendable deed, such as surgery, depends on the overall context of the act, not just its perceptual features.

To test an ethical theory, we need to consider how the theory would apply to a particular case. If the theory has an implication that contradicts our ethical judgments about the case, then this would constitute evidence against the theory. If the theory has an implication that conforms to our ethical judgments about the case, then this would constitute evidence for the theory. For example, consider a very simple ethical theory, "Minimize human mor-tality." To test this theory, we could consider a case where a doctor has an opportunity to take the organs from a dying patient to save the lives of five people. The patient has told that doctor that he does not want to donate his organs because he believes that the human body is sacred. Should the doctor use the organs from the patient, against his wishes, to save the five people, or should he allow the patient to die and preserve his body? The theory implies that the doctor should take the organs, because this would minimize human mortality. If we judge that this would be the wrong thing to do in this situation, then this would constitute evidence against the theory. If enough evidence of this sort is produced, we may reject the theory or develop a different theory that does a better job of handling this and other cases. Over time, our theories can become better at systematizing our ethical judgments. This method for testing (or justifying) ethical theories is known as reflective equilibrium (Rawls 1971, Harman 1977).

While the comparison between scientific and ethical theories is useful, it goes but so far. It is obvious to most people that science and ethics are very different. First, science makes measurable progress. New theories replace old ones, and over time, theories become better at making predictions and explaining evidence. Quantum physics is better at making predictions and explanations than Newton's physics, which was better than Galileo's physics, which was superior to Aristotle's. In ethics, this type of progress has not occurred. One might argue that ethical theories have not gotten much better at systematizing our moral judgments over the years. Instead of making progress, ethical theorists seem to be going around in circles, debating the same old problems and issues over and over again. Moral philosophers and theologians still regard the ethical systems proposed by the ancient philosopher Aristotle, the medieval theologian St. Thomas Aquinas, and the enlightenment philosopher Immanuel Kant to be viable theories. But no scientist today would subscribe to Aristotelian physics, alchemy, phlogiston theory, or Lamarckian evolution. These scientific theories have been shown to be false (or incorrect) and are of historical interest only.

Second, scientists from around the world are able to reach consensus about many different questions and problems. In science, well-confirmed hypotheses and theories eventually become accepted as facts. For example, Nicholas Copernicus's heliocentric model of the solar system was regarded as merely a theory during the Renaissance but was accepted as fact by scientists around the world at the beginning of the Enlightenment. Louis Pasteur's germ theory of disease started out as a theory but is now a widely accepted fact in biomedicine in many countries. James Watson and Francis Crick's model of the structure of DNA was a mere theory in 1953 but is now accepted as fact in countries all over the planet. The same cannot be said about ethics. There are many different ethical questions and problems where there is little consensus, if any, such as abortion, euthanasia, capital punishment, and gun control. Also, different societies have different standards concerning human rights, women's rights, homosexuality, and many other issues. While there may be some global agreement on some very general moral rules, such as "don't kill," "don't lie," and "don't steal," there is still considerable disagreement about specific issues, problems, standards, and concepts (Macklin 1999a).

Third, science leads to technological applications that can be used to achieve practical goals. Galileo's physics was useful for aiming cannon balls and making accurate calendars, Newton's physics was useful for constructing sturdy bridges and sending spaceships to the moon and back, quantum physics has been useful in developing nuclear power and engineering nanoscale materials and devices, the germ theory of disease has been useful in developing vaccinations and establishing hygiene regimens, and so on. While ethics is a useful tool for decision making, it has no technological applications. Ethical theories are not useful for building bridges, designing airplanes, manufacturing pharmaceuticals, and so on.

Fourth, science aspires to be objective or unbiased. Many of the methods of science, such as repetition of experiments, peer review, and random assignment to experimental or control groups, are designed to eliminate or reduce bias. Because science is a human activity, it will never be completely free from political, social, cultural, or economic biases. However, science can and does make measurable progress toward objectivity (Kitcher 1993, Haack 2003, Resnik 2007b). For example, at one time religion had considerable influence over astronomy. In ancient times, astronomical theories developed by the Egyptians and Babylonians conformed to their religious cosmologies. In Renaissance Europe, the Catholic Church tried to suppress Copernican (heliocentric) astronomy. Galileo was placed under house arrest for defending Copernicanism. By the 1700s, however, astronomy had freed itself from the influence of religion. Modern astronomical theories and models are completely secular. During the 1800s, psychology was influenced by racist theories of intelligence, which posited that Europeans are more intelligent than Africans and other races. This view persisted into the mid-twentieth century, but today few psychologists accept such theories of intelligence.

Ethics also aspires to be objective, but it has made far less progress toward objectivity than science has. Many different social, political, religious, cultural, and economic biases influence ethical theories, concepts, principles, judgments, and decisions. As noted above, while there may be some rough agreement on some ethical principles, ethical values vary considerably across different cultures and nationalities and among different religious and ethnic groups. The method of reflective equilibrium, described above, can help to reduce bias in ethics because it involves the examination (and reexamination) of ethical theories and principles. It is conceivable that, over time, this process will increase objectivity by exposing biases and subjecting them to scrutiny. However, there is no guarantee that this will happen. It also possible that some biases will remain entrenched.

Because we view ethical theories as very different from scientific ones, we do not regard any single ethical theory as the "true" or "correct" theory. Many different ethical theories capture different insights about the nature of morality (Hinman 2002). Some theories focus on moral rules; others focus on moral virtues. Some theories emphasize individual rights; others stress the good of society, and so on (Beauchamp and Childress 2001). We therefore encourage readers to consider different theoretical perspectives when analyzing problems. Ethics is similar, in some ways, to a scientific discipline before the emergence of a paradigm (i.e., preparadigm science). Brand-new disciplines in science are often marked by fundamental disagreements about methods, concepts, and theories. Scientists in new disciplines may accept or use many different theories until a dominant paradigm and normal science tradition emerge (Kuhn 1970). For example, psychology before the emergence of Freudian psychoanalysis was a preparadigm science in that many different psychological theories and methods were accepted. Even today there are many different approaches in psychology, such as behaviorism, cognitive psychology, neurobiology, and psychoanalysis.

With these comments about ethical theories in mind, the following theories will be useful for the reader to consider when thinking about questions and problems in the ethics of research.

Kantianism

Kantianism is a theory developed by the German Enlightenment philosopher Immanuel Kant (1724–1804), which has been revised and fine-tuned by modern-day Kantians, such as Christine Korsgaard (1996). The basic insight of Kantianism is that ethical conduct is a matter of choosing to live one's life according to moral principles or rules. The concept of a moral agent plays a central role in Kant's theory: A moral agent is someone who can distinguish between right and wrong and can legislate and obey moral laws. Moral agents (or persons) are autonomous (or self-governing) insofar as they can choose to live according to moral rules. For Kant, the motives of agents (or reasons for action) matter a great deal. One should do the right action for the right reason (Pojman 1995). What is the right thing to do? According to Kant (1753 [1981]), the right thing to do is embodied in a principle known as the categorical imperative (CI), which has several versions. The universality version of CI, which some have argued is simply a more sophisticated version of the Golden Rule, holds that one should act in a way that one's conduct could become a universal law for all people. In making a moral decision, one needs to ask, "What if everybody did this?" According to the respect-for-humanity version of CI, one should treat humanity, whether in one's own person or in other persons, always as an end in itself, never only as a means. The basic insight here is that human beings have inherent (or intrinsic) moral dignity or worth: We should not abuse, manipulate, harm, exploit, or deceive people in order to achieve specific goals. As we discuss later, this concept has important applications in the ethics of human research.

Utilitarianism

English philosopher/reformists Jeremy Bentham (1748–1832) and John Stuart Mill (1806–1873) developed the theory of utilitarianism in the 1800s. The basic insight of utilitarianism is that the right thing to do is to produce the best overall consequences for the most people (Frankena 1973, Pojman 1995). Philosophers have introduced the term "consequentialism" to describe theories, such as utilitarianism, that evaluate actions and policies in terms of their outcomes or consequences (good or bad). "Deontological" theories, on the other hand, judge actions or policies insofar as they conform to rules or principles; these theories do not appeal only to consequences directly. For instance, Kantianism is a deontological theory because it holds that actions are morally correct insofar as they result from moral motives and conform to moral laws. Different utilitarian theorists emphasize different types of consequences. Mill and Bentham thought that the consequences that mattered were happiness and unhappiness. According to Mill's Greatest Happiness Principle, one should produce the greatest balance of happiness/ unhappiness for the most people (Mill 1861 [1979]). Due to problems with defining the term "happiness," some modern utilitarians hold that one should maximize the satisfaction of preferences, welfare, or other values. Different utilitarian theorists stress different ways of evaluating human conduct. For instance, act-utilitarians argue that individual actions should be judged according to their utility, whereas rule-utilitarians believe that we should assess the utility of rules, not actions. A number of different approaches to social problems are similar to utilitarianism in that they address the consequences of actions and policies. Cost–benefit analysis examines economic costs and benefits, and risk-assessment theory addresses risks and benefits. All consequentialist theories, including utilitarianism, depend on empirical evidence relating to the probable outcomes. In this book, we discuss how the utilitarian perspective applies to many important ethical questions in research and science policy.

Natural Law

The natural law approach has a long tradition dating back to the time of the ancient Greeks. The Stoics and Aristotle (384–322 B.C.) both adopted natural law approaches. According to this view, some things, such as life, happiness, health, and pleasure, are naturally good, while other things, such as death, suffering, disease, and pain, are naturally evil (Pojman 1995). Our basic ethical duty is to perform actions that promote or enhance those things that are natural goods and to avoid doing things that result in natural evils. One of the most influential natural law theorists, Thomas Aquinas (1225–1274), developed a theological approach to the natural law theory and argued that the natural law is based on God's will. Most natural laws theorists hold that moral rules are objective, because they are based on natural or divine order, not on human ideas or interests. Natural law theorists, like Kantians, also believe that motives matter in morality. Thus, a concept that plays a key role in natural law theory is the concept of double effect: We may be held morally responsible for the intended or foreseeable effects of our actions but not for the unintended or unforeseeable effects. For example, suppose a physician gives a terminally ill patient morphine in order to relieve his pain and suffering, and the patient dies soon thereafter. The physician would not be condemned for killing the patient if his motive was to relieve suffering, not bring about the patient's death, and he could not have reasonably foreseen that the dose would be lethal: Death would be an unintended effect of the morphine administration. One of the important challenges for the natural law theorist is how to respond to developments in science and technology, such as medicine, genetic engineering and assisted reproduction, which can overturn the natural order. Many things are done in medicine, such as surgery, drug therapy, and assisted reproduction, that are unnatural but still might be morally acceptable.

Virtue Ethics

The virtue ethics approach also has a long history dating to antiquity. Virtue theorists, unlike Kantians and utilitarians, focus on the fourth sense of "ethics" mentioned above, developing good character traits. Their key insight is that ethical conduct has to do with living a life marked by excellence and virtue (Aristotle 330 B.C. [1984], Pojman 1995). One develops morally good character traits by practicing them: A person who acts honestly develops the virtue of honesty, a person who performs courageous acts develops the virtue of courage, and so on. Although virtue theorists do not emphasize the importance of moral duties, we cannot become virtuous if we routinely fail to fulfill our moral obligations or duties. Some of the frequently mentioned virtues include honesty, honor, courage, benevolence, fairness, humility, kindness, fairness, and temperance. Integrity is a kind of meta-virtue: We have the virtue of integrity insofar as our character traits, beliefs, decisions, and actions form a coherent, consistent whole. If we have integrity, our actions reflect our beliefs and attitudes; we "talk the talk" and "walk the walk." Moreover, if we have integrity, we are sincere in that our actions and decisions reflect deeply held convictions. However, because we can develop our beliefs, attitudes, and character traits over the course of our lifetime, integrity is more than simply sticking to our convictions, come what may. Changes in beliefs, attitudes, and traits of character should maintain the integrity of the whole person (Whitbeck 1998). Integrity has become an important concept in research ethics. Although many people use it as simply another word for "honesty," "honor," "ethics," it has its own meaning.

Natural Rights

The natural rights approach emerged with the development of property rights in Europe during the 1600s. The British philosopher John Locke (1632–1704) founded this approach, and it is has been refined by many different theorists in the twentieth century. According to this view, all people have some basic rights to life, liberty, property, freedom of thought and expression, freedom of religion, and so on (Locke 1764 [1980]). These rights are "natural" in that they do not depend on any other duties, obligations, or values. The U.S. Constitution, with its emphasis on rights, reflects the natural rights approach to ethics and politics. Although most theorists agree that it is important to protect individual interests and well-being, there is an inherent tension in ethical and policy analysis between respecting individual rights (or interests) and promoting what is best for society. The harm principle is a widely accepted policy for restricting individual rights (Feinberg 1973). According to this rule, society may restrict individual rights in order to prevent harms (or unreasonable risks of harm) to other people. A more controversial restriction on individual rights is known as paternalism. According to this principle, society may restrict individual rights in order to promote the best interests of individuals. Drug regulations are paternalistic in that they are designed to protect individuals, as well as society, from harm (Feinberg 1973).

Social Contract Theory

Social contract theory began with the English philosopher Thomas Hobbes (1588–1679). It was later developed by the French philosopher Jean-Jacques Rousseau (1712–1778) and helped to inspire the French Revolution. The key insight provided by this theory is that moral standards are conventions or rules that people adopt in forming a just society (Rawls 1971). People accept moral and political rules because they recognize them as mutually advantageous. According to Hobbes, people form a civil society because life without it is "a war of all against all...solitary, poor, nasty, brutish, and short" (Hobbes 1651 [1962], p. 100). We discuss in this book how this Hobbesian insight also applies to science: Many of the rules of scientific ethics are conventions designed to promote effective collaboration and cooperation in research (Merton 1973, Hull 1988, Resnik 1998c). Without social conventions concerning authorship, peer review, publication, and intellectual property, science would be a war of all against all.

Divine Command Theory

The divine command theory can trace its history to the beginning of human civilization. As long as they have worshipped deities, people have believed that they should follow the commands of the deities. Many religious texts, such as the Bible, contain claims about divine commandments as well as stories about sin (disobeying God's commands) and redemption from sin. As we discussed above, many natural law theorists base moral laws on divine order. Many theologians and philosophers other than Thomas Aquinas have also defended the divine command approach to ethics. Although we do not criticize the divine command theory here, we note that many philosophers have challenged the connection between morality and religion. Religion may inspire ethical conduct even if it is not the foundation for moral concepts and principles (Pojman 1995). Moreover, many of the ethical commands that one finds in religious texts, such as Jesus's command to love your neighbor as yourself, can be accepted by people from different religious backgrounds.

ETHICAL PRINCIPLES

In addition to these different theories, moral philosophers and theologians have developed a variety of ethical principles, which can be useful in thinking about ethical questions, problems, and decisions. (We define "principles" as highly general rules.) There are several advantages of using ethical principles to frame ethical questions, problems, and decisions. First, principles are usually easier to understand and apply than are theories because they are not as abstract or complex as theories (Fox and DeMarco 1990). It is much easier to understand and apply a rule like "Don't kill innocent human beings" than Kant's moral theory. Second, many ethical principles have widespread theoretical and intuitive support (Beauchamp and Childress 2001). The principle, "Don't kill innocent human beings" is implied by many different moral theories, including Kantian ethics, rule-utilitarianism, natural law theory, natural rights theory, and virtue ethics. Most people around the world also accept some version of this principle. Ethical principles can also be tested (or justified) according to the method of reflective equilibrium mentioned above. Some influential ethical (or moral) principles are as follows:

1. *Autonomy: Allow rational individuals to make decisions concerning their own affairs and to act on them.* A rational individual is someone who is capable of making an informed, responsible choice. He or she understands the difference between right and wrong and can apply information to different choices, in light of his or her values and goals. In Kantian theory, a rational individual is a moral agent, or someone who can develop and follow moral rules. In Locke's theory, a rational individual is someone who has natural rights. In the legal system, rational individuals are regarded as legally competent. Children, for example, are not treated as legally competent, and mentally disabled adults may be declared incompetent by a judge. Respecting autonomy also includes respecting the previous decisions made by rational individuals, such as preferences for medical care expressed through living wills or other documents.

2. *Nonmaleficence: Do not harm yourself or others.*

3. *Beneficence: Promote your own well-being and the well-being of others.* Two interrelated concepts, "harm" and "benefit," play a key role in applying the principles of nonmaleficence and beneficence. According to a standard view, benefits and harms can be understood in terms of interests: To benefit someone is to promote or advance their interests; to harm someone is to undermine or threaten their interests. An interest is something that any rational person would want or need, such as food, shelter, health, love, self-esteem, and freedom from pain or suffering. So, a visit to the dentist's office may be painful, but it is not really harmful, provided that the dentist performs a service for you that promotes your interests, such as repairing a decayed tooth.

4. *Justice: Treat people fairly.* "Justice" is a complex concept that we do not analyze in detail here, but simply note that there are formal as well as material principles of justice. Formal principles, such as "Treat equals equally, unequals, unequally" and "Give people what they deserve," merely set logical conditions for applying principles that have more definite content, such as "Allocate resources fairly." Some of the approaches to resource distribution include equality ("distribute equally"), need ("distribute according to need"), merit ("distribute according to desert or merit"), chance ("distribute randomly"), and utility ("distribute resources so as to promote utility").

These four ethical principles imply a variety of subsidiary rules, such as respect for personal freedom, property, and privacy, and prohibitions against lying, cheating, killing, maiming, stealing, deceiving, coercing, exploiting, and so on. The subsidiary rules also imply more specific rules. For example, the principle of nonmaleficence implies the rule "don't steal," which implies the rule "don't steal intellectual property."

It is important to note that these ethical principles, as well as the more specific rules derived from them, may conflict in some situations (Ross 1930). For example, the principles of autonomy and beneficence may conflict when a patient wants to refuse necessary medical care. Doctors attending to the patient have an obligation to prevent the patient from harming himself (via the principle of beneficence), but they also have an obligation to respect the patient's autonomy. When two or more ethical principles conflict, one must decide which principle takes precedence in the particular situation. Resolving a conflict of principles depends on a careful assessment of the relevant facts and social, economic, and other circumstances. For example, in deciding whether to respect the patient's autonomy, the doctors need to have more information about the patient's mental and emotional state (is the patient capable of making a sound decision?), the patient's medical condition (what is the patient's prognosis? what are some potential therapies?), the patient's financial circumstances (can the patient not afford treatment?), and the patient's knowledge of his own situation and his values (what is the patient basing his decision on?) (Beauchamp and Childress 2001, Iltis 2000). We say a bit more about conflict resolution later.

Principles for Ethical Conduct in Research

Having described some ethical principles, we now consider some principles pertaining to a particular area of conduct, scientific research. To understand these principles, it is important to distinguish between general ethical principles, discussed above, and special ethical principles. General ethical principles (or morals) apply to all people in society. For example, the rule "be honest" applies to everyone, regardless of their social role or position. Special ethical rules, however, apply only to people who occupy specific social roles, positions, occupations, or professions, such as doctor, lawyer, spouse, or parent (Bayles 1988). The rule "do not fabricate data in research" applies to scientists but not necessarily to other social roles or positions. When people enter a profession, they agree, explicitly or implicitly, to abide by the ethical rules of the profession. In medicine, doctors still take the Hippocratic Oath, and many other professionals swear allegiance to ethical codes. As noted in chapter 1, scientists are also professionals and have ethical rules pertaining to their activities.

In the preceding section, we argued that general ethical theories and principles systematize our judgments about right/wrong, good/bad, and so on. Special ethical principles systematize our judgments of right/wrong (and good/bad) with regard to particular social roles (occupations, positions, or professions). Special ethical principles are not simply an application of general principles to particular social roles: Special principles take into account what is different or unique about social roles. In making judgments of right and wrong pertaining to social roles, we draw on our understanding of the general principles of ethics and our understanding of that social role. For example, in making judgments about honesty in science, we consider the general ethical principle "be honest" in the context of scientific conduct, such as recording, reporting, analyzing, and interpreting data. A principle of honesty in science, therefore, includes elements that are unique to science. As a result, honesty in science may be different from honesty in a different social role. For example, a scientist who exaggerates when reporting his results would be acting unethically, but an artist who loves a certain painting who exaggerates when talking about it might be acting ethically. Scientists are expected to be more honest than the art lover.

In making ethical judgments concerning a particular profession, such as science, we must consider (a) the goals or objectives of the profession, (b) interactions among members of the profession, and (c) interactions between members of the profession and society. For example, we judge that fabricating data is wrong because faking data is contrary to the goals of science, which seeks truth, and faking data violates a commonly accepted general ethical rule: be honest. We can judge stealing ideas without giving proper credit to be wrong in research because stealing ideas undermines the trust that is so important for collaboration and cooperation in scientific research, and stealing ideas is a form of theft, which violates general ethical rules. Violating the autonomy of human research subjects is wrong because it violates human rights, undermines the public's support for science, destroys the trust that research subjects have in science (which is necessary for recruiting subjects), and flouts a general ethical rule: respect autonomy. Finally, sharing of data and results is an ethical obligation in science because it promotes collaboration, cooperation, and trust among researchers, and because sharing is supported by the general ethical rule to help others. Thus, ethical principles of scientific research should be consistent with the general principles of ethics, and they should promote good science (Resnik 1996c, 1998a,c).

Earlier in the chapter we argued that the evidence for general ethical theories and principles consists of our ethical judgments about cases, and those general ethical theories and principles cannot be tested by means of observations we make with our senses or by means of instruments. However, because the ethical principles and rules of scientific research are based, in part, on what constitutes good science, they are also based, at least in part, on empirical evidence. Observations we make about what promotes good scientific practice can be relevant to justifying ethical rules and principles for science. For example, consider the rule, "Researchers should keep records that are accurate, thorough, complete, and well organized." This rule can be justified, in part, by determining whether it promotes good scientific practice. If we observe that scientists who follow the rule are more successful in contributing to new knowledge than those who do not, this would count as evidence in favor of the rule. Good scientists keep good records (Schreier et al. 2006). Consider another example: There has been a dispute in the ethics of research with human subjects about whether investigators should disclose financial interests they have related to a study during the informed consent process. One way to help resolve this dispute is to conduct surveys to find out whether subjects would like to have this information. If the evidence shows that most subjects would want this information, then investigators should provide it (Gray et al. 2007).

We now briefly discuss some principles for ethical conduct in research and some subsidiary rules.

1. *Honesty:* Honestly report data, results, methods and procedures, publication status, research contributions, and potential conflicts of interest. Do not fabricate, falsify, or misrepresent data in scientific communications, including grant proposals, reports, and publications (Pellegrino 1992, Resnik 1996a,b).

2. *Objectivity:* Strive for objectivity in experimental design, data analysis, data interpretation, peer review, personnel decisions, grant writing, expert testimony, and other aspects of research where objectivity is expected or required.

3. *Openness:* Share data, results, ideas, tools, materials, and resources. Be open to criticism and new ideas.

4. *Confidentiality:* Protect confidential communications, such as papers or grants submitted for publication, personnel records, business or military secrets, and records that identify individual research subjects or patients.

5. *Carefulness:* Avoid careless errors and negligence; carefully and critically examine your own work and the work of your peers. Keep good records of research activities, such as data collection, research design, consent forms, and correspondence with agencies or journals. Maintain and improve your own professional competence and expertise through lifelong education and learning; take steps to promote competence in science as a whole.

6. *Respect for colleagues:* Respect colleagues, students, and subordinates. Do not harm colleagues; treat them fairly. Do not discriminate against colleagues on the basis of sex, race, ethnicity, religion, or other characteristics not related to scientific qualifications. Help to educate, train, mentor, and advise the next generation of scientists.

7. *Respect for intellectual property:* Honor patents, copyrights, and other forms of intellectual property. Do not use unpublished data, methods, or results without permission. Give credit where credit is due. Do not plagiarize.

8. *Respect for the law:* Understand and comply with relevant laws and institutional policies.

9. *Respect for research subjects:* Show proper respect and care for animals when using them in research. Do not conduct unnecessary or poorly designed animal experiments. When conducting research on human subjects, minimize harms and risks and maximize benefits; respect human dignity, privacy, and autonomy; take special precautions with vulnerable populations; and distribute fairly the benefits and burdens of research.

10. *Stewardship:* Make good use of human, financial, and technological resources. Take care of materials, tools, samples, and research sites.
11. *Social responsibility:* Promote good social consequences and prevent bad ones through research, consulting, expert testimony, public education, and advocacy.
12. *Freedom:* Research institutions and governments should not interfere with freedom of thought and inquiry.

Many of these principles may seem familiar to readers who have some experience with professional codes of ethics in research (Shamoo and Resnik 2006), government funding requirements, oversight agencies, sponsors, or journal policies. We recognize that there are now many useful sources of ethical guidance for researchers; our principles should complement but not undermine existing ethics codes and policies. Some readers may wonder whether these principles are redundant or unnecessary, because other rules and guidelines have already been stated publicly. However, we think the principles above have several important uses. First, they may cover problems and issues not explicitly covered by existing rules or guidelines. Second, they can be helpful in interpreting or justifying existing rules and guidelines. Third, they apply to new and emerging disciplines or practices that have not yet established ethical codes.

The principles we describe here, like the general ethical principles mentioned above, may conflict in some circumstances. For example, the principles of openness and confidentiality conflict when a researcher receives a request to share unpublished data. When conflicts like this arise, researchers must prioritize principles in light of the relevant facts. We discuss conflict resolution in greater detail below.

These principles may also conflict with the goals or rules of the organizations that employ researchers. For example, researchers who work for private industry or the military may face restrictions on information sharing that conflict with the principle of openness. In these situations, researchers must choose between honoring their professional responsibilities and loyalty to the organization and its goals and rules.

ETHICAL DECISION MAKING

Having described some important ethical theories and principles, we are now prepared to discuss ethical decision making (also known as moral reasoning). Ethical decisions involve choices in which the options are not ethically neutral. That is, the choices have ethical implications. For example, choosing between different flavors of ice cream is probably not an ethical decision, because the choice is a matter of personal preference, with almost no impact on other people. However, purchasing an automobile probably is an ethical decision, because the choice can have a significant impact on other people and the environment. Since many of our choices have some impact on other people, and are not simply a matter of personal preference, many of the choices we make in life have an ethical dimension.

Ethical decisions that are particularly vexing and uncomfortable are known as ethical dilemmas. An ethical dilemma is a situation where two or more options appear to be equally supported by different ethical theories, principles, rules, or values (Fox and DeMarco 1990). As a result, there may be considerable doubt or controversy concerning the right thing to do. Consider the following hypothetical case:

> Mr. Gerd is taken to the emergency room (ER) following an automobile accident. He is unconscious and needs a blood transfusion. There is not enough time to contact his family to give consent for the transfusion. The ER physician attending his case, Dr. Kramer, is planning to give Mr. Gerd blood on the grounds that one can presume that he would consent to emergency care in this situation, when an ER nurse discovers that there is card in Mr. Gerd's wallet stating that he is a Jehovah's Witness and does not want to receive any blood or blood products. However, another physician in the ER, Dr. Vaughn, tells Dr. Kramer that Mr. Gerd may qualify for a research study in which he can receive a type of artificial blood called Polyheme. Polyheme is a hemoglobin product made by recombinant DNA technology. Subjects do not need to consent to the study to be enrolled in the study.

How should Dr. Kramer decide what to do? There are many different ways of making decisions at her disposal: She could consult an astrologer, psychic, a pollster, or a lawyer; she could read tea leaves, flip a coin, or pray; she could look for an answer using Google. An ethical approach to decision making is different from all of these methods. A person who is making an ethical decision should use her judgment and intellect to carefully examine the different options in light of the relevant facts and ethical values. She should consider the interests of all of the affected parties and examine her choice from different points of view. An ethical decision need not be perfect, but it should represent a sincere attempt to do the right thing for the right reason.

Philosophers, ethicists, and other scholars have debated about three distinct approaches to ethical decision making in the last few decades: (a) a top-down, theory-based approach, (b) a bottom-up, case-based approach known as casuistry, and (c) a mid-range, principle-based approach known as principalism (Beauchamp and Childress 2001). According to the top-down approach, to make a decision about what to do in a particular situation, one must appeal to a moral theory and infer consequences from the theory. If the theory says to choose a particular option instead of the alternatives, then one should choose that option and implement it. Ethical decisions should be based on ethical theories, and ethicists should focus their attention on moral theorizing, instead of bothering with the details concerning applications.

The top-down approach has been popular among moral philosophers for many years. While we agree the theories are an important part of ethical analysis and justification, they are often not useful guides to ethical decision making. First, as noted above, there is a perennial controversy concerning moral theories. Different theories may give very different recommendations for dealing with particular ethical dilemmas. On the theory-based approach, one would need to decide which theory is the correct one before even approaching the dilemma, because an incorrect theory might lead to the wrong answer to an ethical question. But in science and many other practical endeavors, people do not have time to wait for philosophers or theologians to decide which theory is the correct one—they need answers now. Second, ethical theories are usually abstract and complex and are therefore difficult to interpret and apply. Most people would have a difficult time interpreting and applying Kant's categorical imperative, for example. People who are making practical choices, such as scientists, need guidelines that are not too difficult to interpret and apply. They need rules and guidelines that are clear and easy to understand (Beauchamp and Childress 2001).

In response to problems with theory-based approaches to ethical reasoning, some philosophers have revived a case-based approach known as casuistry (Johnsen and Toulmin 1988, Strong 2000). According to this method of ethical decision making, one should make decisions about particular cases by comparing those cases to previous cases. If cases are similar in relevant ways, then the decisions that one reaches should be the same. If cases are different, then one should reach different decisions. The method is like the case-based approach used in legal reasoning, in which past cases set precedents for future ones. For example, to decide whether one should exclude five data outliers from a data set, one should compare this situation to previous cases in which the scientific community judged it was ethical to exclude data outliers. If the current situation is similar to those other cases, then excluding the data outliers is ethical and one may exclude them. If the current situation is different from those previous cases, or is similar to cases in which excluding outliers was regarded as unethical, then excluding the five outliers may be unethical (Penslar 1995). The method of casuistry is also known as situational ethics, because matters of right and wrong depend on factors inherent in the particular situation.

The casuist approach offers many useful insights for ethical decision making. First, it emphasizes the importance of understanding and appreciating the facts and circumstances concerning cases. In ethics, the details matter. For example, the difference between justified exclusion of outliers and falsification of a data often depends on the details concerning methodology, analysis, and communication. The difference between plagiarism and proper citation may come down to the placement of quotation marks. Second, the casuist approach emphasizes the importance of learning from the past and other cases. If we are to make any progress in ethics, we must learn from good decisions (and bad ones) (Strong 2000).

However, the casuist approach also has some flaws that hamper its ability to guide ethical decision making. First, the casuist approach has no systematic way of comparing cases (Beauchamp and Childress 2001). We need some method or procedure for determining which features of a case are relevant for ethical analysis, because cases have many features that we do not need to consider. For example, if we compare two cases where authors have excluded data from a publication, what aspects of data exclusion should we focus on? The percentage or amount of data excluded? The type of data excluded? The effect of the data exclusion on the results? To answer questions like these, we need ethical principles, rules, or methods for comparing cases, but the casuist approach does not provide these.

Second, the casuist approach does not offer satisfactory justifications for ethical decisions. People are frequently asked to justify their ethical decisions to colleagues, supervisors, governmental officials, or the public. To justify his or her conduct, a person should be able to do more than explain how she or he based the decision on previous cases—the person should also be able to explain how the decision followed from the acceptance of a rule, principle, or value that transcends those cases (Gibbard 1990). For example, a researcher who wants to defend herself from the charge of plagiarism should be able to do more than say that her conduct is similar to other cases that were not regarded as plagiarism; she should also be able to explain why her conduct does not fit the definition of plagiarism and therefore does not violate any rules against plagiarism.

Some proponents of casuistry have responded to objections like those mentioned above by admitting that casuistic reasoning needs to be supplemented with rules, principles, or values. But making this admission changes the approach from a pure case-based method to one that appears to be principle based. Indeed, there would seem to be very little difference between modified casuistry and principle-based approaches (Iltis 2000).

We favor the principle-based approach for many of the reasons noted above. Ethical principles are less controversial than ethical theories. They are also easier to interpret and apply. Ethical principles provide a framework for comparing different cases. So, the principle-based approach does not have the same problems as the other two approaches. One of the most influential books in bioethics, Beauchamp and Childress's *Principles of Biomedical Ethics* (2001), takes a principle-based approach to ethical problems in medicine and health care. We are following their example by articulating a principle-based approach to ethical problems in scientific research (Shamoo and Resnik 2006a).

The principle-based approach is not flawless, however. Because it straddles the fence between theory-based and case-based approaches, it is susceptible to attacks from both sides. Proponents of theory-based approaches argue that principle-based approaches are nothing but an amalgam of different theories, a hodgepodge. Principle-based approaches have no way of settling conflicts among principles: They lack philosophical unity and coherence (Gert 2007). Proponents of case-based approaches argue that principle-based approaches are too abstract and general to provide sufficient guidance for ethical decision making. Principle-based approaches are not practical enough (Strong 2000).

We acknowledge these problems but think the principle-based approach can overcome them. To the theory-minded critics, we respond that one can unify principles by distinguishing between a profession's core principles and its peripheral principles. Core principles are so fundamental to the profession that one cannot conceive of the profession without those principles. Peripheral principles help members of the profession to coexist with each other and operate in society, but they are not essential to the profession itself. We believe that honesty and objectivity are core principles of scientific research. These principles are so essential to the practice of science that we cannot conceive of science without them. A human activity in which honesty and objectivity are not highly valued would be something other than science, such as literature, art, politics, or religion. The other ethical principles, though very important in scientific research, are not essential to the practice of science. Scientists can be secretive, make careless errors, show disrespect for their colleagues, steal intellectual property, violate the law, and mistreat human subjects and still produce valid results. We might call their work irresponsible, illegal, and unethical, but it would still be science.

When core principles conflict with peripheral ones, core principles should take precedence, unless there are some exceptional circumstances that justify abandoning the core principles. For example, consider a public health researcher who is trying to decide how to communicate his findings about nicotine to the public. Let us suppose that he has discovered that nicotine is effective at treating depression. He is worried that if he reports his results honestly and objectively, some people will use this information as a reason to not quit smoking or to even take up the habit. He considers whether he should distort or manipulate his results in order to protect people from harm. While we acknowledge that it is important to protect the public from harm, we do not think scientists should abandon their core principles to do so. Protecting the public from harm should not come at the expense of honesty and objectivity. Scientists should report the truth, nothing more and nothing less.

To the practical-minded critics, we respond that subsidiary rules can aid in the application of principles. As noted above, general principles imply subsidiary rules and definitions. These rules also imply other rules and definitions, and so on. Eventually, one reaches a level of specificity where people can interpret and apply rules without any need for additional rules or definitions (Richardson 2000). For example, consider the general principle "respect research subjects." This implies a subsidiary rule, "respect human research subjects," which implies "respect the autonomy of human research subjects," which implies "obtain informed consent from human research subjects or their representatives," which implies rules for obtaining consent, such as, "informed consent should not take place in circumstances that are coercive." The key insight is that one needs multiple layers of rules and definitions to interpret and apply ethical principles.

Having made these general comments about ethical decision making, we now describe a method for making ethical decisions. We do not claim to be the originators of this method, because many other writers have described methods very similar to this one (Fox and DeMarco 1990, Weil 1993, Whitbeck 1996, Swazey and Bird 1997, Beauchamp and Childress 2001, Shamoo and Resnik 2006a). Nevertheless, it will be useful to review the method here and make some clarifying comments.

A Method for Ethical Decision Making

To illustrate our method, recall Dr. Kramer's ethical dilemma described above regarding giving Mr. Gerd a blood transfusion. The first step

Dr. Kramer should take is to formulate (or frame) the ethical problem, question, or conundrum (see fig. 2.1). For Dr. Kramer, the problem was, "Should I enroll Mr. Gerd in a study that could save his life?" The next step she should take is to gather relevant information. She needs to know whether Mr. Gerd would be opposed to receiving Polyheme. She needs more information about how Polyheme is manufactured—is it derived from human blood? She also needs more information about the pharmacology of Polyheme and how it affects the body: Is Polyheme safe, effective? How does it interact with other drugs? She needs to know how Mr. Gerd is likely to react when he learns that he has received artificial blood. Dr. Kramer also needs more information about the research study involving Polyheme: What procedures does the study involve? What are the inclusion/exclusion criteria? Does the study have any plans for safety monitoring? Dr. Kramer also needs to identify the

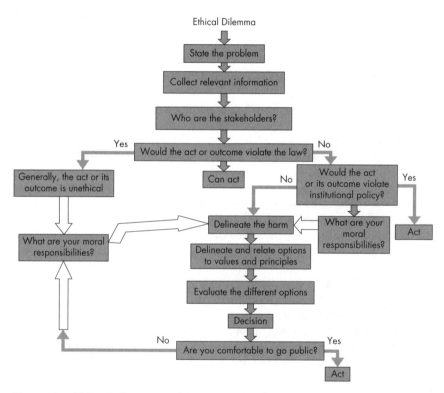

Figure 2.1. Ethical decision-making process. Reprinted with permission of the publisher from: Shamoo, A.E., and Resnik, D.B., 2006, "Ethical Issues for Clinical Research Managers," *Drug Information Journal* 40: 371–383 (listed in the text as Shamoo and Resnik [2006a]). *Note:* We present this diagram as a heuristic aid to the reader. It does not precisely match the all the steps of ethical decision making described in this book.

different people who will be affected by the decision (i.e., the stakeholders), and any laws or institutional policies that may apply to his decision. It is important for Dr. Kramer to gather enough information to make a decision, because ignorance of the relevant facts often leads to poor decision making.

When Dr. Kramer has gathered the information she needs, she can explore her different choices (or options). Her two basic choices are to enroll Mr. Gerd in the study or not enroll Mr. Gerd in the study. But there may be subtle variations on these choices, for example, whether to try to contact a friend, relative, or someone with an understanding of Mr. Gerd's religion, such as a hospital chaplain. Dr. Kramer could also give Mr. Gerd some intravenous fluids and medications to maintain his blood pressure, so she can keep him alive while she is trying to gather more information. Dr. Kramer could decide to not enroll Mr. Gerd in the study but give him human blood. Sometimes it can be difficult for a person to delineate the different options he has in a particular dilemma, because he may get stuck in a particular way of thinking about it (Whitbeck 1996). To avoid this problem, it can be useful to talk to people who have some experience and wisdom to share and can offer a different perspective, such as colleagues or even ethicists. Through discussion and brainstorming, it is sometimes possible to find an option that avoids the horns of the dilemma.

After Dr. Kramer has explored her different options, her next step should be to apply ethical principles (rules or values) to the different options. These may include general ethical principles, such as autonomy and beneficence, as well as special ones, such as objectivity, openness, and respect for research subjects. In the situation that Dr. Kramer faces, respect for dignity and autonomy would imply that Dr. Kramer should not enroll Mr. Gerd in the study, if Mr. Gerd would not want to receive Polyheme. However, beneficence would imply that Dr. Kramer should enroll Mr. Gerd in the study, because this could save his life.

Another principle that might apply to this case is respect for the law. Scientists should comply with laws and institutional policies unless they have a strong justification for violating those rules. Suppose, for example, that the committee that oversees human research at the hospital had approved the Polyheme study with the requirement that investigators obtain consent from the subject or the subject's representative. If this were the case, then enrolling Mr. Gerd in the study would violate the law and institutional policy, and the burden of proof would fall on Dr. Kramer to demonstrate why Mr. Gerd should be enrolled in the study anyway. She might argue, for example, that receiving Polyheme would give Mr. Gerd a chance to live, and it is likely that Mr. Gerd would consent to the study, if he were conscious.

After applying ethical principles to the different options, Dr. Kramer's next step should be to resolve any conflicts among principles. As we have already seen, there may be a conflict between autonomy and beneficence in this situation. To help resolve this conflict, Dr. Kramer may try to acquire more information, explore different options, or even reframe the original question or problem. She should pay special attention to the likely consequences of the different options for the various stakeholders. Will any of the options harm other people or society? Who will benefit most from the different options? Ideally, Dr. Kramer would have the information that she needs before making her choice, but circumstances may force her to make a choice in the face of uncertainty. For example, if Dr. Kramer does not know whether Mr. Gerd would refuse Polyheme, she could err on the side of saving his life and assume that he would consent to receiving Polyheme. If Dr. Kramer has evidence that Mr. Gerd would not want to receive blood and would not want to receive Polyheme (based on his religious beliefs), then Dr. Kramer would have to decide whether she should violate Mr. Gerd's autonomy in order to save his life. She might rationalize this decision by speculating that although Mr. Gerd might be angry when he discovers that his autonomy has been violated, he would at least be alive. In time, he would accept his situation and be grateful to Dr. Kramer for saving his life.

Resolving conflicts among principles is the most challenging part of ethical decision making. Although obtaining more information can usually help to settle conflicts, very often conflicts remain even after one has obtained additional information. When a conflict persists, the person making the decision must decide which principle should take precedence and must set priorities. Above we noted that some people object to the principle-based approach on the grounds that it is unable to resolve conflicts among principles in a coherent way. We responded to this critique by arguing that when core principles of scientific research conflict with peripheral ones, core principles should take precedence, except in unusual circumstances. The situation that Dr. Kramer faces is best understood not as a conflict between scientific principles of ethics, but between general principles of ethics: Dr. Kramer must choose between autonomy and beneficence. To settle conflicts among general principles of ethics, one needs a basis for distinguishing between core and peripheral principles. Core principles are so essential to ethical conduct that we cannot even conceive of ethical conduct without those principles. We consider the principle of autonomy to be the core ethical principle. Autonomy should trump other general ethical principles, except when a person's autonomy threatens other people or society.

Why do we regard autonomy as the core general ethical principle? Giving a complete answer to this question would take us way beyond the scope of this book, but we will provide a brief one here. We conceive of ethical conduct as fundamentally social behavior in which individuals formulate, adopt, modify, and follow rules that apply to all members of the community (Kant 1753 [1981], Rawls 1971, Gert 2007). A moral community consists of individuals who are capable of evaluating their own behavior and the behavior of others. To participate in a moral community, one must be allowed to make choices and act on them. Actions that violate a person's autonomy, therefore, undermine the very foundation of ethics. One should not interfere with a person's autonomy unless that person is making a choice that will harm other members of the community or the community itself. Thus, if Dr. Kramer faced a conflict between autonomy and beneficence that could not be resolved by gathering additional facts or proposing different options, we would recommend that she choose respecting Mr. Gerd's autonomy over saving his life. A note of caution is in order, however. Some people do not view autonomy as a core principle. Some give priority to other ethical principles or values, such as the good of the community (see Etzioni 2004).

After resolving conflicts among ethical principles, Dr. Kramer's next, and final, step should be to make and implement his decision. Implementing a decision may involve a series of steps, over time, to ensure that the option one has chosen becomes effective. For instance, to implement the decision to enroll Mr. Gerd in the Polyheme study, Dr. Kramer would need to contact someone who could begin the enrollment process, such as the investigator or a research nurse. The investigator or nurse would determine whether Mr. Gerd is eligible for the study. If so, Dr. Kramer would sign or complete any necessary forms or documentation, and so on.

Before concluding our discussion of this ethical decision making procedure (Shamoo and Resnik 2006a) (see fig. 2.1), a few comments are in order. First, this procedure is an idealization: Real-world decisions sometimes deviate from this stepwise progression. For example, some may formulate a problem and then start gathering information, only to realize that they need to reformulate the problem; some apply ethical principles and resolve conflicts simultaneously, and so on. Second, this method has limitations. This method requires decision makers to use their reasoning skills to make decisions, but other factors often play a key role in decision making, such as emotion, intuition, and religion. Although we believe that people should try their best to use reasoning to solve ethical dilemmas, we recognize that human reasoning has limitations. If one cannot make a decision after diligently following the method we have outlined here, then it may be appropriate to appeal to some other source of guidance, such as emotion, intuition, religion, or even chance (i.e., flip a coin). Third, using this reasoning method to analyze a choice does not preclude one from bringing other reasoning methods to bear on the choice, such as economic analysis, legal analysis, or rational decision theory. Ethics does not have exclusive dominion over practical choices.

This concludes our first chapter. Although we may not specifically address elsewhere in this book many of the topics included in this chapter, we expect that students and teachers will be able to use the foundational material in thinking about and analyzing the cases, rules, and applications discussed in the remainder of the book.

QUESTIONS FOR DISCUSSION

1. Do you think that most scientists and science students are ethical?
2. When, how, and why are scientists tempted to violate ethical standards in research?
3. What situations in science present the most difficult ethical problems and dilemmas?

4. Do you think researchers should adhere to the same ethical standards that apply to other professions, such as medicine or law? Why or why not?

5. Do you think researchers have ethical duties and responsibilities "over and above" the ethical obligations of ordinary people?

6. Can you think of any principles to add to our list of principles for ethical research conduct? What would they be, and how would they be justified? Do you think our list contains some principles that should be omitted or reworded?

7. Is reasoning the best method for making an ethical decision? Why or why not?

8. Do you think that ethical theories and principles have some bearing on practical choices and decisions? Why or why not?

9. How should one resolve conflicts among ethical principles? Do you agree with our approach to conflict resolution?

10. What ethical principles are most important in society? In science? Why?

3

Data Acquisition and Management

> Proper management of research conduct is essential to achieving reliable results and maintaining the quality, objectivity, and integrity of research data. The different steps of research should be monitored carefully, and research design should include built-in safeguards to ensure the quality and integrity of research data. This chapter addresses ethical conduct in different steps of the research process: hypothesis formation, research design, literature review, data collection, data analysis, data interpretation, publication, and data storage. This chapter also discusses methods that can help assure the quality, objectivity, and integrity of research data, such as good research practices, standard operating procedures, peer review, and data audit.

Scientific research is the systematic attempt to describe, explain, and understand the world. While all three main branches of science—physical science, biomedical science, and social and behavioral science—study different aspects of the natural world, they share some common methods and procedures. These methods and procedures are designed to achieve the goals of science by helping researchers to acquire accurate knowledge and information. Researchers' compliance with scientific methods and procedures can help to minimize falsehoods and biases and maximize truth and objectivity (Shamoo and Annau 1987, Cheny 1993, Kitcher 1993, Resnik 2007b). One pillar of the scientific method is that researchers should subject their theories and hypotheses to rigorous tests (Popper 1959). A test is an attempt to gather empirical evidence (i.e., data) that tends to either confirm or disconfirm a theory or a hypothesis. Ideas that cannot be tested, such as metaphysical theories or ideological claims, are not scientific hypotheses, theories, or facts. Some (but not all) tests involve experiments. In an experiment, a researcher attempts to reduce the number of variables and to control the conditions of a test in order to understand statistical or causal relationships between variables or parameters. For an experiment to be rigorous, a researcher must describe it in enough detail that other researchers can obtain the same results by replicating the experimental conditions (Kirk 1995).

Repeatability is important in experimentation because it confirms that others can carry out the methods and procedures used and attain the same data. Repeatability, or lack thereof, provides substance for public debate and inquiry. Private intuitions, hunches, faith, introspection, or insight can play an important role in generating new ideas to test, but they do not constitute rigorous proof. Therefore, all test results in science, whether from controlled experiments, field observations, surveys, epidemiological studies, computer models, or meta-analyses, should be open to public scrutiny and debate. Peer review, with some limitations, is one of science's most important methods because it promotes the public scrutiny of hypotheses, theories, and test results. Peer review also plays an important preliminary gate-keeping role, ensuring that interpretations of data are self-consistent and consistent with existing literature. In this manner, peer review can contribute to the quality and integrity of published research. Once a hypothesis or theory becomes well established, it may be said to be a "fact," and it is no longer subjected to rigorous tests. For example, the idea that the sun is the center of the solar system is now accepted as a fact, but it was a hypothesis during the time of Copernicus (1542 [1995]). Well-established generalizations, such as Newton's laws of motion and the ideal gas laws, are known as laws of nature (Popper 1959, Hempel 1965, Giere 1991, Resnik 1998c).

Scientific investigators work in different ways to attain new knowledge and have many different motives for conducting research. Most researchers have a deep desire to understand the world, to make new discoveries, and to pursue truth, to the best of their ability. Others want to make an important contribution to the world by improving the human condition or protecting the natural environment. Researchers may also have goals that are less altruistic or noble. Some seek fame, glory, and prestige, and almost all researchers also have strong economic motivations. For most researchers, science is a career and a way to make a living. For example, the U.S. research enterprise consists of more than six million individuals with master's or doctoral degrees, including nearly one million professionals directly engaged in research (Greenberg 2001). Thus, the scientific enterprise is no different from any other business sector of our society, and its performance reflects the values, motives, interests, and shortcomings of our culture (Longino 1990). The failure to understand the selfishness, greed, and bias that is as inherent in science as it is in the other social sectors could lead to unrealistic expectations of the research enterprise and impractical rules and policies.

STEPS IN SCIENTIFIC RESEARCH (OR SCIENTIFIC METHOD)

Plan the Research and Design the Protocol

To develop new knowledge, one must follow numerous steps in planning the research project and designing the protocols. (A protocol is a set of rules, methods, and procedures used to obtain the objectives of a research project.) Each event in the chain of planning and design of the protocol is necessary to ensure that quality, integrity, and objectivity of the data and final results. Every scientist—whether consciously or subconsciously—follows something like the chain of events described below. In the twentieth century, these events became more formalized and rigid, especially in large projects. Although we present these steps in linear order, some of them may occur concurrently, and researchers sometimes return to earlier steps when revising the project and protocol. The following steps outline the processes that usually compose a research project. (For further discussion of the scientific method, see Kitcher 1993, Haack 2003, Resnik 2007b; for a more specific and detailed protocol description for research with human subjects, e.g., clinical trials, see Hulley et al. 2001, Inadomi 2006.)

State the Objectives of the Research Project

The objectives of a research project are the questions that researchers are attempting to answer or the problems they are attempting to solve (Grinnell 1992). For example, if a research project is addressing the toxicity of a particular drug in laboratory mice, then one of its objectives may be to "test toxicity of certain drug in laboratory mice." The questions answered may be "what is the toxicity of the drug?" The knowledge obtained from answering questions such as this could satisfy the needs of society for improving the health and education of its citizens.

Develop Specific Aims for the Project

Specific aims list the particular goals of the project that need to be achieved in order to attain its overall objective(s). The aims may help meet the objectives of the project either wholly or in part. For example, to test toxicity of a certain drug in animals, one must design specific aims to test for lethal doses, toxic doses, and side effects, and one must describe the species and gender of animals, duration of testing, and types of measurements (e.g., blood pathology, temperature, and biopsies).

Propose the Hypotheses

Hypotheses are statements that are designed to answer research questions (Grinnell 1992). One way to test a hypothesis is to put it in the form of a conditional statement, for example, "If A occurs, then B will occur." The antecedent (A) in the conditional specifies the test conditions; the consequent (B) states predicted results (Giere 1991). Suppose we want to know whether a drug inhibits the growth of cancerous tumors. The hypothesis could be "drug A inhibits cancer growth in species S." To test this statement, we can develop a conditional statement that makes predictions about specific results that should occur if the hypothesis is true. For example, the conditional could be "if drug A inhibits cancer growth in species S, then we should be able to find the dose where growth of tumors in the animals used will slow or stop within a month of receiving the dose." If researchers conduct tests with the dose believed to stop tumor growth and it does not stop or slow tumor growth, then these negative results would disconfirm or disprove the hypothesis. But if a large number of animals show reduction in the size of the tumor or a stoppage in growth, then these positive results confirm (or prove) the hypothesis. In this example, the test conditions specify the procedure used for administering the drug to the species, and the outcome (or result) is what happens to the animals. Very often, test conditions include unstated or implied assumptions used in conducting research. For example, the type of syringe used to administer the drug, the diet, and overall health of the population may be unstated test conditions. Sometimes researchers may modify a hypothesis based on negative results, unexpected findings, or difficulties in satisfying the test conditions. However, any modifications to the hypothesis should be clearly stated in the project. In other words, the description of the project and protocol must reflect the realities of what has happened—good, bad, or unexpected—to preserve the integrity, objectivity, and quality of the data.

Conduct a Thorough Literature Search

The literature search step can be the first step in the overall project. This is an important step for the investigator because it can save a great deal of time and money by eliminating a flawed objective or a hypothesis. It can also help researchers to learn whether their projects may make an original or worthwhile contribution or whether they merely repeat previous work or would result in knowledge that has little value. A literature search can also help researchers learn about previously used methods, procedures, and experimental designs and can place the project's experimental design and protocol within the known realities of the subject matter. A thorough literature search can allow researchers to give proper credit to others who have already worked in the area. Failing to acknowledge other relevant work is arrogant and self-serving and is a type of plagiarism or serious bias if one knowingly or unknowingly claims to be the originator of someone else's idea (Shamoo 1992, Resnik 1998a).

It is important to note that an inadequate literature search in clinical research can lead to tragic results. Ellen Roche died while participating in an experiment designed to produce a mild asthma attack in healthy (non-asthmatic) volunteers at Johns Hopkins University. Roche inhaled hexamethonium, a blood pressure medication used in the 1950s and 1960s. Roche developed a cough and breathing difficulties and was put on ventilator. She died from extensive lung damage produced by hexamethonium. An Office of Human Research Protections investigation of Roche's death determined that this tragedy probably could have been avoided if the principal investigator, Alkis Togias, had consulted articles published in the 1950s (and cited in subsequent publications) warning of lung damage due to inhaling the hexamethonium. Togias did a standard PubMed search on hexamethonium and consulted current textbooks, but this literature search did not include references from the 1950s (Savulescu and Spriggs 2002).

Design Experiments or Tests

As mentioned above, each step is important in the chain of research to ensure the integrity, objectivity, and quality of the data and results. Each project usually consists of more than one experimental protocol. The design of experiments (and other tests, e.g., surveys) is one of these crucial steps in preserving the integrity, quality, and objectivity of the research project. It is easy to employ experimental designs that tend to bias the data and results. For example, in the toxicity study mentioned above, using unhealthy or overcrowded mice could affect the data. Another example would be testing a new drug for erectile dysfunction on a population of males younger than 50 while most users would be older than 50. Because no amount of statistical analysis or interpretation can overcome a design flaw, data that result from a flawed design are virtually useless, and using them can be unethical (Irving and Shamoo 1993, Resnik 2000). Generating useless data wastes time, money, and effort, and it can also involve the unnecessary use of human or animal subjects. Sound experimental design is one of the key ethical principles of research with human subjects (Levine 1988, Irving and Shamoo 1993).

Biased designs frequently occur when industries sponsor research projects aimed at promoting their economic interests (Porter 1993, Crossen 1994, Resnik 2007b). For example, biases in research design can occur when a study does not have adequate statistical power to detect a rare adverse effect of a drug, or when a study does not even collect data on a type of adverse effect, such as interference with electrolyte balance. Because scientists have an obligation to be objective in designing experiments, the design of an experiment should not only remove bias but test for it. Some forms of bias involve intentional deception (or research misconduct), such as "cooking" the data (the practice of designing an experiment to produce a desired outcome and not subjecting a hypothesis to a rigorous test). Other types of bias are unintentional—scientists, like all human beings, are susceptible to self-deception (Broad and Wade 1982 [1993]). Because it is not always possible to see the biases in one's own work, it is important to solicit critical feedback from colleagues when designing experiments.

Identify and Describe Methods to Be Used

In this step, researchers identify and describe in detail methods to be used in the project based on their own methods and previous research, existing literature, laboratory manuals, and other sources. Researchers should follow appropriate standards in applying methods and should keep records of what methods they use and how they use them. During initial tests, researchers should use and identify standard (or well-established) methods, but they can modify these methods to suit new experimental applications or testing procedures. It is important for researchers to note changes they make and to state the reasons for them. Furthermore, researchers should not make changes in the middle of a test or experiment, because this will bias or corrupt the data. All accidental changes, such as dropping a test tube, should be noted in the laboratory notebook. If researchers perform tests or experiments to produce data for statistical analysis, the procedures should be carried out in the same exact manner each time. Researchers should not pick and choose among experiments or tests to achieve a desired result. However, they may do so if they recognize a variable inherent in the protocol that was not first recognized in earlier stages of the project. For example, in testing a new drug in humans, researchers may realize that an unanticipated side effect should be recorded and could therefore change the protocol and design a new experiment that measures this side effect. However, researchers should record these decisions and discuss them in detail at the same time and place where the experiments are recorded, derived, or manipulated.

Collect and Record Data

Proper documentation of all aspects of research (e.g., methods, protocols, and data) is crucial to ensuring accountability in research and to keeping a proper paper trail for management and for other future interested parties to authenticate the data. Thorough documentation is also useful for future analysis, verification, and replication by others or investigations of misconduct, error, or other problems (Shamoo 1989, 1991a,b). Detailed and accurate record keeping is essential for proving ownership of intellectual property, such as copyrights and patents. Although this may sound strict to some, we believe that research records can be viewed as quasi-legal documents analogous to medical records, business inventories, or investment accounts.

Raw (or original) data are the records of direct or indirect observations of a test or experiment. Some observations involve the unaided senses, while others involve instruments or devices (Grinnell 1992). For example, when conducting an experiment on rodent maze-learning behavior, the raw data may be the records of the observations made with one's eyes (e.g., the rat completed the maze at a specific time). When testing a metal for electrical conductivity, the raw data would be a record of the output of an ammeter that could be connected to a computer for recording. Raw data, therefore, are those data drawn directly from the experiment or test: data recorded on a laboratory notebook from direct observations, recorder charts, field notes, machine tapes, computer printouts or disks, slides, photographs, and the like. It is important to note that modern researchers are moving away from laboratory notebooks toward computerized record keeping. Electronic records can enhance security and efficiency, but they also have potential problems for manipulation without a paper trail.

Researchers record the raw data in a data notebook or its equivalent, such as a computer disk, computer printout, or instrument output. The data notebook (or other document) is crucial to future review and to test the integrity and quality of the research output. A laboratory data notebook should be bound and the pages numbered consecutively. Loose-leaf notebooks are hazardous and may tempt a beginning researcher or technician to tear off pages with mistakes. All entries in the laboratory notebook should be made legibly with permanent, nonerasable ink. Ideally, entries should also be signed (or initialed) and dated. Researchers should draw a line through a mistaken entry, without making it completely illegible, and should not use correction fluid. Mistakes that can be traced can be valuable in assessing the progress of the project or observing new, unintended findings. All additive information directly relevant to the raw data, such as derived data, tables, calculations, or graphs, should be either done directly in the laboratory notebook or taped thoroughly on an adjacent page in the notebook. If this is not feasible, files can be used; providing clear identification of the data and the page where the data were derived from is essential (National Academy of Sciences 1994, Macrina 2005). Although many researchers take data with them when they change jobs, we strongly recommend that research institutions keep copies of all raw data while allowing individuals to have copies. Some universities follow the example of private industry and treat research data as the property of the institution. Keeping the data within the institution is important so that future interested parties can check the original data against derived data, graphs, or published results.

There are no excuses for not keeping accurate and informative records of methods and data. In the modern age, a large number of principal investigators are far removed in their research operation from these two steps. Therefore, investigators need to exert quality control and insist on proper documentation to assure that the integrity of the data is preserved. Moreover, the existence of an accurate paper trail may provide invaluable data for future discoveries. Some investigators may design detailed standard operating procedures (SOPs) to monitor the research protocol. Other investigators or supervisors may even add a blind quality assurance sample to ensure quality control and integrity of the process.

Some recent studies indicate that academic researchers are not doing a good job of record keeping. In a survey of 1,479 researchers funded by the National Institutes of Health (2007a), Martinson et al. (2005) found that the most prevalent (27.5%) self-reported inappropriate behavior was "inadequate record keeping." Moreover, one in ten had withheld details in publications, used inadequate experimental design, or dropped data. At 90 major research institutions, 38% of research integrity officers reported encountering problems with research records during misconduct inquiries and investigations, which often delayed investigations or made them impossible to complete (Wilson et al. 2007). In a survey conducted at the National Institute of Environmental Health Sciences (NIEHS), 31% of 243 researchers said that they had encountered poor record keeping at the NIEHS (Resnik 2006).

Raw data are usually manipulated (or processed) through many stages, depending on the type of research, before they are presented as graphs, charts, or tables or in a publishable form. As data are processed, the risk of introducing (intentional or unintentional) biases, adjustments, or errors increases (Grinnell 1992, Shamoo 1989, 1991a,b). Thus, it is important to include quality assurance steps to maintain the quality, objectivity, and integrity of derived data—data obtained, calculated, or derived from the raw data. Derived data appear in many forms, most commonly quantitative and qualitative data such as outputs from computer programs or instruments that process data, such as optical scanners, gas chromatographs, or DNA sequencing machines.

In addition to taking good care of raw and derived data, researchers should also act as good stewards of materials that are needed to carry out experiments or conduct research, such as chemical reagents, cells, bacteria, viruses, vectors, transgenic animals, tissue samples, blood samples, and many others. SOPs, computer software, and the like are not data but are an adjunct to the research protocol. However, they are needed to correctly carry out the proposed experiments and thus replicate the results.

Analyze the Data

The analysis of data in modern science involves the application of various statistical techniques, such as correlation, regression, analysis of variance (ANOVA), and chi-square tests. These techniques provide a way of drawing inductive inferences from data and distinguishing any real phenomena or effects from random fluctuations present in the data. A responsible researcher will make every attempt to draw an unbiased inference from a data set. Different disciplines use different statistical techniques, and statistical practices vary a great deal across different disciplines. Most fields have norms or accepted practices for data analysis, and it is prudent for researchers to follow the accepted norms (Resnik 2000). These norms are usually based on two factors: (a) the nature of the variables used (i.e., quantitative, comparative, or qualitative), and (b) assumptions about the population from which the data are drawn (e.g., random distribution, independence, sample size). There is nothing inherently unethical in the use of unconventional norms. It is important, however, to be forthright in clearly stating the method of analysis, why it is being used, and how it differs from others. It is unethical to fail to disclose some important information relevant to the data analysis (Resnik 2000).

Given the complexities of data analysis, it is easy to introduce biases or other errors in the analysis and misrepresent the data (Bailar 1986). The failure to provide an honest and accurate analysis of the data can have as significant an impact on research results as recording data improperly. Moreover, research indicates that statistical errors are fairly common in science (DeMets 1999). Thus, this step is crucial to ensuring the objectivity, integrity, and quality of research. Some aspects of data analysis that raise ethical concerns are excluding outliers, imputing data (i.e., using a statistical method to fill in missing data), editing data, analyzing databases for trends and patterns (or data mining), developing graphical representations of the data, and establishing the statistical and practical significance of the data. While none of these areas of data analysis is inherently deceptive, biased, or unethical, researchers must be sure to follow good statistical practices and honestly describe their statistical methods and assumptions to avoid errors in data analysis (American Statistical Association 1999). Intentionally misrepresenting the data can be regarded as a type of misconduct (Resnik 2000).

Manipulate the Data

When data are published in a journal article, report, or Web page, they are rarely published in raw form and are usually highly processed. Choices of how to present derived data, which portion, why, how, and to whom are all important scientific aspects of data manipulation. For example, a researcher may select a particular set of tables, figures, or spectral analysis, and not others. All of the data (presented in a publication or a report) are part of the supporting documentation of the published material. It is important to keep an adequate and accurate paper trail of data manipulation for future review and potential use (Shamoo 1989). This information is valuable not only to ensure the integrity of the process but also for the additional use of the data. For example, data from several sources may provide a new finding, or a new investigator may find a whole new way to interpret the data missed by the original investigator.

Computer programs, such as Photoshop, can be used to manipulate digital images (e.g., pictures of proteins from gel electrophoresis or cell structures). In some cases, researchers have manipulated images in order to deceptively change the image to obtain produce a desired result. Several journals, including *Science* and *Nature*, have special requirements for the submission of images to the journal for publication (Couzin 2006a). The Office of Research Integrity has special instructions on its Web site for forensic tools to detect fraud in images (Office of Research Integrity 2007b). Researchers should be aware of and use these tools, when necessary. While it is acceptable to use image manipulation technologies to make it easier for researchers to perceive patterns in an image, it is not acceptable to manipulate an image in order to mislead or deceive other researchers. The *Journal of Cell Biology* has adopted the following guidelines, which we endorse:

> No specific feature within an image may be enhanced, obscured, moved, removed, or introduced. The grouping of images from different parts of the same gel, or from different gels, fields, or exposures must be made explicit by the arrangement of the figure (i.e., using dividing lines) and in the text of the figure legend. If dividing lines are not included, they will be added by our production department, and this may result in production delays. Adjustments of brightness, contrast, or color balance are acceptable if they are applied to the whole image and as long as they do not obscure, eliminate, or misrepresent any information present in the original, including backgrounds. Without any background information, it is not possible to see exactly how much of the original gel is actually shown. Non-linear adjustments (e.g., changes to gamma settings) must be disclosed in the figure legend. All digital images in manuscripts accepted for publication will be scrutinized by our production department for any indication of improper manipulation. Questions raised by the production department will be referred to the Editors, who will request the original data from the authors for comparison to the prepared figures. If the original data cannot be produced, the acceptance of the manuscript may be revoked. Cases of deliberate misrepresentation of data will result in revocation of acceptance, and will be reported to the corresponding author's home institution or funding agency. (*Journal of Cell Biology* 2007)

Researchers should also develop the habit of stamping laboratory notebooks and other records to indicate who owns copyrights (or patents) and who is the principle investigator. Entries should be dated with perhaps the signature of a witness. Industry researchers already use such practices to help in future claims regarding issues of intellectual properties, such as copyrights and patents (discussed in chapter 9).

There are no set standards of how and where the data should be stored. Most researchers keep their data practically until they retire, or until they no longer have sufficient space to store them. Recently, the U.S. Public Health Service adopted the same storage time requirement for research data as for the financial records: three years from the time of last report filed to the federal agency. Keeping data for this short period of time is not justifiable, because publication of these data may take several years beyond the time of the last expenditure report. It is our view that data should be stored for at least seven years after the last expenditure report or submitted publication (Shamoo and Teaf 1990). In the event the federal agency is auditing and inquiring about the data, the time is automatically extended to the needed length of time. If the federal agency is investigating certain data, the agency has the right to obtain these data if they are available regardless when they were obtained in the first place. Few universities have provisions for storing data in centralized facilities. Furthermore, granting agencies have not provided the means to store data for future generations. We recommend that research institutions develop computerized data archives and that they require those researchers not working with human subjects to deposit data in these archives on an annual basis. (Researchers have an obligation to destroy some data pertaining to human subject's research if they contain an identifying trail, as discussed in chapter 12) Like any business keeping a centralized inventory of its stock, a research institution should keep a centralized inventory of its research.

Although it is important to store data, research materials, and other records, storage introduces problems of space allocation. Some of these problems can be handled by using technologies to transfer all types of data to digital formats. Data and other records can be stored on CD-ROMs, hard drives, servers, or other media, thus vastly reducing the space problem. However, converting data to digital formats introduces other problems, such as choosing a format that will not soon become obsolete or developing procedures for periodically transferring data to new formats. Most institutions may have trouble finding room for data storage, maintaining technologies, or transferring data to new media. Thus, while data storage is important, specific ethical dilemmas concerning the storage of data must be approached with an eye to other important values and concerns in research (Resnik 1998c). The federal government can and should provide funding to develop resources for data storage, such as databanks or archives. Additionally, research material cannot be easily converted into digital form. For example, cell or tissue samples are physical objects that take up space. Researchers who create banks for storing biological samples will face significant issues concerning space allocation.

Interpret the Data

If all researchers interpreted the data in the same way, science would be a dry and dull profession. But this is not the case. Many important and heated debates in science, such as research on firearm violence, studies of intelligence tests, and studies of global warming, involve disputes about the interpretation of data. Sometimes an important discovery or advance in science occurs as the result of a new interpretation of existing data. Of course, challenging a standard interpretation of the data is risky: Those who challenge the existing paradigm either go down in flames or win the Nobel Prize. Most challenges to the existing paradigm turn out to be wrong. But those few times that the new interpretation is correct can change and advance our knowledge in a revolutionary fashion. For example, Peter Mitchell won the Nobel Prize for his chemiosmotic theory. He advanced the notion that a proton gradient across the mitochondrial membrane is the driving force to synthesize adenosine triphosphate (ATP) from adenosine diphosphate (ADP) and inorganic phosphate. The chemiosmotic theory was originally considered heresy because it contradicted the long-held theory of a phosphorylated intermediate for the synthesis of ATP.

The path of a trailblazer is full of hazards. Most researchers resist new ideas and stick to generally accepted standards, despite their image as being open-minded and liberal. Although revolutions do occur in science, most research conforms to the model of "normal" science—science that falls within accepted standards, traditions, and procedures (Kuhn 1970). It is often the case that researchers who have new interpretations are scoffed at before their ideas are accepted. For example, the idea of continental drift was viewed as ludicrous, as was the idea that a bacterium could cause ulcers. However, if researchers can find new ways of interpreting data, they should be encouraged. And their new interpretations will be more readily accepted (or at least considered) if they properly acknowledge the existing paradigm (Resnik 1994).

Even within the existing paradigm, the interpretation of the same data can take very different pathways, none of which are likely to be unethical. As we discuss in chapter 8, there is an important distinction between misconduct and disagreement. Just because one researcher disagrees with another's interpretation does not mean that one of them is being dishonest. It is especially important for researchers with new interpretations to be even more careful in documenting and leaving a thorough paper trail of their data, so that other researchers will be able to understand their interpretations and not dismiss them as resulting from fraud or error. Ensuring the integrity of research data does not mean straitjacketing the investigator's creativity and latitude in introducing new ideas and interpretation. However, prudence suggests that all interpretations of data should be consistent with the existing knowledge. If the interpretation of new data is inconsistent with existing knowledge, an honest discussion of the differences is in order.

One common ethical problem with data interpretation is what we will call "overreaching." Researchers overreach when they claim that their data are more significant or important than they really are. This problem often occurs with industry-funded pharmaceutical research (Resnik 2007b). For example, suppose that a study shows that a new analgesic medication is 2% more effective at reducing arthritis pain compared to acetaminophen and 4% more effective than aspirin. However, the new medication also increases systolic and diastolic blood pressure by 10% in about 30% of the people who take it. Since its patent has not expired, the new medication will be much more expensive than acetaminophen or aspirin. The researchers would be overreaching if they claimed that the new medication is superior to acetaminophen and aspirin, because the medication brings a marginal improvement in pain relief and has some dangerous side effects. Overreaching can be an ethical problem in clinical research if it causes physicians to prescribe new medications to their patients without considering their higher costs or side effects (Angell 2004).

Put the Results into a Publishable Format

Putting results into a publishable format, such as a paper or conference presentation, is an important part of the research project, because this is how results are disseminated to the wider community of researchers. Research papers should provide readers with an honest, accurate, and thorough description of all the steps of the research project. Researchers should accurately report data and the contributions of other contributors. They should also disclose sources of funding or outside support as well as any potential conflicts of interest (other chapters cover these issues in more depth).

Most journals require authors to divide their papers into specific sections, such as abstract, introduction, materials and methods, results, discussion, and conclusion. The abstract is a short summary of the paper that reports its key findings. Because computer database programs for literature searches usually search for words found in abstracts, it is important for authors to write an accurate and useful abstract. In the introduction section, researchers usually review previous research in the area of the project, describe its aims and objectives, and discuss its importance. In the materials and methods section, authors describe the design of the tests or experiments as well as materials, methods, statistical techniques, and procedures used. In the results section, authors report processed and sometimes raw data as well as the results of the analysis of the data. In the discussion section, the authors may address a number of different issues relating to the research, such as placing the new data within the existing data in the literature; how and why they differ; potential biases, flaws, or shortcomings; difficulties in conducting the research; significance of the research and its relation to other studies; and areas for further research and exploration. The conclusion section summarizes all aspects of the research.

All papers and presentations should be clearly written. If the authors have some help from an editor or writer in preparing their publication, they should acknowledge this contribution.

Publish the Results

We discuss aspects of publication of results in more detail in chapter 7 on publication and peer review. For now, we simply note that researchers have an obligation to disseminate work for the obvious reason that science cannot advance unless researchers report and share results. Dissemination can include publication in peer-reviewed journals, monographs or other books, and Web pages, as well as presentations at professional meetings. The important ethical consideration is that research should be disseminated to colleagues and the public for scrutiny and review. Indeed, researchers who receive grants from the government or private funding agencies are usually required to specify a plan for disseminating their research in the grant proposal and to report to the agency about publications that result from the grant (Grinnell 1992). However, researchers who work for business and industry or the military often sign agreements to not publish results or to withhold publication until they obtain approval from management (Gibbs 1996, Blumenthal 1997). For instance, researchers working for the tobacco industry did not publish their work on nicotine's addictive properties for many years (Resnik 1998a). We explore these issues in chapter 10, as well. As noted in chapter 10, pharmaceutical companies have also suppressed data pertaining to their products. Officials at Merck, for example, knew that Vioxx was associated with cardiovascular problems, but they kept this information from the public for several years (Resnik 2007b).

Replicate the Results

Once the results are published, it is important for other researchers to be able to replicate the results. Although scientists do not frequently repeat each other's tests or experiments, because of an emphasis on original research, it is important that research be repeatable in principle. Repeatability is the primary assurance of the integrity of research data. Moreover, repeated confirmation of results by others lends greater credence that the data are usable, especially those data that can have an impact on the well-being of millions of people. The ability of other investigators to replicate the experiments by following the method in the published report is crucial to the advancement of science. It is important that the published work give sufficient details for others to replicate it. If there is not sufficient journal space to state the experimental details, several other means should be attempted to publish and provide the experimental details, such as mini-print appendixes, archives, and an invitation to the reader to ask for the full report from the investigator.

Share Data and Results

As noted above, openness is a key principle in research ethics. Scientists should share data and results (a) to promote the advancement of knowledge by making information publicly known; (b) to allow criticism and feedback as well as replication; (c) to build and maintain a culture of trust, cooperation, and collaboration among researchers; and (d) to build support from the public by demonstrating openness and trustworthiness. The ideal of openness is considered by many people, including many researchers, to be a fundamental part of research and scholarship. The real world of research does not usually conform to this ideal, however. Although researchers share data within the same team of collaborators working on a common project, they rarely share data with noncollaborators and often do not welcome requests to share data from other researchers in the field, much less people from outside the research community. The resistance to data sharing is especially high among researchers who have concerns about intellectual property, such as potential patents or trade secrets, but resistance is also high among researchers who want to protect their own interests in claiming priority (to be first) for discoveries or publishing original research.

Several recent studies have documented problems with data sharing in the biomedical science. In a survey by Campbell et al. (2002) of academic geneticists concerning their experiences with data withholding, 47% stated that at least one of their requests to share data or research materials related to published research had been denied in the last three years; 28% reported that they had been unable to confirm published research due to refusals to share data or materials; and 12% said that they had denied a request to share data or materials. Of those who refused to share data or materials, 80% said they refused because sharing required too much effort; 64% said they refused to share to protect someone else's ability to publish, and 53% to protect their own ability to publish (Campbell et al. 2002). A survey by Blumenthal et al. (2006) found that 32% of biomedical researchers had engaged in some type of data withholding in the last three years and that data withholding is common in the biomedical sciences.

Although refusals to share data and materials appear to be common, especially in biomedical sciences, some organizations have adopted policies that require researchers to share data and materials following publication. Many government granting agencies, such as the National Institutes of Health (NIH) and the National Science Foundation, encourage or require researchers to share data and materials. The NIH expects intramural and extramural researchers to share data:

> Data sharing promotes many goals of the NIH research endeavor. It is particularly important for unique data that cannot be readily replicated. Data sharing allows scientists to expedite the translation of research results into knowledge, products, and procedures to improve human health. There are many reasons to share data from NIH-supported studies. Sharing data reinforces open scientific inquiry, encourages diversity of analysis and opinion, promotes new research, makes possible the testing of new or alternative hypotheses and methods of analysis, supports studies on data collection methods and measurement, facilitates the education of new researchers, enables the exploration of topics not envisioned by the initial investigators, and permits the creation of new datasets when data from multiple sources are combined. In NIH's view, all data should be considered for data sharing. Data should be made as widely and freely available as possible while safeguarding the privacy of participants, and protecting confidential and proprietary data. (National Institutes of Health 2003)

The NIH also has policies that encourage or require funded researchers to share reagents and model organisms (e.g., transgenic animals). The NIH also requires researchers to state their plans to share data, reagents, or organisms in their grant applications or to explain any proposed restrictions on sharing (National Institutes of Health 1998b, 2003). Also, the NIH has a genomewide association studies (GWAS) policy that establishes an open repository for all GWAS data obtained with NIH funding (National Institutes of Health 2007c,d). There are additional guidelines for many other purposes such as intellectual property and stem cells.

Many scientific journals have also created policies that require researchers to share supporting data or materials as a condition of publication. Many journals have Web sites where researchers can deposit data and other supporting materials that do not appear in the published article. For example, *Science* requires researchers to share data: "*Science* supports the efforts of databases that aggregate published data for the use of the scientific community. Therefore, before publication, large data sets (including microarray data, protein or DNA sequences, and atomic coordinates or electron microscopy maps for macromolecular structures) must be deposited in an approved database and an accession number provided for inclusion in the published paper....After publication, all data necessary to understand, assess, and extend the conclusions of the manuscript must be available to any reader of *Science* (*Science* 2007)." *Science* also requires researchers to share materials following publication:

> After publication, all reasonable requests for materials must be fulfilled. A charge for time and materials involved in the transfer may be made. *Science* must be informed of any restrictions on sharing of materials [Materials Transfer Agreements or patents, e.g.] applying to materials used in the reported research. Any such restrictions should be indicated in the cover letter at the time of submission, and each individual author will be asked to reaffirm this on the Conditions of Acceptance forms that he or she executes at the time the final version of the manuscript is submitted. The nature of the restrictions should be noted in the paper. Unreasonable restrictions may preclude publication. (*Science* 2007)

While the progress of science thrives on sharing information and data and as soon as possible, there are some legitimate reasons to refuse to share data or materials, at least temporarily:

1. *To protect intellectual property claims.* Sometimes investigators are conducting research that may be patentable. Sharing data or other information related to the research prior to submitting a patent application can jeopardize the patent. Thus, researchers may refuse to share data to protect their potential patents. It is important for society to protect patent rights to stimulate invention and private investment in research and development, or R&D (Resnik 1998a, 2007b). We discuss intellectual property issues in more depth in chapter 9.

2. *To protect a researcher's interests in publishing articles from the data or materials.* If a researcher collects data or develops materials for a project, she should not have to share the data or materials until she is ready to publish, since sharing prior to publication may affect her ability to publish. But once a researcher has published, she has an obligation to share. A difficult question arises when a researcher has acquired a large database and hopes to publish a series of papers from the database. Should the researcher be required to share the whole database as soon as she publishes the first paper from it? If she must share the whole database with other investigators, this could jeopardize her

ability to publish other papers from it, since the other investigators might beat her to it. One way of handling this dilemma is to allow a researcher to publish a specific number of papers from her database before releasing the entire database to the public. Another solution is for researchers with databases to collaborate with other researchers when they share data, so that they both can receive publication credit. Difficult questions also can arise with sharing research materials, since sharing materials with others can jeopardize one's prospects of publishing articles based on those materials. Also, if the materials are in limited supply and cannot be recreated, then researchers must decide how to allocate the materials. For example, a blood sample is a limited quantity—once it has been used up, it is gone. To protect their own ability to use the sample in research, investigators need to decide carefully whom to share it with.

3. *To protect a researcher's reputation.* Researchers may not want to share data because they are not ready to present it to the public. They may need to do quality control checks on the data or analyze the data. A researcher may fear that his reputation could be damaged if he publishes data prematurely and there are problems with the data. Charles Darwin waited more than 20 years to publish his theory of natural selection so he could solidify the arguments and evidence in favor of the theory and anticipate objections.

4. *To protect confidential information pertaining to human subjects, personnel files, trade secrets, or national security.* We discuss the confidentiality of human subjects research in more depth in chapter 12. We note at this point, however, that it is often possible to share information pertaining to human subjects research without threatening confidentiality by removing personal identifiers from the data.

5. *To avoid wasting time, effort, and money.* Sometimes it takes a great deal of time, effort, or money to share data or materials with other researchers. There are significant costs with answering requests, shipping materials, taking care of animals, and synthesizing chemicals. One way of dealing with this problem is to deposit data on a public Web site or license a private company to make data or materials available to other researchers. Whenever data or materials are shared, a reasonable fee can be charged to cover the costs of sharing.

6. *To avoid being hassled by industry or political interest groups.* Sometimes industry representatives will request data in order to reanalyze the data or reinterpret the results. For example, if a study finds that exposure to a pesticide increases the risk of Parkinson's disease, the manufacturer of the pesticide might want to acquire the data to reanalyze it or challenge the study. Political interests groups, such as animal rights activists, may also request data or other information to harass or intimidate researchers. While these requests can sometimes be legitimate attempts to advance scientific knowledge, they often are not.

In the United States, if federally funded researchers refuse to share data (or other information), outside parties may still be able to obtain the data under the 1966 Freedom of Information Act (FOIA), as amended in 2002. FOIA allows the public to gain access to recorded information gathered or generated using federal funds, including scientific research records. To gain access to information under FOIA, one must send a request in writing to the head of the appropriate federal agency asking for the records that are sought. One must also specify the records being sought and explain why they are being sought. The agency should respond to this request within 20 days by sending the documents, promising to send the documents within a reasonable time, or explaining why they cannot be sent. The agency may charge a reasonable fee for sending the records sought. There are some exceptions to FOIA: Agencies can refuse to share records pertaining to national security or foreign relations, agency rules or practices, confidential business information, information related to personal privacy, some types of law enforcement records, and information pertaining to the supervision of financial institutions. Federal authorities have determined that some of these exceptions apply to federally funded scientific research. For example, researchers do not have to disclose confidential information pertaining to human subjects. They also do not have to disclose information protected by trade secrecy law, including information pertaining to potential patents (U.S. Department of Justice 2007).

Some scientists have objected to FOIA on the grounds that it could subject them to harassment from people who want to interfere with their work (Macilwain 1999). Although it is important for researchers to be free from harassment from industry representatives, political activists, or other parties, we do not think that researchers who receive public funds can be completely shielded from this threat. It is difficult to know in advance whether any particular request for information would be harassment of researchers. Without having this knowledge in advance, any policy short of answering all requests for data would be arbitrary and possibly biased.

CORRUPTING INFLUENCES IN THE CURRENT RESEARCH ENVIRONMENT

Although following the steps of the scientific method discussed above can help to promote the objectivity, quality, and integrity of research, a number of different psychological, social, economic, political, and institutional influences can undermine the research process. Objectivity, quality, and integrity are ideal standards that may be difficult to achieve in practice. The following influences and trends can interfere with and undermine research results and their application.

Pressure to Produce Results

Researchers who receive government funds face enormous financial and institutional pressures to produce results in order to obtain new funding, to continue receiving funding, or to publish papers. The phrase "publish or perish" accurately describes the life of the academic researcher. Researchers employed by private industry face similar pressures to produce results that provide useful information, although they may not face any pressure to publish. Because researchers in private industry who do not produce results can lose their jobs or their sources of funding, the phrase "produce or perish" is more apt. When the research plan is not going well or is taking longer than expected, researchers can be tempted to compromise the integrity of the research process in order to obtain results, which may result in bias, error, sloppiness, or even fraud (Broad and Wade 1982 [1993], Woolf 1986, Shamoo 1989, National Academy of Sciences 1992, Resnik 1998a). These pressures often fall most severely on junior researchers and graduate students, who may be under pressure to produce results for senior colleagues in order to advance their own careers or retain their jobs (Browing 1995).

Careerism

Research is no longer an avocation or hobby; it is a career. Career advancement usually results from many different types of achievements, such as publications, grants, intellectual property, special appointments, awards, and recognition. While the desire for a career in research can inspire dedication and hard work, it can also lead researchers to violate standards of ethics and integrity to achieve various goals along the path of career advancement (National Academy of Sciences 1992, 1994).

Conflicts of Interest

Conflicts of interest, financial or otherwise, can undermine trustworthiness and objectivity, which are required in virtually all steps of the scientific method, especially study design, data interpretation, and publication (Shamoo 1989, 1992, Krimsky 1996, 2003, Bok 2003, Angell 2004, Resnik 2007b). These issues are discussed in more depth in chapter 10.

Intellectual Property Interests

Researchers in the private or public sector frequently seek intellectual property rights, such as patents or copyrights (Krimsky 1996, 2003). To secure these rights, researchers may be tempted to violate standards of ethics and integrity. Some but not all attempts to secure intellectual property create conflicts of interest.

Complexity

Research projects often involve many different collaborators, institutions, disciplines, research sites, and sources of funding. It may be very difficult for any one person to understand or control an entire research project. This complexity can lead to problems in monitoring data, revising hypotheses or study aims, initiating tests or experiments, and so on. Complexity can also lead to communication problems among different members of the research team, and between the team and institutions, sponsoring organizations, and government agencies (Grinnell 1992).

Remoteness

The principal investigator or the manager of the project sometimes is far removed physically from the project itself and its various collaborators. More important, often the investigator relies on less knowledgeable individuals to carry out daily monitoring and discussions. This remoteness results in the investigator being mentally removed from the day-to-day operations of obtaining research data. Remoteness also results in poor supervision of those directly involved in data acquisition, analysis, and manipulation. According to several studies, poor supervision of subordinates is one of the chief causes of misconduct, error, and bias in research (National Academy of Sciences 1992). Because principal investigators may be so far removed from the day-to-day research operations, trust is a key component of ethical conduct: Principal investigators and managers must be able to trust subordinates, and vice versa.

Self-deception

Although self-criticism and skepticism play an important role in research planning and execution and are part of the ethos of science, scientists, like other people, succumb to self-deception (Broad and Wade 1982 [1993]). There are many different types of self-deception, including observer bias (where one sees what one wants to see), that can affect data collection, research design, and data analysis (Resnik 1998a). The steps of the scientific method are designed to counteract self-deception, but nothing can change the fact that research is conducted by human beings who often believe what they want to believe.

Political and Sociocultural Biases and Pressures

Many research projects address issues that are clouded with political and social controversy. Researchers on different sides of a controversial topic, such as gun control, global warming, tissue engineering, consumer safety, environmental management, homosexuality, or human intelligence, may have different stakes in the results of research. These pressures can undermine objective judgment and decision making in all phases of research (Longino 1990, Crossen 1994, Shrader-Frechette 1994, Resnik 1998a).

THE IMPORTANCE OF GOOD RESEARCH PRACTICES

To deal with corrupting influences and assure the quality, objectivity, and integrity of research data, it is important to promote ethical attitudes and good research practices (GRPs) among the participants in the enterprise (e.g., the principal investigator, technician, supervisor, and management) toward research integrity (Shamoo 2006). An ethical attitude embodies a positive orientation toward ethics, an awareness of ethical issues, and a commitment to ethics in research. A person with an ethical attitude wants to do the right things for the right reasons. GRPs (described below) are rules that researchers can follow to help ensure the quality, objectivity, and integrity of data (Shamoo 1989, 1991a, Shamoo and Davis 1989, Glick and Shamoo

1993, 1994). People in positions of leadership in research can play a key role in developing a culture in which ethical attitudes and GRPs prevail. If principal investigators, managers, corporations, and government agencies demonstrate and tolerate unethical attitudes and poor research practices, then unethical attitudes and poor research practices will prevail. The research culture (attitudes and behaviors) sets the tone for the importance of quality, objectivity, and integrity of data and results.

GRP Aims

The following are aims of GRPs (Shamoo 1991a).

Cardinal Rules

1. *Published data are verifiable:* A paper trail exists documenting the origin of the data.
2. *Published data are reproducible:* Other investigators are able to reproduce the data by following the published procedures.

Commonsense Rule The conclusions drawn from the published research data should be consistent with the data.

Procedures to Achieve GRP Objectives and Aims

The aims of GRPs can be achieved by numerous methods. The investigator should be more concerned with making a sincere effort to accomplish these aims than with following any proposed specific procedure. All proposed procedures should serve the GRP aims and not hinder them. Therefore, any proposed procedure that may result in a contrary outcome should be either modified or discarded.

Quality Control and Quality Assurance

Industries have used quality control and quality assurance processes for years (Shamoo 1991a,b). These concepts are based on common sense and good business practices. Quality control is concerned with developing rules and procedures designed to ensure that products and services during production are of high quality and conform to original specifications. Quality assurance is concerned with reviewing (or testing) the quality of products and services after they have been produced or used (Shamoo 1991a,b).

Unfortunately, these concepts are foreign to researchers, especially to those in academic institutions, but this should not be the case. Many in academic researchers loath the concepts of quality control and quality assurance because they associate them with increased bureaucracy and red tape. But we believe that these concepts can be adapted to the research environment without increasing red tape and bureaucracy and without interfering with the creative process (Shamoo 1989). A recent report by Harvey et al. (2007) found that implementing SOPs for data extraction in a surgical research protocol improved the completeness of data extraction from a range of 14–100% to 95–100%.

The following rules can help establish quality control in research:

1. All equipment used should have been calibrated within the time frame recommended by the manufacturer.
2. All equipment should have its SOP available.
3. All materials consumed during research experiments should be properly labeled with all of the essential information, such as date of manufacture, expiration date, concentration, pH, storage conditions (e.g., refrigeration, freezer, room temperature).
4. Documentation of procedures used to carry out the research protocol should be in its proper place. Documents may include lab notebooks, computer software, or automated instruments.
5. All research data should be promptly recorded, dated, and signed in a permanent manner.
6. Except for some types of human subjects research, all research data and records should be retained for a specified period of time (we recommend at least seven years).

The following rules, which pertain to actions performed after research is conducted, can help promote quality assurance:

1. Require researchers to keep original data at the institution.
2. Develop centralized data archives that can be used to audit or review data.
3. Examine research for potential conflicts of interest and disclose those conflicts to the appropriate parties (see chapter 10 on conflicts of interest).
4. Audit or peer review data every three years. The audit or peer review should be conducted by the research chief or director, an outside researcher within the research institution, or a researcher from outside the research institution.
5. File the results of data audit or peer review with the research institution and the agency sponsoring the research.
6. Verify in peer review/data audit all of the data, from the published data all the way to the original raw data.

The Potential Value of GRP and Peer Review/Data Audit

Implementing the rules listed above can help promote GRPs (Shamoo 1989, Glick and Shamoo 1994) and can also help protect the human and financial investment in research (Shamoo and Davis 1990). Peer review/data audit has some special advantages:

Reduction of Errors Human errors are a part of any human function, and the human factor has a large influence on the quality of data produced. Peer review/data audit can uncover errors both before and after those errors have contributed to large problems. Modern methods of online data processing certainly have reduced errors but have also contributed to the increased volume of data that requires human judgment, either before or after online data processing. Thus, the possibility of error in data analysis remains a significant factor (Resnik 2000).

Reduction of Irregularities Data audit can uncover both unintentional and intentional errors. Instituting peer review/data auditing itself may discourage such irregularities, especially those that are intentional. The amount of data to be fully audited can vary, depending on the preliminary analysis. A strong system of quality assurance can promote cost-effectiveness by ensuring accurate and efficient processing of data as well as reduce the cost of the peer review/data audit.

Reduction of Fraud and Potential for Fraud The existence of a systematic way to examine independently the data, or at least part of the data—or the thought that the data may be examined—may help reduce fraud and the potential for fraud. But peer review/data audit is by no means a foolproof method to eliminate fraud. If a researcher or an organization is intent on deception, there is as much potential for success as for fraudulent financial transactions. Peer reviewers/data auditors should therefore strive to reduce the ability of a person or an organization to succeed in fraud. But at the same time, peer reviewers/data auditors should be aware of their limitations—the system is not perfect.

Protection for the Public, Government, Industry, and Investigators All segments of society benefit from the peer review/data audit because it increases the likelihood of actually achieving worthwhile products. The greatest benefit of peer review/data audit falls to the public, which is the ultimate consumer of all goods and services of industry and government. After all, it is people who must suffer the consequences of poorly tested pharmaceuticals or poorly designed cars. Industry can protect itself from overzealous or dishonest employees as well as unintentional errors. The company's reputation as well as its actual services or products would be enhanced by peer review/data audit. Also, individual investigators can be protected from false accusations, rumors, and gossip.

Improvement of Products and Services Data audit/peer review not only can eliminate error but also can improve the overall quality of a product or service. Toxicological research, for example, is conducted by some in the chemical industry for certain products covered by a regulatory law. Furthermore, management uses these research data for the purpose of protecting the consumers and employees from these potentially toxic substances.

There are broad and general benefits to the consumers of the products and services. Industries such as pharmaceutical, chemical, petrochemical, and pesticide manufacturers, among others, may use and benefit from these procedures. For toxic substances that have specific federal regulations, the National Toxicology Program uses project management systems and quality assurance systems to monitor and evaluate the toxicological studies performed by the industry.

Establishment and Improvement of Future Research and Development Standards   The study and application of peer review/data audit are in their infancy. Peer review/data audit standards that individuals and organizations involved in research data production should adhere to and abide by have yet to be well established. One of the primary objectives of applying peer review/data audit in organizations would be to establish such standards, first by borrowing from the experience of finance and molding it to the world of research. Another objective would be to consistently review and revise these standards as needed.

Improvement of Risk Assessment   Risk assessment is an important tool that helps policy makers to determine regulations regarding such issues as nuclear safety, carcinogens, and pesticides. One of the critical factors involved in risk assessment is the use of the historical database collected from past research. Thus, the accuracy and reliability of these data become crucial for risk assessment. Peer review/data audit, by ensuring the accuracy and reliability of these data, can become an important partner in public health and safety.

Reduction of Liability   Reduction of legal liability is not an immediate or an obvious purpose of peer review/data audit, but it can be a side effect of that process. In the long run, peer review/data audit may contribute to a greater reliability of and confidence in the data in various segments, such as markets, courts, government, and insurance companies. For example, the increased confidence in the data could provide an incentive for insurance companies to reduce liability. Peer review/data audit could also strengthen the hand of industry in defending itself against unfair charges.

This concludes our discussion of data acquisition and management. Topics discussed in subsequent chapters, such as intellectual property and research misconduct, will have direct or indirect connections to the topic discussed in this chapter. In closing, we also mention that managing a large volume of data often requires tools that are often not available to every researcher. There is a need to develop these tools and make them available to all researchers (Anderson et al. 2007). The NIH has recognized this need and has encouraged centralized facilities such as bioinformatics, statistics, and imaging, among many others (National Institutes of Health 2005).

QUESTIONS FOR DISCUSSION

1. What are your thoughts about scientific research? In your opinion, what is the most crucial part of research?
2. How would you list the steps in carrying out of research? Are there some steps you could skip? Why?
3. How would you introduce quality assurance and integrity into your steps for carrying out research?
4. Can you give an example of how data can be "modified" to suit inappropriate goals in steps of research?

5. Give an example of an experimental design that would bias the data.
6. When would you be justified in refusing to share data?
7. How many of these GRPs do people follow (or fail to follow) in your research environment? Why?
8. Can scientific research incorporate quality control and quality assurance methods? Would this stifle creativity, or increase workload?
9. How do you ensure that peer review/data audit is workable, and how would you modify it to accomplish its purpose in your research environment? Can you suggest a whole new system to ensure the quality and integrity of research data?
10. How is a lab notebook like (or not like) a medical record?

CASES FOR DISCUSSION

Case 1

A medical student has a summer job with a faculty mentor at a research university. The student is bright, hard working, and industrious and hopes to publish a paper at the end of the summer. He is the son of a colleague of the mentor at a distant university. The student is working on a cancer cell line that requires three weeks to grow in order to test for the development of a specific antibody. His project plan is to identify the antibody by the end of the summer. The student has written a short paper describing his work.

The mentor went over the raw data and found that some of the data were written on pieces of yellow pads without clearly identifying from which experiment the data came. She also noticed that some of the experiments shown in the paper's table were repeated several times without an explanation as to why. The mentor was not happy about the data or the paper, but she likes the student and does not want to discourage him from a potential career in research.

- What is the primary responsibility of the mentor?
- Should the mentor write a short paper and send it for publication?
- Should the student write a short paper and send it for publication?
- If you were the mentor, what would you do?
- Should the mentor or her representative have paid more attention to the student's work during the course of the summer? Should the student been taught some quality control and/or GRP during the summer?

Case 2

A graduate student at a research university finished her dissertation and graduated with honors. Her mentor gave the continuation of the project to a new graduate student. As usual, the mentor gave the entire laboratory notebook (or computer disk) to the new graduate student, who had to repeat the isolation of the newly discovered chemical entity with high-pressure liquid chromatography (HPLC) in order to follow up the chemical and physical characterization of the new compound.

The new graduate student found that if he follows the exact method described in the laboratory notebooks and published by the previous student, he obtains the new chemical entity not at the same HPLC location as published, but slightly shifted to the left, and there was a different peak at the location stated. However, the new student discovered that if the ionic strength is doubled, he could find the same chemical at the same location in accordance with the previous student's dissertation. The new student discussed with the mentor how he should proceed. The mentor replied, "Why make a fuss about it? Just proceed with your slightly different method and we can move on."

- What are the responsibilities of the new student? Should the new student refuse to accommodate the mentor's request?
- Should the new student have read more thoroughly the relevant laboratory notebooks prior to starting the experiment? Should there have been a paper trail of the error in the laboratory notebook? Do you think the error was intentional, and does it matter?
- If the laboratory notebook does not reveal the error, is it then misconduct? Does it indicate that a better recording of the data would have been helpful?
- Should the mentor have had training and education in responsible conduct of research (RCR)? If the mentor has had the necessary training in RCR, what actions would you suggest?
- Can you propose a reasonable resolution to the problem?

Case 3

A new postdoctoral fellow in a genetic research laboratory must sequence a 4-kDa fragment. After the sequence, he is to prepare a 200-base unit to use as a potential regulator of a DNA-related enzyme. The 4-kDa fragment is suspected to contain the 200-base unit. The sequence of the 200-base unit is already known in the literature, but not as part of the 4-kDa fragment, and not as a potential regulator. The fact that the 200-base unit is known is what gave the mentor the idea that it may have a functional role.

The new postdoctoral fellow tried for three months to sequence the 4-kDa fragment, without success, and so simply proceeded to synthesize the 200-base unit without locating it within the fragment. After two years of research, the 4-kDa fragment appeared to play a key regulatory role in an important discovery, but at this time the mentor learned that the postdoc never sequenced the original 4-kDa fragment. The mentor could never find a "good" record of the attempts to sequence the 4-kDa fragment.

- What impression do you gather about how this mentor runs the laboratory?
- Should there be records of sequence attempts of the 4-kDa fragment?

- How should the mentor proceed?
- If you were the new postdoc, what steps you would take to ensure proper records of your work?

Case 4

A graduate student prepared for her thesis a table showing that a toxic substance inhibits an enzyme's activity by about 20%. She has done only six experiments. The mentor looked at the data and found that one of the data points showed a stimulation of 20% and that this point is the one that skewed the results to a low level of inhibition with a large standard of deviation. The mentor further determined with the student that the outlier is outside the mean by 2.1 times the standard derivation and that it is reasonable not to include it with the rest of the data. This would make the inhibition about 30% and thus make the potential paper more in line with other research results and hence more "respectable." The mentor instructed the student to do so.

- Should the student simply proceed with the mentor's instructions?
- Should the mentor have been more specific regarding what to do with the outlier? In what way?
- Can you propose a resolution? Should the outlier be mentioned in the paper?
- How should this laboratory handle similar issues in the future? Should each laboratory have an agreed-upon SOP for such a statistical issue?

Case 5

A social scientist is conducting an anonymous survey of college students on their opinions on various academic integrity issues. The survey is administered in four different sections of an introduction to sociology. The survey includes 20 questions in which respondents can use a Likert scale to answer various questions: 1 = strongly agree, 2 = agree, 3 = neither agree nor disagree, 4 = disagree, and 5 = strongly disagree. The survey also includes 10 open-ended questions that ask for respondents to state their opinions or attitudes. The social scientist distributes 480 surveys and 320 students respond. A graduate student is helping the social scientist compile the survey data. When examining the surveys, the student encounters some problems. First, it appears that eight surveys are practical jokes. The persons filling out these surveys wrote obscene comments and for many questions added extra numbers to the Likert scale. Although some of the 20 Likert-scale questions in these surveys appear to be usable, others are not. Second, in 35 surveys, the respondents appeared to have misunderstood the instructions on how to use the Likert scale. They answered "5" on questions where it would seem that "1" would be the most logical answer, given their answers to other Likert-scale questions and their written comments. Third, on 29 surveys, the respondents wrote their names on the survey, when they were instructed not to do so.

- How should the researchers deal with these issues with their data?
- Should they try to edit/fix surveys that have problems?
- Should they throw away any surveys? Which ones?
- How might their decisions concerning the disposition of these surveys affect their overall results?

Case 6

A pharmaceutical company conducts five different phase I studies on a new drug to establish its safety in healthy individuals. Three of these studies had a p-value < 0.05, indicating significant results; two had a p-value > 0.05, indicating nonsignificant results. As it so happens, undesirable side effects were observed in one of the studies with the nonsignificant results. None of the studies with significant results had a significant proportion of side effects. The researchers report these results to the U.S. Food and Drug Administration and in a publication, but they do not mention the nonsignificant results.

- Is there an ethical responsibility to report all of the data? Would it make a difference if the subjects were not human (i.e., animals)?
- What are the responsibilities of the researchers to this company, to themselves, and to society?
- Should there be a federal mandate to report all side effects?
- How could the GRP help in this case? Would a quality assurance program be helpful?

Case 7

A graduate student is planning to write a thesis on the affects of exercise in managing diabetes in dogs. He is planning to do a trial with dogs with diabetes, control matched with respect to age, breed, sex, and other factors. One group will receive no extra exercise; another group will receive 30 minutes of exercise a day; and another group will receive two 30-minute periods of exercise per day. He will measure blood sugar levels in all of these dogs, as well as the quantities of drugs used to control diabetes.

The graduate student is also conducting a literature review on the subject as a background to his research. In conducting this review, he searches various computer databases, such as the Science Citation Index and MEDLINE, for the past five years. He gathers many abstracts and papers. For much of the research, he reads only abstracts and not the full papers. Also, he does not include some of the important work on diabetes in dogs that took place more than five years ago.

- Should the graduate student read articles, not just abstracts?
- If he cites an article in a publication or in his thesis, should he read the full article?
- If he cites a book, should he read the full book or only the part that he uses?
- Should the graduate student include articles published more than five years ago?

Case 8

Two physicians in a department of emergency medicine reviewed the medical records of all patients admitted for diabetic crisis to the hospital in the last five years. They correlated several different variables, such as length of hospital stay, survival, cost of care, rehospitalization, age, race, and family history. If they found any statistically significant trends (p-values ≤ 0.05), they planned to publish those results, and they planned to ask a statistician to help them determine if any of their correlations are significant.

- Do you see any potential methodological or ethical problems with this study design?
- Does it conform to the rules described in this chapter?
- Would it be unethical for the researchers to publish their results if they do not make it clear in the paper that they have mined the data for significant trends?
- Would you be more confident in their research if they had a hypothesis prior to conducting the study?

Case 9

A pesticide company conducts a small study (25 human subjects) on the adverse effects associated with one of its products. In the study, the researchers observe the subjects for 48 hours after they ingest a trace amount of the pesticide. The amount of pesticide ingested is barely large enough to show up in a urine test. The researchers measure vital signs and collect blood and urine samples. They perform chemical tests on the blood and urine to measure biomarkers associated with toxicity and damage to various body tissues (e.g., heart, liver, kidney). At the end of the study, the researchers find that the pesticide produced no measurable adverse effects at the dose level administered to the subjects. The company plans to submit data from the study to the U.S. Environmental Protection Agency (EPA) to argue that the EPA should allow pesticide applicators to increase the amount of the chemical they apply to fruit trees. What are some of the statistical, methodological, and ethical issues in this case?

Case 10

Dr. Reno is a junior investigator who has just received her first major NIH grant. She has used NIH funds to create a transgenic mouse model to study depression. The mouse has a genetic defect the leads to an underproduction of serotonin. She has used the model to show how a compound found in an herbal medicine increases serotonin levels in mice and also produces effects associated with normal (i.e., nondepressed) behavior, such as normal levels of exercise and normal sleep patterns. Dr. Reno applies for a patent on this compound with the intent of eventually testing it on human subjects. She also publishes a paper in a high-impact journal describing her work with the mouse model. Almost immediately after the paper appears in print, she receives dozens of requests from researchers who would like to use her transgenic mice in their own research. Dr. Reno is flattered but also overwhelmed. She has barely enough mice for her own work, and she doesn't want to turn her laboratory into a factory for producing mice for someone else's research.

- How should Dr. Reno deal with these requests to share transgenic mice?
- Does she have an obligation to the share the mice?

Case 11

Drs. Kessenbaum and Wilcox are conducting a long-term, observational study of the health of pesticide applicators. The protocol calls for an initial health assessment, including a health history, physical exam, and blood and urine tests. The researchers will also collect a DNA sample from cheek scrapings and collect dust samples from their clothing and hair and underneath their fingernails. After the initial health assessment, the applicators will complete yearly health surveys and undergo a full health assessment every four years. The researchers will follow the subjects for at least 25 years. Their work is funded by the NIH. Drs. Kessenbaum and Wilcox have been conducting their study for 15 years, and they have compiled an impressive database. They have already published more than a dozen papers from the database. Whenever they share data, they require researchers who request it to sign elaborate data-sharing agreements, which spell out clearly how the data will be used. In the past month, they have received some requests to access their database. One request has come from a pesticide company, another has come from a competing research team also studying the health of pesticide applicators, and another has come from a radical environmental group with an anti-pesticide agenda.

- How should Drs. Kessenbaum and Wilcox handle these requests to access their database?
- Is it ethical to require people who request data to sign elaborate data sharing agreements?

4

Mentoring and Collaboration

> Mentoring and collaboration are cornerstones of modern science, but they also raise some ethical problems and dilemmas. This chapter explores a variety of issues related to mentoring and collaboration, including such moral dimensions as proper training, setting examples, trust, accountability, and collegiality. The chapter also considers some reasons that collaborations sometimes fail in research, addresses policies designed to promote collaboration and effective mentoring, and discusses leadership in mentoring, collaboration, and other professional relationships in science.

The institution of mentoring traces its history back to ancient Greece. The word "mentor" comes from the name of a man who was the adviser to King Odysseus and teacher to Telemachus in Homer's *Odyssey*. Mentor provided education and moral guidance to his students. Following this model, a mentor was an older male teacher who had a close relationship with an adolescent young man. Socrates, the father of philosophy, mentored Plato and many other students in Athens. Plato created his own school and mentored many students, including Aristotle, who made important contributions to philosophy, logic, politics, literary theory, and science. Today, mentoring is a very important component of the research enterprise. In academic institutions, mentors not only transmit knowledge and skills to students but also teach attitudes, traditions, values, and other things that cannot be learned in formal courses. Mentoring plays a major role in shaping individuals.

Collaboration was also important in ancient science, but it is even more so today. Although many people still retain the image of the isolated researcher toiling away in the laboratory, modern science is a highly social activity. Archimedes, Galileo, Newton, Harvey, and Mendel managed to do a great deal of work without collaborating significantly, but today's scientists often must work with many different colleagues. Researchers share data, databases, ideas, equipment, computers, methods, reagents, cell lines, research sites, personnel, and many other technical and human resources. Researchers collaborate and have professional relationships with people from different departments, institutions, disciplines, and nationalities. Collaborators may include graduate students, postdoctoral students, junior and senior researchers, basic scientists, and clinical researchers. Some research projects, such as the Human Genome Project, involve thousands of researchers from dozens of disciplines working in many different countries (Grinnell 1992, Macrina 2005). The National Science Foundation established six science and technology centers across the United States with up to $20 million for five years (Mervis 2002). The centers were controversial at first because 143 entrants were narrowed down to six. However, all agree that they have been very successful ventures for collaborative research. The National Institutes of Health has also established centers of excellence to promote collaboration in research.

Successful professional relationships and collaborations in science cannot occur without a high level of cooperation, trust, collegiality, fairness, competence, and accountability. It is important to have competence at all levels of science to facilitate research and engender public confidence (McCallin 2006). Cooperation is essential to collaboration, because collaborators must share data and resources and coordinate activities, experiments, and tests. Trust is important in collaboration because researchers need to trust that their collaborators will keep the agreements, perform according to expectations, will not lie, and so on (Whitbeck 1998). Many different factors can undermine trust, including selfishness, incompetence, negligence, unfairness, careerism, and conflicts of interest.

Collegiality, one of sociologist Robert Merton's (1973) four norms of science, is important in maintaining a social environment that promotes cooperation and trust: Researchers who treat one another as colleagues are more likely to trust one another and to cooperate. The norm of collegiality requires researchers to treat each other with the respect accorded to a friend or ally in pursuit of a common goal. Colleagues not only help one another but also provide constructive criticism. Behaviors that can undermine collegiality include harassment (sexual or otherwise); racial, ethnic, or sexual discrimination; verbal abuse; personal grudges; theft; and jealousy.

Fairness is important in collaboration because collaborators want to ensure that they receive a fair share of the rewards of research, such as authorship or intellectual property rights, and that they are not unfairly burdened with some of the more tedious or unpleasant aspects of research (Resnik 1998a,c). Although research is a cooperative activity, researchers still retain their individual interests, and they expect that those interests will be treated fairly. Plagiarism and undeserved authorship are extreme violations of fairness. Fairness is also very important in other issues relating to collaboration, such as peer review and personnel decisions.

Last but not least, accountability is important in collaboration because when many people work together, it is especially important to know who can be held accountable for the successes or the failures of a project (Rennie et al. 1997). Modern research involves a division of intellectual labor: Different people do different jobs in research, such as designing experiments, gathering data, analyzing data, and writing papers (Kitcher 1993). Many research projects encompass research techniques, methods, and disciplines so different from each other that no one person can be responsible for or even knowledgeable about all the different aspects of the project. For example, the study of a new DNA vaccine may involve enzyme kinetics, X-ray crystallography, reverse transcriptase, electron micrographs, recombinant DNA, polymerase chain reactions, clinical trials, pharmacology, microbiology, immunology, and statistical analysis.

Because so many people are doing different jobs, it is often hard to keep track of all of these different laborers, the standards governing their work, and the products of their labor. Leading a research team can be in some ways like managing a small organization. Like any organization, problems relating to communication and supervision can occur in scientific research. Indeed, many of the problems related to misconduct in science often boil down to poor communication and supervision (Broad and Wade 1982 [1993], LaFollette 1992, National Academy of Sciences 1992).

In any research project, different people need to be held accountable for different parts of the project as well as the whole project itself. Accountability is often confused with responsibility, but we distinguish between these two concepts (Davis 1995b). People are accountable for an action if they have the obligation to give an account of the action. People are responsible for an action if they deserve to be praised or blamed for the action. People may be held accountable even if they are not responsible. For example, if a 14-year-old boy throws a rock through a window, he is responsible for throwing the rock, but his parents may be held accountable for his actions. In a large research project with many collaborators, different people may be held accountable for the project as a whole or its various parts. A person who was not responsible for some aspect of the project, such as recording data, may still be held accountable for the project as a whole. Clearly, accountability and responsibility are both important in research, but it is also important to keep these notions distinct.

RECOMMENDATIONS FOR RESEARCH COLLABORATION

To help prevent ethical problems from arising during collaboration, we recommend that collaborators discuss important research issues before and during their collaborative work:

1. *Extent of the collaboration:* How much of a commitment of time, effort, and money is involved in the collaboration?
2. *Roles and responsibilities:* What is the role of each collaborator? Who is responsible for different parts of the collaboration?
3. *Funding:* How will the project be funded? Who will apply for funding?
4. *Conflicts of interest:* Are there any financial or other interests affecting the research that collaborators should know about?
5. *Resource sharing:* How will data and other resources be managed and shared? Will data be kept confidential? For how long? Will data be placed on a publicly available database?
6. *Dissemination of results:* How and where will the research be published or disseminated? Is there a plan for discussing research results with the media or the public?
7. *Deadlines:* Are there any institutional, funding, or other deadlines relevant to the project?

8. *Regulations:* What regulations pertain to the project?
9. *Authorship and publication:* Who will be an author? Who will receive acknowledgment? When will publication take place? Where will results be published? Who will write the paper? Who will be the corresponding author?
10. *Ownership of data:* Who will own the data? What type of ownership will they have? What portion of the data will they own?
11. *Intellectual property rights:* Who will have intellectual property rights (patents, copyrights)? Will a patent be applied for? Who will apply for the patent?
12. *Data access:* Who will have access to the data? Under what terms?
13. *Closure:* What will bring the collaboration to a formal end?

Collaborative research raises a number of different ethical issues that we cover in other chapters in this book, such as authorship, intellectual property, and interactions between academic institutions and private industry. In this chapter, we focus on only two issues related to collaboration: mentoring and leadership.

MENTORING

Mentoring is a very important part of scientific education and training and is the social foundation of research. Although mentors provide students with knowledge and information as well as wisdom and advice, mentors are role models, so they also teach their students by example. Most people have different mentors at different times for different reasons. A mentor could be a family member, a pastor, a coach, a friend, a teacher, a business leader, a policeman, or anyone a student knows and admires. In science, a mentor is usually a senior researcher who supervises a number of different graduate students. Usually students' graduate advisers or thesis advisers are also their mentors, but many research students obtain mentoring from senior researchers who have no formal advising responsibilities (Weil and Arzbaecher 1997). Many students consider more than one person to be their mentor, but unfortunately, some students have no one whom they would consider a mentor. Students who benefit from mentoring include graduate and undergraduate students, postdoctoral students, and even junior-level professors (National Academy of Sciences 1992, 1997, National Institutes of Health 2000).

Mentors interact with their students in many ways. Some of the most important activities include the following:

- *Teaching students how to do research:* Mentors help students learn the techniques, methods, and traditions of research. They show students how to design experiments, conduct experiments, collect and record data, analyze data, and write up results.
- *Critiquing and supporting students' research and teaching:* Mentors read students' lab notebooks, research protocols, and manuscripts (Macrina 2005) and scrutinize students' research design and data analysis. They

may attend classes that the students teach, read students' evaluations of teaching, or provide feedback on teaching style and technique. Although it is very important for mentors to criticize students, they also need to offer support and encouragement, and they need to carefully tread the line between constructive and destructive criticism. Mentors need to guard against discrimination, favoritism, and excessively high (or low) expectations when critiquing students.

- *Promoting their students' careers:* Mentors help students to enter the job market. They help students find jobs and submit job applications, they write letters of recommendation, and they help students prepare for job interviews (Macrina 2005).

- *Helping students understand the legal, social, and financial aspects of research:* Mentors teach their students about research rules and regulations, such as animal care and use regulations, human experimentation regulations, and biological and radiation safety regulations. They also help students understand the social structure of the research environment, including relationships with colleagues, students, administrators, funding agencies, and the public. They help students understand the funding of research and help them write grant applications and obtain scholarships and fellowships.

- *Teaching students about research ethics:* Mentors play an important role in teaching students about the ethical aspects of research (National Academy of Sciences 1992, Weil and Arzbaecher 1997, Swazey and Bird 1997, Macrina 2005). Such teaching may involve didactic lectures, workshops, and seminars as well as informal discussions. Most important, mentors need to provide students with a good example of how to behave ethically in research, because students learn a great deal of ethics by example.

- *Involvement in students' personal lives:* Although mentors should maintain professional distance from their students, they should not ignore their students' personal lives. For example, if mentors are aware of their students' psychological, medical, or legal problems, mentors should help their students find the proper resources or help. Mentors can listen to their students' problems and support them in difficult circumstances. Even though it is important for mentors to be aware of their students' personal life, judgment and discretion should be used so that the relationships remain professional and do not become too personal. Mentors should avoid becoming too involved in their students' personal lives so that they can maintain a measure of objectivity and fairness.

The list above shows that mentors perform many important duties for their students. Mentors are more than mere teachers: They are also advisers, counselors, and often friends. Because students usually also work for mentors as teaching and research assistants, mentors also serve as employers and supervisors. These different roles may sometimes conflict. For instance, mentors may give students so much work to do that they do not have adequate time for their own research. In these cases, a mentor's role of employer/supervisor conflicts with the role of teacher. Or a mentor may believe that it is in the student's best interests to transfer to a different university to work with someone who has more expertise in that student's chosen area of research, but yet may hesitate to convey this advice to the student if the mentor needs him as a research or teaching assistant.

Steiner et al. (2004) surveyed 139 primary care fellows of the National Research Service Award from 1988 through 1997 regarding their subsequent career development and research productivity. The fellows indicated whether during the fellowship they had no sustained and influential mentorship, influential but not sustained mentorship, or influential and sustained mentorship. Steiner et al. found that those with sustained and influential mentorship were more engaged in research, were publishing more often, were more likely to be a principal investigator on a grant, and were more likely to provide good mentorship to others.

To understand the ethical dimensions of the mentor–student relationship, it is important to realize that mentors have more power, experience, knowledge, and expertise than their students (Weil and Arzbaecher 1997, Macrina 2005). Students also depend on their mentors for education, training, advice, and often employment. Mentors may also depend on students for assistance in teaching or research, but students are far more dependent on mentors than vice versa. Given their minimal power, experience, knowledge, and expertise and the high degree of dependency, students are highly vulnerable. It is very easy for mentors to manipulate, control, or exploit their students, because students often may be unable to prevent or avoid such abuses of power. Thus, the mentor–student relationship resembles other professional relationships where one party is highly vulnerable, such as the doctor–patient relationship and the lawyer–client relationship. These relationships are sometimes called fiduciary relationships because the powerful party is entrusted with protecting the interests of the vulnerable party. The powerful party has a variety of duties toward the vulnerable party, including beneficence, nonmaleficence, confidentiality, respect, and justice (Bayles 1988).

Although it should be fairly obvious that mentors have a variety of ethical duties to their students, ethical problems and dilemmas can still arise in this relationship. Unfortunately, various forms of exploitation are fairly common in mentoring. Mentors sometimes do not protect their students from harm or treat them fairly. For instance, mentors often do not give students proper credit for their work. They may fail to give students acknowledgments in papers or include them as coauthors (Banoub-Baddour and Gien 1991). They may fail to list students as first authors when students make the most important contribution to the research. In some of the more egregious cases, mentors have stolen ideas from their students without giving them any credit at all (Marshall 1999d, 2000b, Dreyfuss 2000). One well-known case of this type of exploitation involved the famous scientist Robert Millikan and his student Harvey Fletcher. Millikan was conducting experiments to measure the minimum electrostatic charge, the charge of an electron. In his experiments, he dropped water droplets through charged plates. By comparing the rate of fall of the droplets without charged plates with their rate of fall through the plates, Millikan would be able to determine the electrostatic force on the droplets and therefore the minimum charges. His experiment was not working well, and Fletcher suggested that Millikan use oil droplets instead of water droplets. Millikan took this advice and ended up winning the Nobel Prize in 1923 for the discovery of the charge of an electron. However, Millikan did not acknowledge Fletcher's contribution in his paper describing these experiments (Holton 1978).

Mentors may also overwork their students by assigning them too many experiments to run, too many papers to grade, too many undergraduate students to tutor, and so on. If students are assigned too much work, they will not have enough time for their own education and research. In recent years, graduate students have formed unions to deal with poor working conditions. Postdoctoral students often face especially demanding and exploitative working conditions. They are usually nontenured researchers who are paid through "soft money," that is, money from research grants. Postdoctoral students are paid much less than regular faculty members even though they have doctoral degrees and often do just as much research or teaching. They also do not receive the usual benefits package (e.g., health insurance), and they have little job security (Barinaga 2000). Although some postdoctoral students enjoy their work, others feel mistreated or exploited. Given their vulnerability, it is very hard for these students to complain about working conditions or about their mentors, because they face the real threat of retaliation. For example, a mentor could refuse to work with the student any longer, recommend that the student be expelled from the program, or encourage his colleagues not to work with the student.

Other examples of ways in which mentors may abuse their students include the following:

- Giving students misinformation or poor advice
- Intimidating or harassing students
- Discriminating against students
- Showing favoritism to one of more students
- Failing to help students advance their careers
- Not recognizing when students are having psychological troubles that require counseling

Given the importance of the mentor–student relationship for scientific research, and the kinds of problems that routinely arise, many universities and professional organizations have developed programs and policies aimed at improving mentoring (National Academy of Sciences 1997, National Institutes of Health 2000). Some of these policies include the following:

1. *Rewarding mentors for effective mentoring:* Most universities do not emphasize or even consider mentoring skills when they review

faculty for hiring and promotion, but this needs to change if we want to improve mentoring (Djerassi 1999).

2. *Providing mentors with enough time for mentoring:* Professors who do not have adequate time for mentoring will do a poor job of mentoring. Professors who have heavy mentoring responsibilities should be released from other administrative or teaching obligations.

3. *Developing clear rules concerning workloads, teaching duties, research opportunities, authorship, time commitments, and intellectual property:* Many of the problems that occur in mentoring are due to poor communication. Communication can be improved by clearly defining expectations and obligations (Macrina 2005).

4. *Establishing procedures and channels for evaluating mentoring and for allowing students and mentors to voice their grievances.*

5. *Ensuring that students who "blow the whistle" on mentors are protected:* A whistle-blower is someone who reports unethical or illegal conduct. Whistle-blowers often face retaliation. To avoid this, whistle-blowers must be protected. We discuss this recommendation again in chapter 8 on scientific misconduct.

6. *Promoting a psychologically safe work environment:* Students and mentors both need to have an environment that is free from sexual, religious, ethnic, and other forms of harassment (National Academy of Sciences 1992). Sexual harassment is unethical and can also be illegal. Although most researchers agree on the need to protect students and others from sexual harassment, there are disputes about the definition of sexual harassment as well as the proper response to sexual harassment (Swisher 1995). For further discussion, see Resnik (1998a).

7. *Promoting a nondiscriminatory work environment:* Racial, ethnic, sexual, religious, and other types of discrimination are also unethical and often illegal. Women have for many years labored under the yoke of sex discrimination in science. Although women have made significant gains in some sciences, such as anthropology, biology, and medicine, women are still vastly underrepresented in engineering and physical science. Racial and ethnic discrimination continue to be a problem in science as more minorities enter the workplace (Johnson 1993, Manning 1998). Although African Americans have historically been the most frequent victims of discrimination, Asian Americans also experience discrimination (Lawler 2000). Scientists should be judged by the quality of their research, education, and character, not by the color of their skin, their national origin, their religious views, or their gender. Effective mentoring cannot take place when discrimination affects the laboratory (for further discussion, see Resnik 1998a).

8. *Promoting a diverse workforce in research:* Because mentors serve as role models as well as advisers and friends, one could argue that it is important to promote diversity in science in order to enhance

mentoring and education. Science students have different gender, racial, ethnic, and religious characteristics. The scientific workforce should reflect this diversity so that students can benefit from having role models with whom they can identify (Mervis 1999, Holden 2000). An excellent way to promote the effective mentoring of women in science is to hire and promote more women scientists (Etzkowitz et al. 1994), which will also encourage more women to enter science. This same "diversity" argument also applies to racial and ethnic diversity, which raises the question of affirmative action in science: Should hiring and promotion of scientists be decided based on racial or ethnic features of a person? This is a complex legal, moral, and political question that we do not explore in depth here. We favor a weak form of affirmative action that increases the diversity of the workforce without compromising quality. Racial, ethnic, and gender considerations should be treated as one factor among many that can be used to enhance diversity. Other factors might include geography, socioeconomic status, and life experiences. Affirmative action should not be used to promote incompetence or tokenism (Resnik 2005).

ETHICAL LEADERSHIP

Leadership is an important aspect of mentoring, collaboration, laboratory management, and research administration. The leading psychologist/leadership scholar Fred Fiedler (1967) studied the effectiveness of leaders who are task motivators as compared with relationship-based motivators. The task motivator emphasizes to his subordinates the importance of the task at hand to the organization. The relationship-based motivator emphasizes the importance of building personal relationship with his subordinates in order to achieve the task. Fiedler found that the task motivator was more effective in four out of eight leadership characteristics and that the relationship-based motivator was more effective in the other four leadership characteristics. The contingency model of leadership was born from combining the several types of leadership as thus described by Fiedler (1967). Vroom and Jago (2007, p. 18) define leadership as "a process of motivating people to work together collaboratively to accomplish great things." These two authors dismiss the need for the moral component in the definition of leadership. Wong (2006) found that the most common characteristics of scientists who became leaders of research in key industries are determination, drive and diligence, passion, broad experience, flexibility, and inspiration. Other characteristics of leadership include vision, initiative, realism, and ethics (Koestenbaum 1991, Gardner 1995).

We think that ethics is one of the most important characteristic of leadership. Many leaders who lacked ethics, such as Hitler, Genghis Khan, Stalin, and Pol Pot, have wreaked havoc on our civilization. Virtuous or ethical leaders, such as Gandhi, Roosevelt, Martin Luther King, and Mother Teresa, have brought much good to society. Ethical leadership does not imply perfection—even ethical leaders make mistakes. But ethical leaders respond to mistakes by admitting them, learning from them, and taking steps to amend them, not by denying them or covering them up. Placing more emphasis on ethics in scientific leadership will be good for science and society. We therefore adapt Vroom and Jago's (2007) definition of leadership for our own:

> Leadership is a process of motivating people ethically to work together collaboratively to accomplish a given task.

Most research endeavors are accomplished through the exercise of leadership by a researcher/manager/boss/leader. If ethical conduct is important in research, then all of the stakeholders in the research, including individual scientists as well as various organizations, have an obligation to promote ethical conduct and ethical decision making (Swazey and Bird 1997, Resnik 1998c, National Academy of Sciences 1992). Although it is important to promulgate and enforce standards of conduct in research, promoting ethical conduct requires much more than investigating and adjudicating claims of misconduct. Ethics is not police work. By far the most effective way to promote ethical conduct in research is to teach students at all levels about ethics and integrity in research (Bird 1993, Swazey 1993, Swazey and Bird 1997). Education in research ethics can help students develop moral sensitivity (an awareness of ethical issues and concerns), moral reasoning (the ability to make moral decisions), and moral commitment (the willingness do to the right thing even at some personal cost) (Pimple 1995). Although courses, seminars, workshops, and lectures on research ethics can be an effective way of teaching ethics in research, empirical research suggests that role modeling and leadership play a very important role in learning ethical conduct. That is, all of us, including scientists and other professionals, learn professional norms and virtues, in part, by example (Kuhn 1970; see also Aristotle ca. 330 B.C. [1984], Kopelman 1999).

Scientists who model ethical behavior for their students and colleagues exhibit moral or ethical leadership. Ethical leadership promotes ethical conduct by setting an example and a general tone or attitude in an organization. Organizational leaders can promote an ethical corporate culture or unethical corporate culture, depending on their demeanor. If members of an organization (e.g., students, researchers, employees) can see that ethics is important to the leaders of the organization and that the leaders take ethical issues seriously, then members will also value ethical conduct and take ethics seriously (Fraedrich 1991). However, if leaders do not emphasize ethics, or if they engage in unethical practices, such as dishonesty, deception, negligence, and law breaking, then members of their organization will follow their example.

Today, many large corporations have adopted ethics statements that emphasize the importance of ethical leadership (Murphy 1998). Although all researchers should teach ethics by example, it is especially important for researchers who have important leadership or management positions, in academia and private industry, to lead by example and affirm their commitment to ethics in research. Individuals as well as institutions have a responsibility to promote ethics by developing policies and procedures and promoting effective leadership (Berger and Gert 1997). Thus, department chairs, laboratory and center directors, section heads, vice presidents, deans, and presidents should all lead by example. By doing so, they can affect the behavior of students, colleagues, and employees and promote an ethical culture. Good leaders inspire learning, creativity, and a sense of duty to society and our civilization. We discuss many of these policies in various chapters of our book.

QUESTIONS FOR DISCUSSION

1. How would you describe the current state of collaboration in research?
2. Why is collaboration important for research?
3. Can you think of any examples, from your own experience or from the experiences of colleagues, where good collaboration was lacking?
4. How would you react if someone you do not know wants a sample of research material or data?
5. Can you describe how you are being mentored? Are there modifications you could suggest?
6. What are the qualities (or virtues) of a good mentor?
7. What are the qualities of a good leader?

CASES FOR DISCUSSION

Case 1

A postdoctoral fellow got into a severe conflict with her mentor. Her mentor provided her salary from his grant resources, and she was working on one of his primary projects. She found another job and took all three laboratory notebooks with her when she left. The mentor was very angry when he found out. He asked her to return the lab notebooks immediately or she would be accused of theft. He claimed the lab notebooks belonged to him and to the university, but he invited her to copy the books for her use. She returned the notebooks after making copies.

Two years later, the mentor learned that she had published a paper without mentioning his name anywhere, but that his grant was acknowledged. What should the mentor do?

Case 2

A graduate student worked for a year with an adviser on replicating a new small protein. He spent part of the year developing the methodology before conducting the replications. However, the graduate student did not like his adviser and moved to a different adviser within the same department, who happened to be the director of the graduate program in that department. The student's new research program was in a different area from the previous work.

A year later, the student learned that a subsequent student of his former mentor had used his method for replicating the protein in subsequent research, and that they were writing a paper without listing him as a coauthor. He protested but was told that the new graduate student had to do the whole thing all over again and that they were not using his data. The student argued that the new technique used to collect the data was a novel technique developed by him and not available in the open literature. The student's former adviser, after meeting with everyone including the director, reluctantly agreed to publish at a later date a small technical paper on the technique naming the student as coauthor. The first paper will still appear, much sooner, and without his name. The student agreed, under protest, but he knew his life would be difficult if he insisted on a different outcome.

- What would you have done under these circumstances?
- Should the first adviser have done what he did?
- What should the new student have done, and what should he do now? The director?

Case 3

Dr. Barnes has been telling dirty jokes in his anatomy and physiology class for as long as anyone can remember. He is a very friendly, outgoing teacher, gets excellent evaluations on his teaching, and has won several teaching awards. Dr. Barnes has three teaching assistants who help with various aspects of the course.

During one of the labs for the course, one of his teaching assistants, Heather, overhears two female students complaining about his jokes. They find his jokes both rude and offensive. Heather talks to the two students. She says that she also found his jokes rude and offensive at first, but that she got used to them and they don't bother her anymore. They say that they do not want to get used to his jokes and that they are planning to talk to the dean of students. What should Heather do?

Case 4

Dr. Trotter is a molecular geneticist who applies Darwin's principle "survival of the fittest" to his laboratory environment. Each year, he hires two new postdoctoral students for one year. He assigns them both to work on the same experiment. Whoever finishes the work first, with reproducible results, will get to be an author on a paper; the loser will not. He runs several such contests during the year. At the end of the year, the postdoctoral student who has the best results will be hired for a three-year position, and the loser will be terminated. What do you think about this Dr. Trotter's policy? Is it ethical?

Case 5

Dr. Stabler is the dean of a prestigious medical school. In response to a scandal involving misconduct in a privately funded clinical trial conducted on campus, the medical school has decided to revamp its conflict of interest policies. A faculty committee has proposed that faculty are not allowed to own more than $5,000 in stock in any particular pharmaceutical or biotechnology company or receive more than $5,000 in consulting fees or honoraria from any particular pharmaceutical or biotechnology company in a calendar year. Faculty must disclose their financial relationships with companies each year. The committee also recommended tougher standards for administrators: Deans and department heads should own no stock and receive no consulting fees or honoraria from pharmaceutical or biotechnology companies. When this proposed policy was discussed at a faculty meeting, Dr. Stabler strongly opposed tougher standards for administrators, and it did not pass. The faculty adopted a policy that treats all faculty equally with respect to conflicts of interest. Do you think Dr. Stabler exhibited ethical leadership?

5

Collaboration between Academia and Private Industry

This chapter examines some of the ethical dilemmas and issues arising from relationships between higher learning institutions and private industry, including conflicts of interest, research bias, suppression of research, secrecy, and the threat to academic values, such as openness, objectivity, freedom of inquiry, and the pursuit of knowledge for its own sake. The chapter also provides an overview of the historical, social, and economic aspects of the academic–industry interface and addresses some policies for ensuring that this relationship benefits researchers, universities, industry, and society.

THE DEVELOPING RELATIONSHIP BETWEEN SCIENCE AND INDUSTRY

The petroleum industry was hatched in a very modern symbiosis of business acumen and scientific ingenuity. In the 1850s, George Bissell, a Dartmouth College graduate in his early thirties who had enjoyed a checkered career as a reporter, Greek professor, school principal, and lawyer, had the inspired intuition that the rock oil plentiful in western Pennsylvania was more likely than coal oil to yield a first-rate illuminant. To test this novel proposition, he organized the Pennsylvania Rock-Oil Company, leasing land along Oil Creek, a tributary of the Allegheny River, and sending a specimen of local oil to be analyzed by one of the most renowned chemists of the day, Professor Benjamin Silliman, Jr. of Yale. In his landmark 1855 report, Silliman vindicated Bissell's hunch that this oil could be distilled to produce a fine illuminant, plus a host of other useful products. Now the Pennsylvania Rock-Oil Company faced a single, seemingly insurmountable obstacle: how to find sizable quantities of petroleum to turn Professor Silliman's findings into spendable cash. (Chernow 1998, p. 74)

Although the times have changed and much more money is at stake, nineteenth-century scientist-entrepreneurs such as Bissell and Silliman have much in common with their contemporary counterparts, such as Craig Venter, the former head of Celera Genomics Corporation (for further discussion, see chapter 14). The ingredients are the same: A well-known scientist from a well-known institution invents a product or develops a process that can be used by large numbers of people and hopes to make a lot of money from it. In Venter's case, he became frustrated with academic science and launched his own company and research program. Venter's key innovation was the "shotgun" approach to genome sequencing, which stands in sharp contrast to the traditional, and slower, "clone-by-clone" method. Venter has raised several hundred million dollars in venture capital to use this approach to sequence mouse, fruit fly, and human genomes. The company sequenced the human genome for about $300 million. The shotgun approach breaks

81

apart an entire genome, uses automated sequencing machines to sequence different fragments, and then uses supercomputers to put all the parts back together again (Marshall 1999b). Although Venter and academic researchers have clashed over issues related to data access, release, and control, Celera Genomics and the publicly funded Human Genome Project (HGP) collaborated to produce a complete sequence of the human genome, which was published in *Science* and *Nature* on February 16, 2001 (Venter et al. 2001; see also Robert 2001). However, clashes between the two groups have remained, with some HGP researchers accusing Celera of relying on the HGP's data and methods (Wade 2001). Although the company has yet to turn a profit, its earnings have increased steadily and it has sold more than $1 billion in stock. Because most of the human genome is now available on the Internet, Celera, located in Rockville, Maryland, plans to make most of its money from selling genomics information services (Celera Genomics 2008).

As discussed in more depth in chapter 9, the collaboration between the scientist James Watt and the entrepreneur Mathew Bolton on the development of the steam engine in the late 1700s set the example for future science–industry ventures. The modern industrial laboratory was developed in mid-nineteenth-century Germany, when the dye industry hired chemists to invent and produce synthetic dyes. The development of the petroleum industry, as exemplified by the contributions of Bissell and Silliman, followed a similar pattern. Business leaders began to realize the economic value of applying scientific methods and concepts to solving technical and practical problems. Science became a commodity (Dickson 1988). Thus, the first marriages between science and industry took place outside of the university setting, in private laboratories, businesses, or research centers.

This is still a very important aspect of the science–industry relationship. Pharmaceutical companies, biotechnology companies, food companies, seed companies, software companies, chemical companies, automotive companies, and electronics companies all conduct a great deal of research outside of the university setting. Private industry funds approximately 60% of the research and development (R&D) in the United States, while the federal government funds about 35% (Resnik 2007b); 70% of all R&D in the United States is conducted in private venues, followed by federal laboratories and universities, each at about 15% (Resnik 2007b).

An important and at that time unusual marriage between science and private industry began in the 1920s at the University of Wisconsin, when Professor Henry Steenbock discovered that he could activate vitamin D by irradiating food. He patented his discovery, hoping to benefit his own university and ensure quality of the manufacturing process (Bowie 1994). The Board of Regents of the University of Wisconsin rejected Steenbock's offer to share in the profits of the invention. A group headed by some of the school's alumni started a nonprofit foundation, the Wisconsin Alumni Research Foundation (WARF), to provide additional funding for R&D. WARF was responsible for thousands of patent disclosures and hundreds of license agreements, and had an income of millions of dollars (Bowie 1994).

The foundation gave 15% of invention royalties to the inventor and 85% to the foundation in order to disburse research grants to university faculties. Although WARF was formally not part of the University of Wisconsin, it was for all practical purposes a part of the university, because university officials controlled the foundation, and its functions were comingled with university functions. To make money, WARF attempted to obtain monopolistic control over vitamin D synthesis and negotiated and enforced narrow license agreements that prevented poor people from having access to vitamin D. For example, WARF claimed that its patent applied to vitamin D produced by sunlight, and WARF refused to license other companies to use its vitamin D irradiation process to develop other food products, such as oleomargarine (Bowie 1994).

To understand the significance of the change in universities represented by the patenting of research conducted at universities, one must place universities in their social and historical context. Since ancient times, there were schools of higher learning, such as Plato's Academy or Aristotle's Lyceum. There were also schools that provided students with training for priesthood. But these schools were highly exclusive and did not enroll many students. The birth of universities took place during the thirteenth and fourteenth centuries in Europe. During the Crusades of the early 1100s, Europeans reclaimed territory that had been occupied by the Islamic Empire during the middle ages. They discovered hundreds of ancient Greek, Roman, and Islamic texts on a variety of subjects, including law, medicine, biology, physics, mathematics, philosophy, logic, optics, history, and alchemy. Much of this knowledge consisted of ancient texts, such as Ptolemy's *Almagest* and most of Aristotle's scientific works, which Europeans had "lost" during the Dark Ages but that had been preserved for posterity by Islamic scholars. The "new" knowledge also included original works by Islamic scientists, such as Avicenna, Alhazen, and Averroes.

Because these texts were written in Arabic, Europeans had to translate them into Latin or the vernacular. Once these new works became available, European students were eager to learn them. Because these students were not familiar with these texts, they hired scholars to help them read and interpret them. The first university was a law school established in Bologna, Italy, in 1158. The students hired the teachers, set the rules, established residencies, and procured books (Burke 1995). A school of theology was established in Paris in 1200. During the next century, universities were established in Germany, England, Spain, and the Netherlands. The first subjects taught in universities included civil law, cannon law, theology, medicine, and "the arts" (i.e., geometry, music, astronomy, and mathematics). Other scientific subjects, such as biology, chemistry, and physics, were not taught in universities until much later (Goldstein 1980).

Education was the primary reason that universities were established. Universities did not even begin to emphasize research until the 1500s, when the University of Padua in Italy became known as a haven for scholarly exchange, free thought, and intellectual development. The "modern" university system, consisting of professors, instructors, and doctoral, master's, and undergraduate students, did not emerge until the nineteenth century in Germany. The "commercialized" university, that is, a university with strong intellectual property interests and ties to private industry, is a relatively recent development. For hundreds of years, universities emphasized academic norms such as openness, free inquiry, and the pursuit of knowledge for its own sake. Business norms, such as secrecy, directed inquiry, and the pursuit of knowledge for the sake of profit, simply were not part of the university culture until the twentieth century (Bok 2003).

Before World War II, the U.S. government did not invest heavily in research. Indeed, in the 1930s Bell Laboratories conducted more basic research than any university (Dickson 1988). During the war, the government began funding large-scale research programs, such as the Manhattan Project and the development of radar, jet airplanes, and computers. After World War II, it became clear to many people involved in government and in science policy that science and technology helped the Allies win the war and that continued investment in R&D would be a key pillar in national security and defense (Dickson 1988, Guston 2000). Federal investments in R&D grew from less than $1 billion in 1947 (or less than 0.3% of the gross national product) to $3.45 billion in 1957 (or 0.8% of the gross national product). A crucial turning point occurred in 1957, when the Soviet Union launched Sputnik, convincing many politicians that America was in grave danger of losing its scientific and technological edge over the Russians. From that period onward, federal funding of R&D grew steadily at 5–10% a year, and the government made more extensive use of scientific advisers. Federal support of R&D climbed to $26 billion in 1965, then leveled off from 1966 through 1976, and then climbed again to $40 billion in 1985 (Dickson 1988). During this period, the federal government became the leading sponsor of R&D in the United States, but this pattern changed in the 1980s, as private industry increased its investments in R&D. In 1980, private industry and the federal government each sponsored about 48% of research, but the percentage has shifted, and private industry is now once again the leading sponsor of R&D (Jaffe 1996).

After World War II, universities and colleges began to proliferate and expand around the United States, fueled by an increasing number of students seeking higher education. The GI Bill allowed soldiers returning from the war to obtain an education during the 1940s and 1950s, and the baby boom generation increased enrollments during the 1960s and 1970s. Throughout the United States, state governments helped fund expansion of the higher education system, and the federal government contributed by funding R&D. However, many state colleges and universities began facing financial difficulties in the 1970s due to increasing enrollments but insufficient or shrinking state funds (Bowie 1994). In response, these institutions sought to increase revenues from traditional sources, such as tuition, fees, and donations from alumni. Universities (and to a lesser extent colleges) also began to try to obtain additional revenues from sources related to R&D, such as grants, contracts, entrepreneurial activities, and intellectual property rights. These very same economic pressures exist today and help explain why universities have become increasingly involved in commercial activities (Bowie 1994, Bok 2003). At the same time, many academic leaders have sought to expand their institutions and enhance their economic strength and overall status by building ties to the private sector. Public and private institutions have both taken a strong interest since the 1970s in increasing revenues related to research activities.

Important initiatives implemented in the 1970s included the University Cooperative Research Projects Program and the Industry/University Cooperative Research Centers Program, administered by the National Science Foundation (NSF). These efforts created dozens of centers to enhance multidisciplinary areas of research in all of the sciences. During the 1970s and 1980s, universities began more actively pursing patents and developed technology transfer offices, a development that was encouraged by legislation, such the Bayh-Dole Act of 1980 and the Technology Transfer Act of 1986 (discussed in chapter 9) (Dickson 1988).

The twentieth century has also seen several different eras in private R&D investments. For example, during the 1940s, companies invested heavily in the development of vacuum tubes and television. During the 1950s and 1960s, companies sponsored a great deal of R&D on transistors, semiconductors, and pharmaceuticals. The 1970s and 1980s saw massive R&D investments in integrated circuits and computer hardware and software. By the 1980s, biotechnology became the rage (Teitelman 1994). The molecular biology revolution that began in the 1970s was the catalyst for the private industry investment in biotechnology. For industry to tap the scientific basis of and develop this new field, it realized it needed collaboration with university-based researchers. It began in the early 1980s when the West German chemical company Hoechst GA invested $70 million in a joint venture with Massachusetts General Hospital (MGH), affiliated with Harvard Medical School (Bowie 1994). MGH was to hold all patent rights of any discovery, but Hoechst would have first rights to any commercial products.

The true first major corporate investment in a university came in 1974, when Monsanto and Harvard Medical School entered into a 12-year joint venture, whereby Harvard received $23 million for research and Monsanto received all of the patent rights from the venture. Also, Monsanto received a first look at research results prior to publication. The university formed a committee to "protect academic freedom" and the "public good" (Culliton 1977) to review publication delays. The Harvard/Monsanto agreement was kept confidential until 1985, reflecting the university's discomfort at that time with collaboration with industry.

In 1981, Edwin Whitehead, president of Technicon, a medical diagnostics company, made a $120 million contribution to establish the Whitehead Institute at the Massachusetts Institute of Technology (MIT). The mission of the institute was to conduct biomedical research with potential applications in medicine and biotechnology. Whitehead himself retained control of faculty appointments at the institute. Faculty also held appointments at MIT. When the institute was formed, faculty at MIT objected to the commercial ties between the university and private corporations and the potential conflicts of interest (COIs). Despite these concerns, researchers at the institute have made important contributions to basic and applied biomedical science (Teitelman 1994).

In 1982, Monsanto negotiated an agreement with Washington University in St. Louis, Missouri. Monsanto gave $23.5 million to Washington University Medical School for research in the area of biotechnology. The agreement stipulated that royalties from patents and licenses would be split three ways among the university, the medical school, and the principal investigator's laboratory, but none would be paid directly to the investigator (Bowie 1994). It was claimed that excluding investigators from receiving royalties would avoid COIs among the faculty members, but this arrangement intensified the university's COI, because this would increase the university's financial interests. Monsanto received first rights of refusal to patents and the right to review publications. In 1994, the Scripps Research Institute entered a $20 million per year agreement with the Swiss pharmaceutical company Sandoz. Under this arrangement, Sandoz funded research conducted at Scripps in exchange for first-refusal rights to almost half of Scripp's research results (Beardsley 1994).

The next example of corporate investment into a university illustrates how the instability of the corporate model can affect academic life. In the mid-1980s, the Italian drug company FIDIA planned to give Georgetown University $3 million a year for 20 years (a total of $60 million) for research on neuroscience, to be housed in an institute within the university (Bowie 1994). The institute would primarily be devoted to basic research. FIDIA presumably would have received in return a strong connection to the best and brightest in this field, credibility, and the first rights of refusal on patents. Unfortunately, in 1993, FIDIA filed for bankruptcy in Italy, and its contributions to Georgetown were reduced to a few million dollars. The university had to downscale its efforts to build its program and hire investigators.

In response to the rapid growth of the biotechnology and university–industry partnerships in the field, *Scientific American*, not known as a radical journal, published an article titled "Big-Time Biology" (Beardsley 1994). The article raised serious concerns regarding the rapid growth of biological sciences into an industry and how it is affecting university culture. An example to illustrate the modern complex relationship between private gains, universities, and researchers (Ready 1999) involves Dennis Selkoe, a researcher at Harvard Medical School who founded California-based Athena Neurosciences. The company is based on Selkoe's own research and has already earned him millions of dollars. Athena Neurosciences manufactures a controversial blood test for Alzheimer's disease. Without disclosing his relationship to the company, he was one of the endorsers of a National Institutes of Health (NIH) report for the new Alzheimer's disease test (National Institutes of Health 1998a). A Harvard panel concluded that he did not violate the university's COI policy, but he did agree to publish a disclosure statement in the same journal that published his original work. Apparently, Selkoe routinely wrote about Alzheimer's disease without reporting his affiliation with Athena. He also had a $50,000 per year consulting contract and served on the board of directors. As another example, in 1997 the University of California at Irvine found out that its cancer researchers had failed to inform the U.S. Food and Drug Administration (FDA) of a side effect of a drug they had developed. These researchers had investments in the company that hoped to sell the drug (Ready 1999). The university shut down the laboratory.

The amount of university research funding from industry continues to rise, and universities themselves are engaged in building their own corporations by joint ventures and equity arrangements. In 1997, U.S. corporations spent $1.7 billion on university research (Shenk 1999). The survey data available from 1994 indicate that 90% of companies involved in life sciences had relationships with university research; 59% of those corporations supported university research directly, at about $1.5 billion, accounting for 11.7% of all R&D funding at universities (Blumenthal et al. 1996a). The survey further indicates that more than 60% of companies involved in university research received patents and products from the data generated. According to Dustira (1992), industry support to R&D in the biomedical field rose from 31% in 1980 to 46% in 1990, while at the same time the contributions of the NIH dropped from 40% to 32%.

The Association of University Technology Managers (AUTM), representing more than 300 university and research institutions, claimed that in 1998 alone, 364 startup companies resulted from technology transfer activities within these institutions. The total number of startup companies since 1980 now exceeds 2,500 (Press and Washburn 2000). AUTM also claims that these technology transfer activities generated $34 billion in 1998, supporting 280,000 American jobs. If one looks only at the global impact of the biotechnology industry, it is clear that this industry is becoming a new force in the world's economy, supporting millions of high-paying jobs and promoting economic activity, investment, and development (Enriquez 1998). The political economy of the twenty-first century will undoubtedly be shaped by biotechnology (Carey et al. 1997).

This brief review illustrates the many ways that university–industry relationships can affect research and educational activities, and the benefits and risks associated with blurring the lines between industry and academia. Individual researchers may use contracts or grants from private companies to conduct research, hold conferences or workshops, or develop educational materials. They may also hold stock or equity in companies, share patents with companies, or receive honoraria, salary, or fees from companies. Institutions engage in similar activities: They receive grants and contracts, hold stock and equity, hold patents, and receive fees. In addition, institutions also receive contributions to their endowments and from companies and form joint ventures. As Derek Bok (2003, p. 2), president of Harvard

University for two decades, said: "What is new about today's commercial practices is not their existence but their unprecedented size and scope."

ACADEMIC VALUES VERSUS BUSINESS VALUES

Numerous tensions have arisen in the marriage between business and academia. Universities and businesses traditionally have emphasized different values or norms (Dickson 1988, Bowie 1994, Davis 1995a, Rule and Shamoo 1997, Resnik 1999a,b, 2007b). Below are some examples:

- The main goals of institutions of higher learning are to educate students; to advance human knowledge through research, scholarship, and creative activities; and to conduct public service. Private, for-profit corporations, on the other hand, aim to maximize profits and to produce goods and services. Nonprofit corporations also seek to produce goods and services, and they seek financial growth and stability.
- Higher learning institutions emphasize openness and the free exchange of data and ideas. Private corporations, on the other hand, often use secrecy to protect confidential business information and proprietary interests.
- Higher learning institutions emphasize academic freedom, free speech, and free thought. Private corporations, on the other hand, conduct research for specific purposes and may impose restrictions on public communications to protect the interests of the corporation.
- Higher learning institutions emphasize honesty and objectivity in research. Private corporations may endorse and promote these values in principle, but also as a means of complying with the law, for enhancing market share, or for ensuring the quality of goods and services.
- Higher learning institutions emphasize knowledge for its own sake; corporations emphasize the utilization of knowledge for the sake of profit or other practical goals.
- In capitalistic countries, private corporations compete with different companies on the free market. To compete on the market, corporations must produce high-quality goods and services; invest wisely in technology, equity, infrastructure, and R&D; and make effective use of human, financial, and technical resources. Higher learning institutions, on the other hand, have not traditionally competed with each other in the free market in the same way that corporations have. Many of these institutions are subsidized by government funds and have monopolistic control over local markets.
- Private corporations have obligations to stockholders, customers, employees, the government, and the community. Higher learning institutions have obligations to students, faculty, staff, alumni, the government, and the community.

Can this marriage work, or is it doomed to failure? Will private interests corrupt universities? To address these questions, it will be useful to review some of the benefits and harms of collaborations between academic institutions and private corporations.

POSSIBLE BENEFITS OF COLLABORATION BETWEEN ACADEMIA AND PRIVATE INDUSTRY

Benefits to Academic Institutions

It should be apparent to the reader that universities and colleges benefit financially from relationships with industry. Private corporations and private interests provide billions of dollars for universities to conduct research; to develop courses, educational materials, and curricula; to hire faculty and staff; to build laboratories and facilities; to acquire tools and instruments; to hold workshops and conferences; and even to enroll students. Academic institutions also benefit from the knowledge and expertise obtained from the private sector, which can be useful in generating ideas for research or education. For example, a professor teaching a course in bioengineering may consult with leaders in the private sector to help prepare her students for real-world problems. Knowledge about private industry's plans for new products can be especially useful in clinical research. For example, academic researchers can investigate the molecular mechanisms of a new drug being tested by a pharmaceutical company. Collaborations with industry play an important role in the NIH's Roadmap for Medical Research, which emphasizes the importance of translating basic discoveries made in the laboratory into medical applications (National Institutes of Health 2008). Additionally, many institutions also appoint people from the private sector to boards and committees in order to obtain advice in preparing students for careers in business, industry, or the various professions. Industry leaders often provide key advice on research and education as consultants to colleges and universities.

Benefits to Industry

Industry benefits from having access to top-notch university facilities, equipment, and researchers. Although many companies already hire their own excellent researchers and have state-of-the-art facilities and equipment, universities also still have economically valuable resources. Although research has become increasingly privatized in the last two decades, many of the world's top researchers are still employed by universities, not by private companies. Corporations may also benefit from good publicity and visibility in their association with universities or colleges.

Benefits to Society

Society benefits from this private investment in academic institutions, which have helped to expand and support universities, have created higher paying jobs, and have contributed to economic development. About 8 million people in the United States are employed in R&D, and approximately 6% of the U.S. economy is directly related to R&D (Resnik 2007b). Society also benefits from the goods and services produced by academic–industry collaborations, such as new drugs, medical devices, and information technologies. And society benefits from the increases in productivity resulting from academic–business collaborations. The increased productivity is evident from a survey indicating that industry-supported university research produced four times more patents per dollar than did industry research and increased the rate of the technology transfer from laboratory to useful products (Blumenthal et al. 1986, 1996a,b). Francisco J. Ayala, a prominent biologist and once a member of a presidential committee on science and technology, stated that biotechnology is "contributing to the economic welfare of the nation, and it has by and large benefited academic progress. The benefits have far outstripped harms that may have occurred" (quoted in Beardsley 1994, p. 93).

POSSIBLE HARMS OF COLLABORATION BETWEEN ACADEMIA AND PRIVATE INDUSTRY

Individual Conflicts of Interest

Chapter 10 discusses COIs in more detail, but we briefly explore the topic here. We focus mostly on the COIs for individual researchers arising from their relationships to industry. Some of these include funding or research through contracts or grants, ownership of stock or equity, intellectual property rights, honoraria, consulting agreements, fees for enrolling patients in clinical trials, and gifts from industry. All of these COIs can undermine the researcher's primary obligations to the university, to his or her profession, and to society by compromising the objectivity of research. Even if COIs do not actually create biases in research, they can still undermine trust by creating the appearance of bias. Thus, it is important to address apparent COIs, as well.

Institutional Conflicts of Interest

It is now widely recognized that research institutions can also have COIs that undermine objectivity and trust. An institution of higher education may have obligations to students, faculty, alumni, and the community that are affected by its relationships with industry. These COIs can affect the institution's decisions, policies, and actions and can undermine trust in the institution (Resnik and Shamoo 2002). Institutional COIs can affect many different university activities, including research, education, and compliance with regulations.

Increased Secrecy in Academia

Problems with data sharing and suppression of research related to relationships with industry have been discussed above. Private corporations view research as proprietary information, and they often prohibit researchers from discussing data, ideas, or results prior to publication (Rosenberg 1996). The work by Blumenthal et al. (1996a,b) shows that researchers are under some pressure to not share data and to delay or stop publication in order to protect intellectual property or honor obligations to private companies. Betty Dong's case, discussed below, illustrates how companies may attempt to suppress unfavorable results. This climate of secrecy can have a detrimental impact on students, as well, who may become involved in secret research that they cannot discuss with other students or professors. As mentioned several times above, openness is vital to scientific progress and is one of the most important academic norms. Some fear secrecy will have broad effects on research and academic life (Bok 2003, Washburn 2006). On the other hand, some argue that academic researchers may influence industry to promote greater openness in industry-sponsored research (Davis 1991). So far, it does not appear that the academic researchers are winning the battle for openness. Moreover, many researchers may not feel that they can criticize the industry's secrecy policies because they do not want to jeopardize their own funding or risk the wrath of university administrators. Some commentators have also pointed out that researchers have reasons to keep data and results confidential even if they are not receiving funds from industry. In any case, the conflict between secrecy and openness in research bears watching. We need to find ways to promote openness while protecting private interests. In discussing cases of how secrecy has prevailed to the determent of research progress, Steven Rosenberg, then the chief of surgery at the National Cancer Institute, remarked, "This is the real dark side of science" (quoted in Press and Washburn 2000, p. 42).

Diversion of University Resources from Basic to Applied R&D

Although many universities emphasize applied research as a part of their mission, basic research should not be neglected in order to achieve short-term economic or practical goals (Bowie 1994, Zolla-Parker 1994, Brown 2000). If corporations can influence academic appointments, they can increase the percentage of faculty who conduct applied research at a university, and faculty interests can have a trickle-down effect on graduate and postdoctoral students. As corporate funds continue to flow into a university, it may lose its commitment to basic research and knowledge for its own sake and may become more interested in applications and profits. The curriculum can also become transformed from one that emphasizes education in critical thinking and the liberal arts to one that emphasizes the training of students for careers in business, industry, or the professions. Talented faculty who might normally be involved in teaching may be encouraged to conduct research so the university can increase its revenues from contracts and grants (Brown 2000). Corporate dollars can transform an academic institution into a private laboratory and technical training school.

Research Bias

Because private corporations are interested in producing and marketing specific products to enhance the profit, market share, and stock value of the corporations, they may influence the process of research to ensure "favorable" outcomes (Crossen 1994, Resnik 1999a, 2001e, 2007b). For example, if a study is not supporting a conclusion that is favorable to the company, the company can withdraw funding. Once a study is completed, a company may try to prevent unfavorable results from being published. A company may also attempt to manipulate or misrepresent data to produce a favorable result, or discredit research that is "unfavorable" to its products. Although research bias can occur in the public or private sector, the economic incentives to conduct biased research are stronger in the private sector and therefore pose a hazard for university–industry relationships. Moreover, the penalties for bias, such as loss of funding or tenure or a damaged reputation, are potentially more damaging to academic researchers than to private researchers or corporations.

For an example of research bias related to industry, consider the controversy over calcium-channel blockers. In 1995, Psaty et al. published results of a population-based case–control study of hypertensive patients and the risk of myocardial infarction associated with calcium-channel blockers versus beta-blockers versus diuretics. With 335 cases and 1,395 controls, they found that the risk of myocardial infarction was increased by 60% among users of calcium-channel blockers with or without diuretics. With 384 cases and 1,108 controls, they found that the use of calcium-channel blockers was associated with about a 60% increase in the risk for myocardial infarction compared with beta-blockers. Because the study was supported by federal funds, several pharmaceutical companies who manufacture beta-blockers requested all of the documents, including original data, under the Freedom of Information Act (Deyo et al. 1997). The companies hired several academic consultants to analyze the research, and the consultants wrote scathing critiques of the study. The companies even attempted to interfere with the publication of the article (Deyo et al. 1997).

Following the controversy of calcium-channel blockers, Stelfox et al. (1998) conducted a survey of literature published from March 1995 through September 1996 on the subject. They demonstrated a strong association between authors supporting the safety of calcium-channel blockers and their financial ties with pharmaceutical companies. Stelfox et al. classified the authors into three categories: critical, neutral, and supportive of the safety of calcium-channel blockers. They found that 96% of supportive authors had ties with the manufacturer, compared with 60% and 37% of neutral and critical authors, respectively. Earlier, Davidson (1986) obtained similar results that demonstrated a link between industry funding sources and favorable outcomes for new therapies.

Subsequent research has demonstrated a strong connection between sources of funding and research results. Cho and Bero (1996) and Cho (1998) found that the overwhelming percentage of articles with drug-company support (90%) favor the company's drug, compared with those without drug-company support (79%). Friedberg et al. (1999) reported that only 5% of new cancer drug studies sponsored by the company developing the new drug reached unfavorable conclusions, compared with 38% of studies on the same drugs sponsored by other organizations. In the dispute over second-hand smoke, articles sponsored by the tobacco industry indicated that second-hand smoke has no ill health effects, whereas studies sponsored by other organizations supported the opposite conclusion (Bero et al. 1994). The tobacco industry also blocked research (Hilts 1994) and the publication of unfavorable results (Wadman 1996). Rochon et al. (1994) found that the industry-sponsored research on nonsteroidal anti-inflammatory drugs (NSAIDs) favored these medications as opposed to control drugs.

In 1995, the Boots Company made Betty Dong, clinical pharmacologist at the University of California, San Francisco, withdraw a paper on drugs used to treat hypothyroidism that had been accepted for publication in the *Journal of the American Medical Association* (*JAMA*). Boots had funded her research in the hopes that she could show that its product, Synthroid, was superior to three generic alternatives. Synthroid is a brand name for levothyroxine, which was manufactured in 1987 by Flint Laboratories and later bought by the Boots pharmaceutical company and then by the Knoll Pharmaceutical Company. The sale of Synthroid generated nearly $500 million dollars annually (Weiss 1997). To Boots's surprise, Dong showed that several generic drugs were just as safe and effective as Synthroid and also less expensive. Boots responded by trying to change her report, and the company spent several years trying to discredit her research. Because Dong had signed an agreement with Boots not to publish the results without written consent from the company, Boots invoked this clause in the contract to prevent Dong from publishing the paper in *JAMA* (Wadman 1996). In the meantime, the company published a more favorable article in another journal (Altman 1997, Weiss 1997, Shenk 1999). To avoid a lawsuit, Dong withdrew her submission from *JAMA*. Two years later, the *New England Journal of Medicine* published her findings.

Shenk (1999) recites several cases where corporate sponsors of research derailed the publication of adverse findings: Nancy Olivieri, a University of Toronto researcher, who was threatened by the Canadian drug manufacturer Apotex if she published her results on drug L1; David Kern, a Brown University researcher, who was pressured by a local company and his own university to not publish his data on lung disease related to the company's plant; Curt Furberg, a hypertension researcher, resigned from a research program on calcium-channel blockers funded by Sandoz in order to avoid tempering his data. Bodenheimer (2000) discusses concerns about the relationship between academia and the pharmaceutical industry and addresses problems with the use of ghostwriters by industry. As mentioned in chapter 6, some companies hire professional writers to write research manuscripts but list physicians or scientists as authors. (We discuss the issue of bias again in chapter 10.)

CONCLUSION: MANAGING THE UNIVERSITY–INDUSTRY RELATIONSHIP

To summarize, there are benefits as well as risks in university–industry relationships. We may have appeared to have stressed the negative aspects of this relationship (Krimsky 2003, Washburn 2006), but we would like to temper this impression. As noted above, there are many benefits of the university–industry interface. Moreover, it is hard to imagine how the modern research enterprise could proceed without some type of collaboration between academia and industry. For the last few hundred years, ties between universities and private corporations have been growing stronger as research has increased in its practical relevance, economic and commercial value, complexity, and cost. Changes in the financial and economic aspects of research have transformed universities from institutions of that focus on education and "pure" research to institutions that also have commercial ties and interests and conduct "applied" research. Although it is important to preserve and promote values traditionally associated with academic institutions, such as openness, objectivity, freedom of inquiry, and the pursuit of knowledge for its own sake, it is not realistic to expect that we can return to an earlier era. Instead of romanticizing about the "good old days" or bemoaning the "corrupt" nature of the current system, we should develop practical policies for ensuring that university–industry relationships benefit both parties as well as society (Resnik 2007b).

Based on the discussion in this chapter as well as preceding and subsequent ones, we recommend the following guidelines for university–industry relationships (for further discussion, see Rule and Shamoo 1997, Resnik and Shamoo 2002, Resnik 2007b):

1. *Develop and implement strategies for managing individual and institutional COIs:* Disclosure is the first step in dealing with COIs, but as Krimsky et al. (1996, Krimsky 1996) have found, many researchers do not disclose COIs. Some authors, such as Cho (1998), believe that disclosure is not sufficient. Individuals can take steps to avoid COIs by obtaining multiple sources of funding for research, refusing to hold stock or equity in companies that sponsor research, not taking honoraria or consulting fees, not accepting patient recruitment fees over and above administrative costs, refusing certain types of gifts, and so on. We also believe that disclosure may sometimes be insufficient. However, because it is not realistic to expect that universities or individuals can avoid all COIs in research, decisions to prohibit or avoid a COI must be made on a case-by-case basis by considering the strength of the COI as well as the benefits and harms created by the situation. Despite this, the research community should consider prohibiting certain financial relationships such as owning equity in small corporations while conducting its clinical trials. Because the federal government does not set very restrictive COI rules for obtaining grants, universities have a variety of COI policies (McCrary et al. 2000). Harvard Medical School has some of the strictest guidelines: Researchers can have no more than $30,000 in stocks in corporation where they are involved in studying their products. Also, they cannot receive more than $20,000 in royalties

or consulting fees from companies that sponsor their research (Harvard University 2004). Prior to 2004, Harvard contemplated drastically relaxing its COI rules at a time of great concern about COIs, which brought a raft of criticism (Angell 2000, Mangan 2000, O'Harrow 2000). Later, Harvard backed down and announced that it will not change its COI rules (O'Harrow 2000). The Harvard example illustrates the powerful lure of industry money even for a highly endowed and reputable institution. Cho (1997) argues that the first moral obligation of a researcher is to the integrity of research, and the obligation to financial interest should be secondary.

2. *Develop COI committees to manage COIs:* Most research institutions use existing administrative hierarchies (e.g., department chair, section head, director, or dean) for managing COIs of its employees. The U.S. Public Health Service (PHS) has recommended that universities also appointment committees to manage COIs, especially those that require more than mere disclosure (Public Health Service 2000a). Both the PHS and Moses and Martin (2001) also suggest that each university develop a committee to manage its own COIs as an institution.

3. *Where possible, take steps to dilute COIs, such as placing investments in blind trusts, pursuing multiple sources of funding, and developing separate research institutes:* Although these steps will not eliminate COIs, they may help minimize their impact.

4. *Ensure that research conducted at universities is not suppressed:* Time and again, problems have arisen when private corporations have suppressed research. Although companies should be able to control research conducted in private laboratories, and they should have some input into the publication of any research they sponsor, research conducted at universities should not be barred from the public domain. One way that universities can take an active approach to this problem is require that researchers who conduct their work in university facilities do not sign contracts granting companies the right to stop publication of the results.

5. *Be especially wary of involving students in secret research conducted for industry:* Students enroll in universities for education and career advancement and should not be exploited as a cheap source of labor for commercial research. Because students should be allowed to discuss their work with their colleagues and teachers, no student master's or doctoral thesis should involve secret research. Some involvement in private research can help students learn and advance their careers, but great care should be taken in involving them in this work.

6. *Develop intellectual property policies that protect the rights of researchers as well as the interests of the university, industry, and the public, and provide researchers with proper legal counsel so they will be more willing to share data while they are filing for a patent:* Agreements between universities and industry, such as cooperative research and development

agreements, should address concerns about data sharing and management related to intellectual property.

7. *Use contracts and agreements to set the terms for university–industry collaborations, such as intellectual property rights, publication, sharing of data and resources, research oversight, auditing, and use of funds.*

Some other management strategies include selective recusal from parts of the design, additional oversight of recruiting of research subjects and adverse event reporting, and special approval of consulting and other arrangements.

QUESTIONS FOR DISCUSSION

1. What do you think of the development of university–industry relationships in the twentieth century?
2. What kind of university–industry relationship would you like to see in the twenty-first century?
3. Give some examples from your own experience of university–industry relationships that were acceptable and not acceptable to you.
4. Are there or should there be different moral limits in industry, academia, or society?
5. Can you give some examples of benefits to academia from industry, and vice versa, that are not listed in this chapter?
6. What proposals do you have for managing university–industry relationships?

CASES FOR DISCUSSION

Case 1

A biotechnology company is giving $2 million for three years to a researcher at a university to test his new diagnostic device on patients for an FDA-approved protocol. The researcher is one of the few experts in the field. The researcher and his department chair are both on the company's scientific advisory board. They both receive payments of $5,000/year for this service.

- What are your concerns about this arrangement?
- Should a COI committee be involved?
- How would you manage the issue, if any management is needed?
- Would you answer differently if both had equity interest n the company?

Case 2

A large pharmaceutical company awards a $6 million contract for three years to a researcher at a university to conduct trials on patients. The company's proposed conditions are as follows:

1. Six-month delay in publication of research outcomes
2. First rights of patenting and licensing of the discoveries

3. The researcher serves on its scientific advisory board
4. The researcher will conduct the research

- What is the role of the university in this arrangement?
- Should the university accept the terms as they are offered or make a counteroffer? What should the counteroffer be?
- How would you manage the conflict, if one exists?

Case 3

A large pharmaceutical company pays a researcher at a university consulting fees of $10,000 per year, and once or twice a year an honoraria of about $3,000 to give a talk. The same researcher also serves on one of the FDA advisory councils.

- Do you have any concerns? What are they?
- How should these concerns, if any, be managed?

Case 4

A fifth-year graduate student about to receive her doctoral degree discovers through gossip at a national meeting that her research has been tapped by a company. After talking to her mentor, she learns he has been discussing patents and obtaining a contract to develop the discovery into a patent. The student was told that all this would occur in the future and that she has nothing to do with it. She was also told that her data will earn her the PhD she sought and that is what the work was for.

- Do you see any problem?
- Does the student have any grievance toward her mentor?
- What should the student do, if anything?
- How you would have dealt with this situation if you were the mentor?

Case 5

Five large industrial plants near a university are planning to give $12 million each to endow a Center for Environmental Science (CES). The total gift to the university will be $60 million. As a condition of the gift, the companies require the following:

1. The CES will have a governing board consisting of 10 members of academia and industry; five members of the board will represent the companies giving the gift.
2. The governing board will have the power to appoint a director of the CES and will also approve all faculty appointments to the CES.
3. The governing board will also have the power to approve course development activities, conferences, guest lectureships, and research projects sponsored by the CES.

- Do you see any problems with this arrangement?
- Should the university accept this gift?

6

Authorship

Authorship is a prized commodity in science, because most of the tangible rewards of academic research are based on a person's publication record. Since the twentieth century, the number of authors per scientific paper has been steadily increasing. Authorship disputes are also becoming much more common. This chapter explores some of ethical issues and problems involving authorship and describes some criteria and guidelines for authorship.

An author has been traditionally defined as someone who creates a literary text, such as a novel, a history book, or a newspaper editorial. In science, an author is someone who creates a scientific text, such as an article published in a journal, a poster presented at a meeting, a chapter in a book, or a monograph. For many years, determining authorship was a fairly simple process in science because most scientific works had, at most, a few authors, who were all involved in conducting the research and writing the text. For example, Robert Boyle was the author of *The Sceptical Chemist*, Isaac Newton was the author of *Principia*, Charles Darwin was the author of *The Origin of Species*, Sigmund Freud was the author of *Civilization and Its Discontents*, and James Watson and Francis Crick authored a paper describing the structure of DNA.

Determining authorship became more complicated in the second half of the twentieth century, as scientists began working in larger research groups and collaborating with other researchers in other institutes and disciplines (LaFollette 1992, McLellin 1995). Collaboration increased because research problems became more complex, demanding, expensive, and interdisciplinary. A division of labor arose as people worked on different parts of research projects. Instead of a single person designing, conducting, and analyzing an experiment and writing up results, many people became involved in different parts of the process. Today, an author listed on a paper may not be responsible for writing the paper, other than reading the final draft of the manuscript.

The number of authors on scientific papers has steadily increased. In biomedical research, the average number of authors per paper increased from 1.7 in 1960 to 3.1 in 1990 (Drenth 1996). It is not at all uncommon for a paper published in biomedicine to have five or more authors, and multiple authorship is now the norm (Rennie et al. 1997). The average number of authors for all fields rose from 1.67 in 1960 to 2.58 in 1980 (Broad 1981). In the 1980s, the average number of authors per article in medical research rose to 4.5 (LaFollette 1992). In physics, papers may have 50–500 authors. In the life sciences, 37 papers were published in 1994 that had more than 100 authors, compared with almost none with 100 authors in the 1980s (Regaldo 1995).

Although some disciplines in the humanities and the social sciences have bucked this trend, multiple authorship is also becoming more common in humanistic disciplines, such as bioethics (Borry et al. 2006).

This rise in the number of authors per paper has been accompanied by a rise in disputes about authorship (Wilcox 1998). Such disputes were the most common ethical problem reported by respondents to a survey of researchers at the National Institute of Environmental Health Sciences: 35.6% of respondents had experienced or knew about authorship disputes, and 26.7% had experienced or knew about undeserved authorship (Resnik 2006). A survey of 3,247 scientists conducted by Martinson et al. (2005) found that 10% of respondents admitted to inappropriately assigning authorship in the past three years. Of the misconduct allegations that the Office of Research Integrity (ORI) received in 1997, 27% were authorship disputes that had been improperly interpreted as misconduct (Scheetz 1999). In many of these cases, a person who did not receive authorship on a paper claimed that he had been the victim of plagiarism.

The reason for authorship disputes in science is not hard to understand: People fight about what they care about, and scientists generally care about authorship. Most of the tangible rewards in academic science, such as tenure and promotion, contracts and grants, honors and prizes, and status and prestige, are based on a person's publication record. Scientists are rewarded for the quality and quantity of the publications. Scientific databases, such as the Science Citation Index, keep track of author citation records, which measure how many times an author is cited by other authors. One can increase one's citation score by publishing high-impact papers and by publishing many papers. The phrases "publish or perish" and "authorship is a meal ticket" reflect the grim realities of academic life (LaFollette 1992). Many believe that this emphasis on quantity has undermined the quality of published work (Chubin and Hackett 1990). Publication is important to a lesser extent in the private sector. Private corporations may not care whether their employees publish, and they may even discourage publication. However, one's publication record may still be important in obtaining employment in the private sector.

UNETHICAL PRACTICES RELATED TO AUTHORSHIP AND CITATION

Researchers, ethicists, and scholars have documented a number of ethically questionable practices related to authorship:

- *Gift authorship:* A person is listed as an author as a personal or professional favor. Some researchers have even developed reciprocal arrangements for listing each other as coauthors (LaFollette 1992).
- *Honorary authorship:* A person is listed as an author as a sign of respect or gratitude. Some laboratory directors have insisted that they be listed as an author on every publication that their lab produces (Jones 2000).

- *Prestige authorship:* A person with a high degree of prestige or notoriety is listed as an author in order to give the publication more visibility or impact (LaFollette 1992).
- *Ghost authorship:* A ghostwriter is used to write a manuscript, who may not be involved in planning or conducting the research. In extreme cases, even the researchers are ghosts. Pharmaceutical companies have asked physicians to be listed as authors on papers in order to give the papers prestige or status. These "authors" have been involved in no aspects of planning or conducting the research or writing up the results—they are authors in name only (Jones 2000, Flanagin et al. 1998). Ghost authorship is common in industry-sponsored clinical trials (Gøtzsche et al. 2007).

In addition to the problems noted above, many people have been denied authorship when they deserved it. There are several reasons for preventing someone else from receiving credit for intellectual or other contributions. First, the quest for priority has been and always will be a key aspect of research: Scientists want to be the first to discover a new phenomenon or propose a new theory (Merton 1973). Many bitter battles in science have been fought over issues of priority, including the international incident over the discovery of the HIV virus involving Robert Gallo and Luc Montagnier described in case 1 below (Cohen 1994). A researcher may be tempted to not list someone as an author or not acknowledge that person's contribution in order to achieve an edge in the contest for priority. Second, concerns about priority and credit play a key role in intellectual property rights, such as patents. To solidify their intellectual property claims, researchers may attempt to deny authorship as well as inventorship to people who have contributed to a new discovery or invention. Although we focus on authorship in this chapter, much of our discussion also applies to questions about inventions and patents. Just as a person may be listed as a coauthor on a publication, they may also be listed as a co-inventor on a patent application (Dreyfuss 2000). Third, although multiple authorship is becoming the norm, researchers may want to hold the number of authors to a minimum to prevent authorship from being diluted. For instance, many journals cite papers with more than two or three authors as "[first author's name] et al." If a paper has three authors, the second and third authors may have a strong interest in not allowing someone to be added as a fourth author, in order keep their names visible.

Extreme examples of failing to grant proper authorship or credit include the following:

- Plagiarism, in which a person is not given proper credit or acknowledgment for his or her work. According to standard definition, plagiarism is wrongfully representing someone else's ideas, words, images, inventions, or other creative works as one's own. Plagiarism has been and continues to be a common issue in misconduct cases investigated by the U.S. government (Steneck 1999). There are many different forms of plagiarism, not all of which are as extreme as copying a paper word for word (LaFollette 1992, Goode 1993). Although most cases of leaving someone's name off a manuscript amount to authorship disputes rather than plagiarism, scientists should be mindful that they can face a charge of plagiarism if they do not give credit or acknowledgment to people who make original and meaningful contributions to the research.

- Self-plagiarism, in which researchers fail to acknowledge, to disclose, or to obtain permission to use material from their own previous publications. Although self-plagiarism is not as serious an ethical transgression as plagiarism, it is still an ethical problem. Although it does not involve the theft of someone else's ideas or creative works, self-plagiarism is a poor citation practice that may also involve copyright violations (LaFollette 1992). The best practice is to provide proper reference and obtain permissions (if necessary) when reusing one's previously published work.
- "Citation amnesia," in which a person fails to cite important works in the field. As noted in chapter 3, conducting a thorough and fair review of the background literature is one of the most important steps of scientific research. It also can be useful in ensuring that people who make important contributions to the literature receive proper recognition. For many, citation is almost as important as authorship. Those who fail to cite important contributors present the reader with inaccurate and biased information and unfairly deny others the rewards of research and scholarship (LaFollette 1992, Resnik 1998c).

CRITERIA FOR AUTHORSHIP

In light of these problems related to credit allocation in science, it is imperative that researchers discuss issues related to authorship and establish standards for authorship in research. In the past, many journals have used as the standard for authorship whether the person has made a "significant contribution" to the project through planning, conceptualization, or research design; providing, collecting, or recording data; analyzing or interpreting data; or writing and editing the manuscript (LaFollette 1992). We believe that this popular approach is vague and may not promote accountability, because it would allow a person to be listed as an author as a result of planning a project, even though this person might not know how the project was actually implemented. Likewise, a person may become involved in the project at the very end and not have an important role in planning the study.

Our basic ethical principle is that authorship and accountability should go hand in hand: People should be listed as an author on a paper only if they (a) have made a significant intellectual contribution to the paper; (b) are prepared to defend and explain the paper and its results; and (c) have read and reviewed the paper (Rennie et al. 1997, Resnik 1997). An author need not be responsible for every aspect of the paper, but he or she must be accountable for the paper as a whole. Deciding who should be listed as an author involves deciding who can be held accountable for the whole project. People may contribute to a project in important ways without being considered an author. Important contributors can be recognized in an acknowledgments section of a publication.

Our approach is very similar to the one adopted by the International Committee of Medical Journal Editors (2007, p. 3): "Authorship credit should be based on 1) substantial contributions to conception and design, or acquisition of data, or analysis and interpretation of data; 2) drafting the article or revising it critically for important intellectual content; and 3) final approval of the version to be published. Authors should meet conditions 1, 2, and 3."

There are at least three reasons for keeping a tight connection between authorship and accountability. First, granting authorship in the absence of accountability can lead to the irresponsible authorship practices. Second, if there is a problem with the paper, such as an error, omission, or potential misconduct, it is important to be able to hold people accountable. In many research misconduct cases, authors on the paper have tried to avoid culpability by claiming that it was someone else's fault and that they knew nothing about the problem (LaFollette 1992). Third, linking authorship and accountability promotes justice and fairness in research (Resnik 1997). It is unfair to deny authorship to someone who is accountable for the paper. Conversely, it is also unfair to grant authorship to someone who cannot be held accountable for the paper.

To foster accountability in authorship, we further recommend that research articles have a section at the end describing the different roles and responsibilities of the authors and contributors to a project (Rennie et al. 1997, Resnik 1997). For example, an article might say, "Joe Smith was involved in conception, data interpretation and drafting the manuscript; Karen Shaw helped with data collection and analysis, and editing the manuscript," and so on. A few journals, such as the *Journal of the American Medical Association* (2007) and *Nature* (2007), require that articles submitted for publication include this type of information. Accountability can sometimes be difficult to pin down, because it is difficult to give a complete accounting of every piece of data when a large number of researchers from different disciplines collaborate on a project. For example, when using some technologies in research, such as X-ray crystallography, different collaborators may work with only a part of the data, and no single person my be responsible for all the data. Whole-genome analysis studies and multisite clinical trials present similar concerns. A precise description of each author's contribution may help to resolve potential problems such as these.

To make sense of authorship issues, it is important to carefully describe the different roles and responsibilities in research, because these will determine who should or should not be listed as an author or as a contributor, as well as the order of the listing. Although research roles vary considerably across disciplines and even among different projects within the same discipline, people may contribute to a project in basic ways, based on our discussion of the process of research:

1. Defining problems
2. Proposing hypotheses
3. Summarizing the background literature
4. Designing experiments
5. Developing the methodology
6. Collecting and recording data
7. Providing data
8. Managing data
9. Analyzing data
10. Interpreting results
11. Assisting in technical aspects of research
12. Assisting in logistical aspects of research
13. Applying for a grant/obtaining funding
14. Drafting and editing manuscripts

One person may be involved in many of these different aspects of research, although different people may perform different roles. For instance, in academic research, the principal investigator (PI) is usually a professor who proposes the hypothesis, designs the experiments, develops the methodology, writes the grant proposal, and writes the manuscript. However, the PI may use undergraduate or graduate students or postdoctoral fellows to collect and record data, review the literature, or edit the manuscript. Students and fellows may also provide assistance in other aspects of research, as well. Lab assistants (or technicians) may assist with technical aspects the experiments and are not usually listed as authors. However, lab assistants could be listed as authors if they make an important intellectual contribution in addition to their technical expertise (Resnik 1997). Statisticians may help analyze data and interpret results. Other senior colleagues, such as the laboratory director, may also help the PI by providing data from the research or by helping with experimental design, methodology, or grant writing. The laboratory director is usually a senior researcher who may or may not be the PI. The director may supervise many different projects and many different PIs.

In an academic setting, the PI is supposed to be in charge of the research, but other lines of authority and responsibility are sometimes defined (Rose and Fischer 1995). In private industry, the lines of authority and responsibility are usually more clearly defined, because private industry works on a business model rather than an academic model (Bowie 1994). The laboratory director is in charge of many different projects and many different investigators, who may be assisted by technicians or other scientists. The laboratory director reports to the company's upper level managers. Decisions about which projects to pursue are based on market considerations, not on intellectual considerations. In the military, lines of authority and responsibility are delineated even more strictly, and decisions about funding are based on strategic considerations (Dickson 1988).

ADDITIONAL GUIDANCE ON AUTHORSHIP

In addition to the fundamental guidance on authorship discussed above, we propose the following guidelines to promote accountability, trust, and fairness in authorship:

1. For corporate authorship, at least one person should be listed as an author. Institutions, committees, associations, task forces, and so on, are listed as sole author in a growing number of publications. Although these types of collaborations are vital to research, they can raise serious problems with accountability if they do not identify one or more people who can defend the whole paper. If a publication lists only a corporate entity as an author, then reviewers, readers, and the lay public have no idea who to hold accountable or responsible for the project (International Committee of Medical Journal Editors 2007).

2. The order of authors should also reflect the importance of each person's role in the paper. Every publication should have a person who is the accountable author—who bears the most responsibility for the article. In some disciplines, the accountable author is listed first, and in others, last. The corresponding author should be the person who is responsible for drafting the paper, submitting it for publication, and responding to comments from reviewers. The accountable author and the corresponding author are usually the same person, but they need not be. Some writers (e.g., Diguisto 1994) have proposed quantitative methods for determining authorship order, but we do not find these methods especially useful, because judgments about importance are, in large part, value judgments and cannot be easily quantified (American Psychological Association 1992).

3. In general, all articles that result from a graduate student's thesis project should list the student as first author. In general, graduate students should not be assigned projects where there is very little chance of publication or where publication may be many years away. We include these recommendations to avoid some of the problems of exploitation of subordinates that unfortunately occur in research, which we discuss in more depth under the topic of mentoring in chapter 4.

4. To the extent that it is possible, papers should clearly state the contributions of different authors and should acknowledge all contributors. This policy encourages both accountability and fairness in that it provides readers with some idea of who is most responsible for different aspects of the paper (Nature 1999). Some writers have argued that it is time to do away with the outmoded category of "author" in scientific research and replace it with a different system of credit allocation that simply details how different people have contributed to the project (Rennie et al. 1997, Resnik 1997). We are sympathetic to this proposal but also recognize the difficulty of overcoming

long-standing traditions and institutional personnel policies that place a heavy emphasis on authorship.

5. Many of these guidelines also apply to the concept of inventorship. Inventors, like authors, receive credit for their work and can be held accountable. It is also important to promote accountability, trust, and fairness when collaborations result in an invention (Dreyfuss 2000).

CONCLUSION

Although authorship is not as morally weighty as other topics in scientific research, such as misconduct, conflicts of interest, or protection of human subjects, it is a topic that has great significance for scientists, students, and trainees because of its implications for career advancement. Furthermore, because authorship disputes are often linked to other ethical problems in research, such as exploitation of subordinates, bias, and even plagiarism, authorship has been and will continue to be an important topic in research ethics.

CASES FOR DISCUSSION

Case 1: The Dispute over the Discovery of HIV

One of the reasons that authorship is so important in science is that it helps to establish the priority of discoveries and inventions. Scientists are rewarded for being the first to make a discovery, propose a theory, or develop an invention (Merton 1973). Scientific journals are usually interested in publishing research that is new and original. The patent system (discussed in more detail in chapter 6) awards patents based on the originality of the invention. Many episodes in the history of science illustrate the importance of priority, including the dispute between Newton and Leibniz over the discovery of calculus, the agreement between Charles Darwin and Alfred Russell Wallace concerning publication of the theory of natural selection, and the race between Watson and Crick, Linus Pauling, Maurice Wilkins, and Rosalind Franklin to discover the structure of DNA.

In the 1980s, a bitter priority dispute erupted concerning the discovery of the HIV virus and patents on blood tests for the virus. Two prominent researchers, Luc Montagnier, from the Pasteur Institute in France, and Robert Gallo, from the National Institutes of Health (NIH), were both trying to isolate the virus that causes AIDS, and Montagnier named his strain LAV. Montagnier cultured infected immune system cells (T-lymphocytes) from an AIDS patient known as L.A.I. Montagnier also had blood samples in his laboratory from an AIDS patient known as B.R.U. At the same time, Gallo was attempting to culture T-lymphocytes infected with HIV in his laboratory. Gallo named the virus he was working with HTLV-IIIb. Both researchers believed they were working with different viral strains, and they exchanged samples. There was a spirit of cooperation among the different research groups, at least at first (Hilts 1991).

In 1983, Montagnier was preparing to submit a paper reporting his discovery of the B.R.U. strain to *Nature*. Montagnier received a call from Gallo, who told him that he was preparing to submit a paper to *Science* on the HTLV-IIIb virus. Gallo convinced Montagnier to switch journals, so that both papers would appear in *Science*. Montagnier sent Gallo his paper to submit to *Science*, but he forgot to write an abstract. Gallo wrote the abstract for Montagnier, which stated that Montagnier had discovered an HTLV virus strain, which would confirm Gallo's work on HTLV. Following publication of the two papers, Gallo received top billing for the discovery of HIV and Montagnier was largely ignored (Crewdson 2003). In addition to publishing papers claiming to discover HIV, both researchers sought to patent a test for the virus.

The researchers were stunned when genetic tests indicated that their viral strains were nearly identical (Hilts 1991). Montagnier charged Gallo with misconduct, arguing that Gallo had deliberately taken B.R.U. samples and claimed them as his own. This accusation led to a misconduct investigation by the NIH and ORI. A U.S. House of Representatives committee that oversees the NIH, chaired by John Dingell (D-Mich.), also looked into the case. Gallo was eventually cleared of these allegations, though Dingell accused the NIH of trying to cover up fraudulent research.

The French and U.S. governments fought a legal battle concerning patents on the HIV test. The two governments eventually reached an out-of-court settlement, signed by the presidents of both countries. Under the terms of the agreement, Montagnier and Gallo were acknowledged as codiscovers of HIV. The two governments would also split the royalties from patents on HIV tests, with 80% of the money allocated toward HIV/AIDS research (Hilts 1992). After investigating the different strains of HIV and examining their laboratory practices carefully, the researchers also agreed that the most likely explanation of the similarity between their strains was due to contamination of both strains by the virus from the patient known as L.A.I. According to this explanation, the virulent L.A.I. virus somehow infected B.R.U. samples in Montagnier's laboratory. Montagnier sent samples of the B.R.U. to Gallo. Gallo began working with the samples sent by Montagnier, and the L.A.I. virus also infected the other samples that Gallo had in his laboratory. Although they did not know it at the time, researchers in both laboratories isolated the exact same strain of the virus, because the L.A.I. virus contaminated both strains—LAV and HTLV-IIIb—in their different laboratories (Hilts 1991). Montagnier shared the 2008 Nobel Prize with two others, not including Gallo.

The dispute between Gallo and Montagnier had a negative impact on HIV/AIDS research and on the public's acceptance of the hypothesis that HIV causes AIDS. Leaders in some African nations openly questioned whether HIV causes AIDS, and some still do.

- Why do you think that Gallo and Montagnier had such a bitter dispute over scientific priority?

- What factors do you think contributed to the controversy?
- Do you agree with the way the case was handled by the scientists and the respective governments?

Case 2

A graduate student was studying at a university where his faculty mentor is also a full-time employee of the affiliated hospital. He and his mentor discovered a major environmental airborne toxin, and the media covered the story. The mentor then died suddenly of a heart attack. They had a manuscript already written and about to be submitted for publication in a prestigious journal. The chair of the department, who worked in a closely related area, became the new student's mentor. She took the manuscript and put her name as the senior author (last) and eliminated the name of the deceased. She had no involvement in the project prior to the death of the former mentor. The paper was published after minor revision.

- What should the student do?
- What should the mentor do?

Case 3

Dr. Thomas published a review article on recent developments in mitochondrial genetics. In the article, he mentioned his own work prominently but failed to mention the work of several other researchers who have made key contributions to the field. Dr. Thomas is clearly one of the most important researchers in the field, and he has done some important work. But he also has some long-running feuds with some of the other prominent researchers in the field, based in part on fundamental disagreements about the role of the mitochondrial DNA in human evolution. In his review article, in several instances he cited his own work instead of giving proper credit to colleagues who had published their articles before his.

- Is this plagiarism?
- Is it unethical?

Case 4

Drs. Clearheart and Wing are studying socioeconomic factors related to incarceration. They recently read a paper published by Drs. Langford and Sulu on the relationship between juvenile incarceration and the mother's age at birth. Clearheart and Wing believe that they may be able to incorporate the supporting data obtained by Langford and Sulu in their analysis, so they send them an e-mail asking them to share the data used to support their publication. (Due to space constraints, the journal published some of the data, but not all of the supporting data.) Langford and Sulu agree to send Clearheart and Wing the supporting data provided that Clearheart and Wing list them as authors on any publications that use the data. Dr. Wing is uncomfortable with agreeing to this request, because he believes that one should do more than just provide data to be an author. Dr. Clearheart wants to honor their request. "Our publication will be much stronger," he says, "if we incorporate their data into our analysis."

- What should they do?
- Is it ethical to name someone as an author for providing data?

Case 5

Three economists, Stroud, Jones, and Wicket, published a paper describing a mathematical model for representing collective bargaining. The following year, Stroud and two other authors, Weinberg and Smith, published a paper that applied the model to a collective bargaining dispute between a teacher's union and a school district. The paper included two paragraphs in the methods section that were identical, word for word, to two paragraphs in the methods section of the paper published by Stroud, Jones, and Wicket. The paper by Stroud, Weinberg, and Smith did not mention or even cite the paper by Stroud, Jones, and Wicket.

- Is this unethical?
- Is this plagiarism? Self-plagiarism? A possible copyright violation?

Case 6

Dr. Gumshoe is a family physician specialist conducting a clinical trial on a new treatment for foot wounds for people with diabetes. The treatment is a cream she developed. Although all of the compounds in the cream have been approved by the U.S. Food and Drug Administration (FDA), the treatment itself has not. The FDA has classified it as an investigational combination therapy. Dr. Gumshoe has completed the clinical trial, which has demonstrated that the new treatment is twice as effective as current therapy for diabetic foot wounds. She is getting ready to publish a paper and needs to decide (a) who should be an author and (b) the order of authorship. The following people have worked on the project:

Dr. Gumshoe developed the cream, designed the experiments, interpreted the data, and wrote the paper.

Dr. Wainwright is a pharmacist who helped Dr. Gumshoe develop the cream, interpret the data, and edit the paper.

Dr. Sabrunama is a biostatistician who helped to design the experiments and analyze and interpret the data and who read the paper but did not edit it for content.

Ms. Stetson is a nurse who provided treatment to patients in the study and collected data and who read the paper but did not edit it for content.

Ms. Williams is a medical student who provided treatment to patients in the study and collected data and who read the paper but did not edit it for content.

Mr. Gumshoe is Dr. Gumshoe's husband. He has been providing legal assistance to Dr. Gumshoe concerning her FDA and institutional review board applications and patents on the treatment. He read the paper and edited it for style and grammar.

Mr. Jensen is a pharmacy technician who works for Dr. Wainwright who helped to prepare the treatment.

Dr. Chu is a colleague of Dr. Gumshoe's. He has discussed the project with her several times over lunch. He encouraged her to initiate the clinical trial and has given her critical feedback.

Dr. Rogers is a diabetic foot specialist. He has provided Dr. Gumshoe with tissue samples and data that she has used in her study.

- Who should be an author?
- Who should receive an acknowledgment?
- What should be the authorship order?

7

Publication and Peer Review

This chapter provides a historical overview of scientific publication and peer review and describes the current practices of scientific journals and granting agencies. It also examines a number of different ethical issues and concerns that arise in publication and peer review, such as quality control, confidentiality, fairness, bias, electronic publication, wasteful publication, duplicate publication, publishing controversial research, and editorial independence. The chapter also addresses the ethical responsibilities of reviewers and concludes with a discussion of the relationship between researchers and the media.

A BRIEF HISTORY OF SCIENTIFIC PUBLICATION AND PEER REVIEW

Throughout history, advances in communication technologies have helped to accelerate the progress of science (Lucky 2000). Written language was the first important innovation in communication that helped promote the growth of science. The Egyptians used hieroglyphics as early as 3000 B.C., and by 1700 B.C. the Phoenicians had developed an alphabet that became the basis of the Roman alphabet. With the invention of writing, human beings were able to record their observations and events as well as their ideas. The Egyptians and Babylonians, for example, made detailed observations of the movements of constellations, planets, and the moon in the night sky, as well as the position of the sun in daytime. Ancient Greek and Roman scientists communicated mostly through direct conversations and occasionally through letters. Philosopher-scientists, such as Pythagoras, Hippocrates, Plato, Aristotle, Euclid, Hero, Ptolemy, and Archimedes, discussed their ideas with students and colleagues in their respective schools, academies, and lyceums. Although these scientists also published some influential books, such as Euclid's *Elements*, Plato's *Republic*, and Ptolemy's *Almagest*, books were very rare because they had to be copied by hand on papyrus rolls.

Egypt, especially the city of Alexandria, was the cultural province of Greece and later Rome. The Roman Empire built the largest network of roads and bridges in the world, which increased commercial and scientific communication between the Far East and Middle East and the West. From about 40 B.C. to 640 A.D., most of the world's recorded scientific knowledge rested in the great library of Alexandria. Invading forces burned the library three times, in 269 A.D., 415 A.D., and 640 A.D. Each time the library burned, scholars rushed to save books from being lost forever. To people in the modern world, the idea of there being only a single copy of a book that, if lost, is lost forever is almost inconceivable (Ronan 1982).

Aside from the development of written language, the invention of the printing press during the 1400s by the German goldsmith Johannes Gutenberg was the single most important event in the history of scientific communication. The Chinese had developed paper in the second century A.D. By the ninth century, they had printed books using block print and woodcuttings. These inventions were brought to the Western world through trade with Arabic/Islamic countries. Gutenberg added to these established printing technologies a key innovation, movable type, which allowed the printer to quickly change the letters of the printing template. In 1450 Gutenberg established a printing shop and began to print the first works using his press, which included sacred texts such as the Bible (1454) and prayer books, grammar books, and bureaucratic documents. The selection of books soon expanded to include guidebooks, maps, how-to books, calendars, and currency exchange tables. In 1543, two influential and widely distributed scientific works also appeared in print, Andreas Vesalius's *On the Fabric of the Human Body* and Nicolas Copernicus's *The Revolutions of the Heavenly Bodies*. During the sixteenth century many important scientific and technical books, often originally written in Latin or Greek, were translated into vernacular and printed (Burke 1995).

The printing press increased the rapidity and quantity of scientific communication like no other invention before or since (Lucky 2000). It also helped transform Europe from a medieval to a modern culture and thus helped stimulate the Scientific Revolution. The printing press helped transform Europe from a society based on oral tradition and memory to a literate society based on permanent, shared records and writing. People learned to read and developed a passion for books and learning. The Catholic Church lost control of the interpretation of the Bible when this work was translated into the vernacular, and people began fashioning their own interpretations of what they read, an event that helped spur the Protestant Reformation. In 1517, Martin Luther printed copies of his Ninety-Five Theses criticizing the church and nailed them to a bulletin board in his own church in Wittenberg, Germany. As others printed copies, Luther's complaints about indulgences, corruption, and paganism spread all over Europe in less than a month (Burke 1995).

By the 1600s, scientific books were common items in libraries throughout the world. But because it took several years to research, write, and publish a book, more rapid communication was needed, as well. Scientists continued to correspond with letters, of course. Indeed, the letters of Descartes, Galileo, Newton, Boyle, and Harvey are considered to be key documents in the history of science. An important step toward more rapid and more public communication took place in 1665, when the world's first two scientific journals, *The Philosophical Transactions of the Royal Society of London* and the *Journal des Sçavans*, were first published. The first scientific association, the Royal Society of London, was a private corporation formed in 1662 to support scientific research and the exchange of ideas. Its journal is still published today. The world's second scientific society, the Paris Academy of Sciences, was formed in 1666 as an organization sponsored by the French government. By the 1800s, many other scientific associations and scientific journals had arisen (Meadows 1992).

The advantage of scientific journals is that they provided rapid communication of ideas, data, and results. Journals can also print a high volume of material. However, journals face the problem of quality control. The Royal Society addressed this issue in 1752, when it started evaluating and reviewing manuscripts submitted to its journal (Kronic 1990). The Royal Society took this step because its members became concerned about the quality of papers that were published in the journal. Some of the papers it published had included highly speculative and rambling essays, as well as works of fiction (LaFollette 1992). Soon other journals followed this example, and the peer-review system began to take shape.

Other important innovations in communication technologies included the telegraph (1837), transatlantic telegraph (1858), telephone (1876), phonograph (1877), radio (1894), and television (1925–1933). Developments in transportation also enhanced scientific communication, because printed materials still must be transported in order to disseminate them widely. Important developments in transportation included the steam locomotive (1804), transcontinental railways (1869), the automobile (1859–1867), and the airplane (1903) (Williams 1987).

After the printing press, the computer (1946) is arguably the second most important invention relating to scientific communication. Computers allow researchers to collect, store, transmit, process, and analyze information in digital form. Computers are also the key technology in the Internet, which combines computers and other communication technologies, such as telephone lines, cable lines, and satellite transmitters, to form an information network. The Internet allows researchers to publish scientific information instantaneously and distribute it to an unlimited audience via electronic mail, discussion boards, and Web pages. The Internet also allows scientists to instantly search scientific databases for articles and information (Lucky 2000). It is like having a library, newspaper, shopping catalog, magazine, billboard, museum, atlas, theater, and radio station all accessible through a computer terminal (Graham 1999). Although the Internet is a monumental leap forward in the rapidity and quantity of scientific communication, it also poses tremendous problems relating to quality control, because anyone can publish anything on the Internet (Resnik 1998a).

THE CURRENT PEER-REVIEW SYSTEM

Although peer review had its origins in scientific publishing in the eighteenth century, it was not institutionalized until the twentieth century, when it began to be used to legitimize specific projects that required the expenditure of large sums of public funds (Burnham 1990). In 1937, peer review for awarding grants from the National Cancer Institute (NCI) was made into a public law. Today, all major federal funding agencies, such as the National Institutes of Health (NIH), the National Science Foundation (NSF), the Department of Energy (DOE), the Environmental Protection Agency (EPA), and the National Endowment for the Humanities (NEH), use peer review to make decisions on funding awards.

Peer review brought together two potentially conflicting concepts—expertise and objectivity (Shamoo 1993). The need for peer review of public funds arose in order to provide a mechanism of quality control and to prevent favoritism, the "old boy network," and fraud in allocating public money. The system was implemented in order to make objective decisions pertaining to the quality of grant proposals. Peer review arose in scientific publication in order to evaluate the quality of manuscripts submitted for publication and to provide a quality control mechanism for science. Another advantage of peer review is that it is less centralized and bureaucratic than other processes one might use to fund research (LaFollette 1994a). Scientists, not politicians or bureaucrats, control peer review. On the other hand, peer review also allows for a certain amount of government oversight, and thus it provides a venue for public participation in the setting of funding priorities (Jasanoff 1990).

There are numerous types of peer-review processes developed for variety of purposes. In this chapter, we consider three types of peer review of government contracts and grants, scientific publications, and tenure and promotion decisions.

Government Contracts and Grants

Grants are the most important area where peer review is used. Government grants provide public money to fund a large portion of research across the country. This is one area of peer review under direct public scrutiny where citizens can raise issues of fairness, equity, justice, and public accountability with the public. In addition to public funding agencies, many private organizations, such as private charities or pharmaceutical and biotechnology companies, sponsor research through grants to investigators or institutions.

A contract, on the other hand, is a specific formal agreement between the investigator and a government or private agency. A contract stipulates the conditions to be met, and both parties must agree to these conditions. A contract may or may not go through peer review. Even when contracts undergo peer review, it is selective and does not conform to the usual "independent" peer review used to evaluate grant proposals. Because contracts do not conform to the standards set by independent peer review, one may question the current practices in granting contracts in science, but we do not explore that issue here. We concentrate here on the peer-review standards for grants.

In the peer-review process for government agencies, the first step is the selection of peers and definition of "peers." The U.S. General Accounting Office (GAO; the forerunner of the current Government Accountability Office) report of peer review provides a thorough evaluation of this process. The selection of peer reviewers differs among the NIH, the NSF, and the NEH (U.S. General Accounting Office 1994). Further, the hierarchy of peer-review structure and its impact also differ at the three agencies. The NIH has an initial peer review at the proposal's study section level (it has more than 100 study sections). These study sections belong officially to the Center for Scientific Review (CSR), which reports directly to the NIH director and not to each institute within the NIH. The independence of the CSR from actual program managers at each institute preserves the integrity of the review process. However, in special cases, institutes can construct their own study sections, especially for large program projects and center grants. Study sections usually consist of 18–20 members. Their decisions are not final, because NIH advisory councils have the final authority. However, the councils usually follow the study section ratings. The executive secretary of each study section is the one who recommends new members for the study section, at the approval of senior administrators. The executive secretary keeps a large list of potential study section members, based on discipline, knowledge, gender, age, and so forth. The executive secretary is usually a career staff member of the NIH with an MD or PhD degree. The NIH uses nearly 20,000 reviewers annually (National Institutes of Health 2007b).

At the NSF, most of the initial peer review occurs electronically, via a secure Web site, followed by a meeting of a panel of 8–12 members who review the comments on the Web site. The panels are headed by NSF program directors, who are NSF career employees with advanced degrees. Each program director has the authority to select panel members and send materials to reviewers. Each program director also keeps a large list of potential reviewers based on discipline, publications, gender, race, age, and knowledge. NSF uses over 60,000 reviewers annually (U.S. General Accounting Office 1994).

At the NEH there is only one peer-review system, which is based on standing panels of about five members, who are selected from a large list based on qualification similar to the criteria used by the NIH and NSF. At times, NEH program managers solicit evaluation from external reviewers to supplement the panel's deliberations. Annually, the NEH uses about 1,000 scholars for this review process (U.S. General Accounting Office 1994).

In evaluating proposals, reviewers at government agencies consider and address a number of different criteria:

1. The scientific or scholarly significance of the proposal
2. The proposed methodology
3. The qualifications of the principal investigator (PI) and other participants
4. Prior research or data that support the proposal
5. The level of institutional support and resources
6. The appropriateness of the budget request
7. Dissemination plans
8. Compliance with federal regulations
9. The proposal's potential social impact

Reviewers take these factors into account and rate proposals based on a relative point scale. The NIH uses comparative merit scores based on the scores of several previous review cycles. These comparative scores are given as percentile scores from lowest (0 percentile) to highest (100 percentile). The NSF allows each program to develop its own scoring method, but each program uses the same scoring method for all proposals it reviews. NSF criteria for scoring are similar to those of NIH, except there is a larger variation in scoring since most scoring is done by mail. For both the NIH and NSF, there is a good relationship between ratings and funding decisions. However, funding decisions are also greatly influenced by the program or project managers. The NSF's program directors have much greater ability to bypass the scores in order to expand opportunities for women, minorities, and geographic areas and to initiate what they think are new and innovative projects. The NEH panels review proposals beforehand, deliberate at their meeting, and then score each proposal. The NEH's criteria for scoring are similar to those of the NIH and the NSF.

Private foundations and private corporations follow whatever review process they deem suitable for their purpose, and many use methods that evaluate proposals on ethnic, political, or religious grounds. Many private foundations, such as the such as the Howard Hughes Medical Institute and the American Heart Association, use methods for awarding funds very similar to the peer-review mechanisms employed by government agencies. However, there is very little public scrutiny of their decision-making process other than what little is prescribed by the Internal Revenue Service (IRS) for nonprofit status. Although these foundations are philanthropic and public service institutions that contribute to the advancement of knowledge, most are indirectly subsidized by taxpayers because they are classified by the IRS as nonprofit (501(c)(3)) organizations, which are tax-exempt. One could argue that these organizations should be held more publicly accountable if they receive tax breaks.

Most private and public organizations that awards research grants now use secure Web sites, e-mail, and other electronic communications for submitting, reviewing, and overseeing grants proposals. It is likely that advances in information technology will continue to help improve the efficiency of the process.

Journals

Although the concept of peer review for publication has its origins in the eighteenth century, in the twentieth century peer review evolved into a method for controlling quality of publications as well as providing a stamp of approval by the established peers in a given field. The peer review of manuscripts varies from journal to journal, although the usual procedure has the editor-in-chief (or an associate if it is a large publication) who selects two reviewers, often experts in the field, to review each manuscript. A few editorial offices of journals that receive a large volume of manuscripts, such as *Science, Nature,* and *New England Journal of Medicine*, screen papers for quality and suitability beforehand to select only a small percentage of manuscripts to send to reviewers. Acceptance rates at journals vary from less than 5% at the high-impact publications, such as *Science* and *Nature*, to close to 90%.

Reviewers are given guidelines on reviewing the paper, but in reality reviewers are like jurors in courts: They can ignore the guidelines and base their opinion on whatever factors they deem to be appropriate. Most reviewers, however, evaluate manuscripts based on the following criteria:

1. Appropriateness of the topic for the journal
2. Originality or significance of the research
3. Strength of the conclusions, results, or interpretations as supported by the data or evidence
4. The validity of the research methods and the research design, given the research goals or aims
5. The quality of the writing

Reviewers may recommend that a manuscript be

- Accepted as is
- Accepted following minor revisions
- Accepted following major revisions
- Rejected, but encouraged to resubmit with revisions
- Rejected, not encouraged to resubmit

Usually, if both reviewers highly recommend the acceptance or the rejection of the manuscript, the editor-in-chief follows that recommendation. However, when the reviewers are lukewarm in acceptance or rejection of the manuscript, or if the reviewers conflict in opinion, the editor usually has some latitude in making the decision. Some editors may simply accept or reject a paper, whereas others may send it to a third reviewer for deciding between two conflicting opinions. Of course, editors may also decide, in some rare cases, that the reviewers are completely mistaken and accept a paper that reviewers have rejected, or vice versa (LaFollette 1992).

High-profile cases of misconduct in published research, such as the Hwang stem cell case discussed in chapter 1, have caused some journals, such as *Science* and *Nature*, to rethink their misconduct policies. *Science*, for example, has decided to give additional scrutiny to high-impact papers. Despite the increased attention to issues of scientific misconduct, a survey (Redman and Merz 2006) found that few of the top 50 highest impact journals have adopted policies for dealing with misconduct.

Promotion and Tenure

In public and private academic and research institutions, personnel decisions such as appointment, promotion, and tenure are based on committee recommendations. The criteria followed in making these personnel decisions vary considerably among community colleges, small colleges, local universities, research universities, industrial research organizations, and government laboratories. Although small colleges and community colleges emphasize a person's contributions to teaching, advising, or service to the institution, most of the other institutions place considerable emphasis on a person's research record, that is, the quantity and quality of publications and presentations, and the person's success in obtaining contracts or grants. The committee members as well as relevant administrators are in essence conducting an internal peer review of the candidate's past performance and potential contributions. The committees also solicit reviews of publications and contracts/grants from established researchers/scholars from outside the institution. The input of internal and external peer reviewers is crucial to the personnel decisions. If the committee recommends a candidate for tenure, for example, then other administrators and committees will usually follow this recommendation. Because most institutions emphasize research over teaching, advising, and service, individuals who are seeking tenure or promotion often face difficult choices in balancing their commitments in these different areas.

The discussion of mentoring in chapter 4 notes that one way to improve mentoring is to ensure that researchers are adequately rewarded for high-quality mentoring. We recommend that mentoring play a key role in personnel decisions. Many commentators have also noted that the pressure to publish contributes to a variety of ethical problems in science, such as plagiarism, fraud, undeserved authorship, and careless errors (National Academy of Sciences 1992). To address these pressures to publish, it is important to consider ways of reforming the tenure and promotion system. One key assumption that needs to be examined is the idea that, for publication, "more is better." The emphasis on the quantity of publications, as opposed to quality, is a key factor in the pressure to publish. To address this issue, some universities have decided to review only a select number of publications and to focus on the quality of the candidate's work.

PROBLEMS WITH PEER REVIEW

The idea of peer review makes a great deal of sense. Indeed, peer review is regarded by many philosophers of science as a key pillar of the scientific method because it promotes objectivity and repeatability (Kitcher 1993, Haack 2003). Science can be "self-correcting" because scientists review and criticize research methods, designs, and conclusions and repeat experiments and tests. Indeed, it is hard to imagine how science could make progress without something like a peer-review system (Abby et al. 1994). The peer-review system has come under severe criticism in recent years. In this section, we discuss some ethical and epistemological problems with peer review. Our negative comments about some aspects of the system should not be taken as a condemnation of the system as a whole or of individuals who serve as reviewers. On the contrary, the peer-review system is key to scientific progress, and the overwhelming majority of individuals involved in peer review strive to do their very best.

Bias and Lack of Reliability

Numerous scholars have provided evidence that peer reviewers have significant bias and low agreement of opinions on the same proposal or potential publication (Cole et al. 1978, Cole and Cole 1981, Chubin and Hackett 1990, Bower 1991, Oxman et al. 1991, Roy 1993, Shamoo 1993, 1994a, Fletcher and Fletcher 1997). Some of the biases that can affect peer review include theoretical, conceptual, and methodological disagreements; professional rivalries; institutional biases; and personal feuds (Hull 1988, Godlee 2000). There is considerable evidence that the peer-review process may not catch even simple mistakes and that it is certainly not effective at detecting plagiarism and fraud (Peters and Ceci 1982, LaFollette 1992, Goodman 1994). From our own experience, we have come across many published papers that still have obvious and nontrivial mistakes.

Peer review is not good at detecting fraud (data falsification or fabrication) partly because reviewers usually do not have access to all the materials they would need to detect fraud, such as the original data, protocols, and standard operating procedures. Some journals have begun to use information technology to detect fraud. For example, statisticians have developed computer programs that can determine whether a data set has probably been fabricated or falsified. A fabricated or falsified data set will usually lack the random variation that is found in a genuine data set, because human beings tend to follow patterns when they choose numbers. However, a very astute perpetrator of fraud could evade these programs with a program that introduces random variation into fabricated or falsified data! One computer program that is impossible to evade can detect plagiarism. The program compares submitted papers with published papers for degrees of similarity. If two papers are very similar, or they contact similar sections, they can be flagged as potential plagiarism.

Because editors and grant review panels may reject papers or research proposals based on one negative review, biases can profoundly affect the peer-review process. As noted above, most scientific journals assign only two or three reviewers to a given paper. Although a peer-review panel includes many reviewers, usually only a few people are assigned as primary reviewers for a given proposal. Other panel members will not read the proposal very closely and will not be asked to make extensive comments. Because these panels rely heavily on primary reviewers, it may take only a single biased primary reviewer to eliminate a grant proposal.

This problem is especially pronounced in the realm of controversial research (Barber 1961, Chalmers et al. 1990, Godlee 2000). Controversial research does not fit neatly into well-established research traditions, norms, or paradigms. It is what the famous historian of science Thomas Kuhn (1970) dubbed "revolutionary science" (in contrast with "normal science"). Research can be controversial for a number of reasons: It may be highly creative or innovative, it may challenge previously held theories, or it may be interdisciplinary. Interdisciplinary work provides an additional challenge for reviewers because reviewers may come from different disciplines with different standards for review, and no single reviewer may have all the education or training required to give an overall evaluation of an interdisciplinary project.

Because reviewers often are established researchers with theoretical and professional commitments, they may be very resistant to new, original, or highly innovative ideas or ideas that challenge their own work. Historians of science have understood this phenomenon for many years. For example, Kuhn (1970) notes that quantum mechanics was not fully accepted, even by Einstein, until the old generation of Newtonian physicists died off. Although research carried out within specific disciplinary parameters (i.e., "normal" science) plays a very important role in the overall progress of science, the most important advances in science occur through controversial or "revolutionary" science (Kuhn 1970, Shamoo 1994a). History provides us with many examples of important theories that were resisted and ridiculed by established researchers, such as Gregor Mendel's laws of inheritance, Barbara McLintock's gene jumping hypothesis, Peter Mitchell's chemiosmotic theory, and Alfred Wegener's continental drift hypothesis.

The debate over cold fusion in many ways fits this model, although it is too soon to tell whether cold fusion researchers will ultimately be vindicated (Fleischmann 2000). Critics of cold fusion have lambasted this research as either fraudulent or sloppy, while proponents have charged that critics are close-minded and dogmatic. As a result of this controversy, it has been difficult to conduct peer-reviewed work on cold fusion, because mainstream physics journals select reviewers with strong biases against cold fusion. This has not stopped this research program from going forward, however. Cold fusion researchers have established their own journals and societies in response to these rebukes from the "hot" fusion community.

To provide objective and reliable assessments of controversial research, journal editors and review panel leaders should be willing to do what it takes to "open the doors" to new and novel work. If they close these doors, then they are exerting a form of censorship that is not especially helpful to science or to society. What it takes to open the door to controversial research may vary from case to case, but we suggest that editors and review panel leaders should always try to understand controversial research within its context. For instance, if an editor recognizes that a paper is likely to be controversial, he or she should not automatically reject the paper based on one negative review but should seek other reviews and give the paper a sympathetic reading.

The issue of publishing controversial research reveals an important flaw in the idea of quality control: If a journal tries to ensure that all articles meet specified standards related to quality, then it may not publish some controversial (but good and important) studies. One way to ensure that controversial ideas are not ignored is to publish a larger quantity of articles with less control over quality. Thus, the scientific community faces a dilemma of quality versus quantity. A partial solution to this dilemma is to increase the number of journals as well as the number of ways to disseminate information, to allow outlets for controversial studies. The advent of Internet publishing has increased the quantity of published material without a proportional increase in the cost of publication, which allows researchers to publish more data and results. It will still be difficult to get controversial work published in the top, peer-reviewed journals, but controversial work can still be published in mid-range to lower range journals, in Web-based journals, and so forth (Bingham 2000).

Fairness in Journal Review

As discussed above, the decisions of individual reviewers, editors, and panel leaders can control the outcomes of peer review. A great deal often depends on who is selected to review an article or proposal and on how the reviewers' assessments are interpreted. The system is designed in such a way that it is very easy for reviewers to affect outcomes in order help someone they know and like or to hinder someone they don't know or like. For example, an editor's biased choice of reviewers can have an impact on the review outcome.

Many factors can give unfair advantages (or disadvantages) to the authors of manuscripts. Factors such as the author's name and institutional affiliation affect the judgments of editors and reviewers, who are more likely to give favorable reviews to well-respected authors from prestigious institutions than to unknown authors from less prestigious institutions (LaFollette 1992, Garfunkel et al. 1994). Most scientific journals currently use a single-blind system, where authors do not know the identity of reviewers, but reviewers know the identity of the authors. One way that journals have attempted to deal with this problem is to use a double-blind peer-review system, where reviewers do not know the identity of authors.

There has been some debate about the strengths and weaknesses of double-blinded peer review. Several studies have shown that often the reviewers can still identify authors even when they are not given the authors' names or institutional affiliations (van Rooyen et al. 1998). The effects of blinding on the quality of peer review are not clear. Some studies have shown that masking the identities of authors improves peer review (McNutt et al. 1990, Blank 1991), while other studies indicate that it does not (Justice et al. 1998, van Rooyen et al. 1998). Laband and Piette (1994) examined 1,051 full articles published in 1984 in 28 economic journals. They then compared the citations of these articles in the following five years, 1985–1989. They found that papers from journals using double-blinded peer review were cited more often than were papers from journals using single-blinded review. These results suggest that the quality of published papers that have double-blind reviews is better on average than that of papers published that have single-blind review. However, the results may be because the journals that use double-blind review are better than or more prestigious than journals that use single-blind review. Even if the evidence does not show that masking the identities of authors improves the quality of peer review, one can still argue that this practice is important in promoting confidence, fairness, and integrity of peer review.

Many journals ask authors to suggest potential reviewers or to name individuals who should not serve as reviewers. The journals may use these reviewers, select their own, or use some combination of author-suggested and journal reviewers. The Internet journal *BioMed Central* (*BMC*) requires authors to suggest four reviewers. The editor can chose from his roster and from the author-suggested reviewers. All reviews are published alongside the paper, if accepted. A study by Wager et al. (2006) found that in author-suggested reviewers were more likely to recommend acceptance than were editor-chosen reviewers. However, excluding some differences in the overall recommendation, both sets of reviewers offered similar recommendations for improvement and gave reviews of similar quality.

Some writers have argued that open review (unblinded, unmasked review) would best promote integrity, fairness, and responsibility in peer review, because the current system allows reviewers to hide behind the mask of peer review (DeBakey 1990). In an open system, authors and reviewers will both know each other's identities. Some journals, such as *BMJ*, have experimented with open review. Others, such as the Public Library of Science (PLoS) journals, use an open system. The main argument for open review is that it should improve the quality and integrity of review. If reviewers know that their identities can be discovered and made public, they will be more likely to give careful, responsible, and ethical reviews (Davidoff 1998, Godlee 2002). However, one can argue that if reviewer identities are not anonymous, then reviewers may be less likely to be critical or unbiased, because they may fear repercussions from authors (LaFollette 1992). It might be especially difficult, for example, for a younger researcher to offer candid criticism to an established researcher, because the younger researcher might fear for repercussions to his career. Also, some experts may not want to serve as reviewers in an unmasked system. Some studies have shown the unmasking the reviewers may not improve the quality of peer review (van Rooyen et al. 1998).

Fairness in Grant Review

When it comes to grant review, there is evidence that people who know members of the review panels and who know how the process works are much more likely to have their grants approved than are people who lack this inside information (Shamoo 1993). Some studies have shown that the peer-review system may give favorable treatment to researchers from large or more prestigious institutions and that it may disfavor researchers from certain geographic areas. Men also get higher grant awards than do women (Marshall 1997, Agnew 1999a,b). Thus, something like an "old boys network" exists in peer review (Chubin and Hackett 1990, Glantz and Bero 1994, Armstrong 1997, Marshall 1997, Godlee 2000).

Fairness has become such an important issue in peer review partly because scientific resources and rewards, such as grants and publications, are scarce. When rewards are scarce, questions about how to distribute rewards fairly are paramount. Not every paper will be published in a top journal, and not every grant will be funded. Indeed, many top journals have acceptance rates of 10% or less (LaFollette 1992). After nearly doubling in five years, the NIH budget has leveled off and has begun to decline when inflation is taken into account. The NIH, which at one time funded about 30% of submitted proposals, now funds scarcely more than 20% of the proposals it receives (Gross et al. 1999, Malakoff and Marshall 1999, National Institutes of Health 2007a,b). Other agencies, such as the NSF, have not kept pace with NIH in recent years and tend to fund an even lower percentage of proposals. Some administrations with strong ideological views have attempted to influence the selection of peer reviewers (Ferber 2002).

The selection of peers to review grant proposals can profoundly affect the review process. Glantz and Bero (1994) showed that the professional interests of reviewers play a critical role in their level of enthusiasm for a grant proposal, which affects its score. The overall effect of reviewer selection is that it gives an unfair advantage to applicants who know the professional interests of potential reviewers. These applicants are likely to be those people who know the potential reviewers—the "old boys network."

The requirement of preliminary data can create an unfair disadvantage to beginning investigators or those who lack sufficient funding. Even though no federal agencies officially require preliminary data, the GAO report on peer review (U.S. General Accounting Office 1994) includes testimony from investigators who said that preliminary data are an essential component of a new grant application. Even more troubling for new investigators is where to obtain funding for their preliminary data. If investigators already have funding for another project, they must funnel some of that funding to obtain preliminary data for a new project. There are very few resources to produce truly preliminary data. Therefore, this policy pushes investigators into "theft" of funding from one project to another, an unethical zone of behavior that could manifest elsewhere. The fact that there is little or no funding for preliminary data also encourages investigators to "stretch the truth" on grant applications by overestimating the significance of their preliminary work (Grinnell 1992, LaFollette 1992). Recent survey data confirms such claims (Anderson et al. 2007).

Those who are involved in the peer-review process, especially members of advisory councils, can also gain unfair "insider information" relating to how the process works, what types of proposals agencies would like to fund, what it takes to write convincing applications, and so on. In securities trading, financial institutions or individuals can go to jail for using "insider information" for economic gains. In science, those with insider information are not punished but are free to use it and gain greater chances for funding for their projects. For example, evidence shows that NIH's members of the advisory councils (past, present, and nominated) are twice as likely to have their research grant applications funded compared with the rest of the scientific investigators, despite the fact that their merit scores by the study sections were no different (Shamoo 1993). According to the GAO (U.S. General Accounting Office 1994), 97% of NSF applicants who were successful in obtaining funding had been reviewers in the last five years. One may partly explain these results by arguing that insiders are more likely to obtain funding because they have better expertise, a better understanding of peer review, or better knowledge of research or methods. While we do not deny this point, we doubt that these factors can completely explain these results. We believe it is likely that funding decisions are often influenced by access to the right people with the relevant insider information.

Failure to Control the Quality of Published Work

There is evidence that peer review often does not improve the quality of published work. Some studies have shown that peer review may not be a very effective gate-keeping mechanism, because most articles rejected by more prestigious journals are eventually published elsewhere (Lock 1991). However, as noted above, this is not entirely a bad thing, because it ensures that controversial research is published. Studies have also shown that peer review is effective in improving the quality of published articles even if does not make articles perfect (Goodman 1994, Callahan 1998). As mentioned above, peer review is not an effective means detecting research misconduct. Peer review fails to stop publication of many papers with fabricated, falsified, or erroneous data.

IMPROVING THE PEER-REVIEW SYSTEM

Although the peer-review system is far from perfect, it does work very well for the vast majority of papers and grants proposals. Moreover, it is hard to imagine any reasonable alternative to peer review. Without peer review, researchers would have no way to control the quality of articles or funded research or to promote objective, reliable research—there would be no way to separate the wheat from the chaff. From its inception in 1752, peer review was never meant to be infallible. As Knoll (1990, p. 1331) states: "We tend to forget that just because peer review reviews scientific work does not mean that it is itself a scientific process."

On the other hand, the fact that we have no reasonable alternatives to peer review does not mean that we cannot or should not make an effort to improve the current system. It is important to conduct more research on how the current system works (or fails to work) and to experiment with different peer-review practices. To its credit, the NIH is considering ways to reform its system of grant review, such as awarding longer grant periods, reviewing grants retrospectively, inviting public participation in study sections, relaxing pressures to produce preliminary data, and finding ways to fund controversial research (Zurer 1993, Marshall 1997, National Institutes of Health 2007b).

Journals have also considered ways to reform the peer-review system. As mentioned above, some journals have experimented with open (unblinded) review. The jury is still out on open review. Another recent innovation by a number of journals is the concept of "target articles with open commentary." In this practice, authors submit an article for publication, which is peer reviewed. If the editors decide to publish the article and think it will generate interesting discussion, they then solicit or invite commentaries to be published along with the article. They may also use e-mail discussion boards to allow further debate (Bingham 2000). The advantage of this system is that it combines the virtues of quality control (blinded peer review) with the virtues of openness, diversity, and accessibility. We encourage journals and researchers to experiment with other innovations in peer review.

A challenge for journals and granting agencies is to find ways to give a fair review to creative and controversial projects or papers, as discussed above. Because science often makes progress through creative or unconventional leaps, it is important to test innovative or controversial ideas. Because a few of these ideas may bear fruit, editors and granting agencies may be willing to take some risks in order to reap potentially high rewards. Because these controversial projects may use unorthodox methods, it may be necessary to modify the normal peer-review process somewhat to accommodate these proposals. For example, granting agencies could loosen their informal requirements of prior work as well as their demands to produce results within certain deadlines. To ensure that research is still of high quality, agencies could perform more frequent site visits and data audits.

ETHICAL DUTIES IN PEER REVIEW

This is anecdotal evidence that reviewers have acted unethically during the peer-review process by violating confidentiality, stealing ideas, and so on (Dalton 2001, Lawrence 2003). The Committee on Publication Ethics (2007) has developed some ethical guidelines for reviewers and editors, and various granting agencies have their own rules and policies for peer review. In this section, we briefly describe some ethical duties of editors, panel directors, reviewers, and others involved in peer review and give a brief rationale for these duties.

Confidentiality

Everyone involved in the peer-review process should maintain the confidentiality of materials being reviewed. Papers submitted for publication or grant proposals often contain data, results, ideas, and methods that have not been previously published. This material belongs to those who are submitting the paper or proposal, and it is privileged information. If a reviewer discusses a paper with an outside party, a paper still in the review process, then that outside party could use that information to his or her advantage, for example, to steal the ideas or results. The whole system of peer review would collapse if those who submit papers or proposals could not trust that their work will remain confidential (LaFollette 1992, Godlee 2000). Most granting agencies and journals now require reviewers to agree to treat all materials that they review as confidential, and most even require reviewers to destroy papers or proposals after they complete their review.

Respect for Intellectual Property

The ideas, data, methods, results, and other aspects of a paper or proposal submitted for review should be treated as the intellectual property of the authors or PIs. Those involved in the process should therefore not use any of this property without the explicit permission of the authors or PIs (Godlee 2000). Many scientists can attest to having their ideas stolen during peer review. One common scenario is as follows: Researchers submit a paper to a journal. The journal takes a long time to make a decision on the paper and then finally rejects the paper. In the meantime, another research team publishes a paper on the exact same problem using identical methods with almost identical results. Is this sheer coincidence, or theft? Most researchers will not be able to prove that their ideas have been stolen, but they will remain suspicious. However, many researchers are so wary of this problem that they will omit important information from papers in order to prevent others from replicating their work before it is published (Grinnell 1992, LaFollette 1992). It almost goes without saying that the peer-review system could not function if authors or PIs could not trust that their work would be protected from theft. Why would anyone submit a paper or proposal if they thought someone else would steal their work?

Addressing Conflicts of Interest

If reviewers have a personal, professional, or financial interest that may undermine their ability to give a fair and unbiased review, then they should declare that conflict to the relevant parties, such as editors or panel leaders (Shamoo 1994a, Resnik 1998a, Godlee 2000). One very common type of conflict of interest in peer review is when the reviewers have relationships with authors or PIs, who may be current or former students, colleagues at the same institution, or bitter rivals. Another common type of conflict of interest is when the reviewers have financial interests related to the work they are reviewing, such as stock in the company sponsoring the research or a competitor. Although it is important to disclose these conflicts, some should be avoided altogether, such as reviewing the work of a current or former student or a colleague at the same institution. Many granting agencies and journals also now require reviewers to disclose conflicts or interest. Chapter 10 discusses conflicts on interest in depth.

Punctuality

Reviewers should complete their reviews within the stated deadlines, especially for papers submitted for publication. If reviewers cannot complete the review by the deadline, they should not accept the assignment and may recommend someone else (LaFollette 1992). Scientific research occurs at a very rapid pace. Slow reviews can have adverse effects on the careers of researchers: One research team may be "beaten to the punch" by a different research team while they are waiting for their work to be reviewed. In research related to public policy or current events, much of the material in a paper submitted for publication may go out of date if the paper is not published on time. Researchers may fail to get tenure, promotion, or a good rating if they have some major papers that have not been reviewed in a timely fashion. Although the length of review varies from journal to journal, most have a decision within six months of submission, and many have a decision in less than three months. With the use of e-mail and electronic submission, review can be even more rapid. Authors should not hesitate to contact editors if they are concerned about the status of a manuscript or its stage in the review process.

Professionalism

Reviewers should conduct careful, thorough, critical, and responsible reviews of papers or proposals. Reviewers should not offer to review a paper or proposal if they lack the expertise to make an informed judgment about its quality. In writing comments, reviewers should avoid insults, personal attacks, and other unprofessional remarks and make every effort to provide authors or PIs with useful comments for improving the manuscript or proposal (LaFollette 1992, Resnik 1998a).

Although these ethical rules may seem obvious, they are not always observed. Unfortunately, many researchers have had to deal with unprofessional reviews. Although senior researchers can develop some understanding of the peer-review process and learn to accept harsh and irresponsible criticism, beginning researchers and graduate students may have an especially difficult time dealing with unprofessional reviews of their work. Insults can undermine a person's self-confidence and trust in his or her colleagues. Low-quality reviews are often of little use to the author or PI. Who can learn anything from a review of a manuscript that says "reject" and does not even offer a reason for rejecting the manuscript? Worse yet are reviews of manuscripts that are so poor that it is clear the reviewer did not understood or even read the manuscript.

OTHER ISSUES IN SCIENTIFIC PUBLICATION

Wasteful and Duplicative Publication

There is a tendency to publish papers according to the least publishable unit (LPU). Many commentators have criticized this practice as wasteful and irresponsible (Huth 2000, Jones 2000). We concur with this assessment and encourage researchers not to divide substantial papers into LPUs. However, we recognize that there are sometimes advantages to dividing larger papers into smaller parts, because a large paper may cover too many topics or take up too much room in a journal. Authors should divide papers to improve the quality and clarity of their arguments and analysis.

The practice of duplicative publication has also been criticized (Huth 2000). For example, some researchers have published the exact same paper in different journals without telling the editors. Others have published papers substantially similar to papers they have published elsewhere (LaFollette 1992). Duplicative publication is unethical because it is wasteful and deceptive: Most journals expect that papers submitted for publication have not already been published, even as a different but substantially similar work. Some journals now ask authors to certify that the manuscript is original and has not been previously published. Some types of duplicative publication are warranted, however, because they may serve important educational purposes, for example, when papers are published in a different language or for a completely different audience. If an older paper has historical significance, it may be appropriately republished so that the current generation of researchers can become familiar with it. In all of these cases where duplicative publication is appropriate, the editors should clearly reference the original work.

Multiple Submissions

Most journals forbid multiple submissions, and most require authors to certify that their manuscripts are not under submission elsewhere (LaFollette 1992). There seem to be two reasons for this practice. The first is to save resources: If a journal is going to go to the trouble to review a paper, then it does not want the paper snatched by another journal. The second is to avoid disputes among journals: Journal editors do not want to have to negotiate with other journals or authors for the rights to publish papers. On the other hand, the rest of the publishing world does not follow this exclusive submission policy. Law journals, poetry journals, magazines, newspapers, and commercial and academic presses allow multiple submissions. Indeed, one could argue that the peer-review process would be improved if journals did not require exclusive submissions, because this would force journals to compete for papers and to improve the quality and punctuality of the review process. Established researchers who can afford to wait can often choose among competing journals, even if they only submit a paper to one journal at a time. If researchers submit a paper to a journal and decide that they do not want to make changes recommended by the reviewers, then they can submit the paper to another journal.

Editorial Independence

Many journals are sponsored by professional organizations. For example, the American Association for the Advancement of Science sponsors *Science*, the American Medical Association sponsors the *Journal of the American Medical Association* (*JAMA*), and the Federation of American Societies for Experimental Biology sponsors the *FASEB Journal*. Other journals are sponsored by government or para-governmental organizations, such as the *Proceedings of the National Academy of Sciences of the USA* (*PNAS*) and the *Journal of the National Cancer Institute*. Some journals are sponsored by private think tanks, such as *The Hastings Center Report* and the *Cato Policy Report*. Other journals are sponsored by religious, political, or business organizations. Some journals are completely independent and have no sponsors.

For those journals that have official sponsors, problems relating to editorial independence sometimes arise. In one recent example, it appears that the American Medical Association fired *JAMA*'s editor George Lundberg after he decided to publish a controversial study on sexual practices and perceptions among adolescents (Anonymous 1999). We strongly support the idea of editorial independence as vital to the peer-review process and the progress and integrity of research. However, we also recognize that private organizations have a right to control the materials they publish. A Jewish organization, for example, has no obligation to publish anti-Semitic writings. One way to settle these conflicts is for organizations that sponsor specific publications to publicly state the editorial goals of those publications. This will allow readers and prospective authors to understand a journal's particular editorial bent and to plan accordingly. If authors realize that their work goes against that stated aims of the journal, they may submit their work elsewhere.

Controversial Research

We have already discussed questions related to reviewing controversial research, but ethical dilemmas can also arise even when the reviewers agree on the merits of a controversial paper. The editors may still have some reservations about publishing the paper even when the reviewers recommend publication as a result of the potential social or political implications of the research. For example, in 1998 *Science* and *PNAS* both published papers on human embryonic stem cells. Many people oppose research on embryonic stem cells because it involves the destruction of human embryos, but these journals decided to publish these papers to promote progress in these important new areas of research and to help stimulate public debate on the topic (Miller and Bloom 1998). Other areas of research that have generated a great deal of public debate (and at times acrimony) include cloning, human intelligence, genetic factors in crime, sexuality, and global warming. Although we recognize that editors must consider the social, security, or political implications of articles, their primary obligation is to publish new findings and stimulate debate, not to protect society from "harmful" or "dangerous" research. Only a few decades ago Alfred Kinsey and his colleagues had difficulty publishing their research on human sexuality because of its controversial nature. Indeed, much of this research was conducted with private money and published with private support.

Research Raising Security Issues

The specter of terrorism looms over the twenty-first century. Terrorists have killed thousands of people, destroyed major buildings such as the World Trade Center in New York, and disrupted commerce and trade. Terrorists groups have used biological and chemical weapons and continue to seek other weapons of mass destruction, including nuclear weapons. Given the significance of this threat to national and international security, it is incumbent on all people, including scientists, to help prevent, discourage, mitigate, or respond to terrorism. In recent years, several publications in scientific journals have raised some security concerns. These have included a paper on how to make pox viruses more virulent, a paper on how to use mail-ordered materials to manufacture a polio virus, and a paper proposing a mathematical model for infecting the U.S. milk supply with botulism (Resnik and Shamoo 2005). These and other publications have contained information that could potentially be used by terrorists, criminals, or others to launch attacks on people and infrastructure.

Security issues related to the publication of scientific research are not new. Since World War II, research with pertaining to biological, chemical, and nuclear weapons has been classified. Governments have sought to pull a veil of secrecy over many types of research, including research on cryptography, missile guidance, and missile defense. While there is real need to keep some research secret to protect society from harm, this conflicts with the ethics of scientific openness. As we mention several times in this book, the sharing of information (data, ideas, and methods) is essential to advancement of science. Restrictions on publication, such as treating some research as classified or censoring some types of research, undermine scientific progress. Additionally, the sharing of information is important for achieving practical goals, such as preventing diseases, promoting public health, and protecting against threats to security. The problem is that scientific research almost always has multiple uses: Research on retroviruses could be used to understand viral infections, develop vaccines, create a biological weapon, or prepare for a defense against a biological weapon (Couzin 2002). A policy designed to stop the flow of information used to make biological weapons could also impede basic biological research or the development of medical interventions. Researchers and government officials must use discretion and good judgment when considering or implementing restrictions on the sharing of scientific information.

Any policy on the sharing of scientific information with security implications should consider the following points.

Classified versus Unclassified Research   In the United States, research is considered to be unclassified unless it is classified by a particular government agency with the authority to classify information. If information is classified, then access to it is restricted to people who have a need to know the information and have the appropriate security clearance. Most of the classified research conducted in the United States is sponsored by a government agency with a specific security agenda, such as the Department of Defense, the Department of Homeland Security, or the National Security Agency. Although the government has the authority to classify academic research or privately funded research, since the 1980s the policy of the U.S. government has been to not interfere with research that would normally be open to the public, such as academic research. We endorse this policy: Only in extraordinary circumstances should academic research be changed into to classified research.

Self-Regulation versus Government Regulation   The current system in the United States allows scientists to control the publication of unclassified research. Another alternative to this is for the government to control information that is unclassified. For example, a government panel could oversee the publication of scientific information with national security implications and, if necessary, takes steps to control it, such as censorship, limited publication, or classification. We think that this policy would be unwise for several reasons. First, the scope of the problem is so large that no single government agency or panel could handle it. Thousands of scientific papers are published every day in thousands of scientific journals. The government does not have the resources to examine each paper, but scientific journals do have the resources. Second, allowing the government to oversee the publication of scientific information by journals could easily lead to political manipulation and other abuses of power in the name of "national security." Lyndon Johnson, Richard Nixon, Ronald Reagan, George H.W. Bush, George W. Bush, and many other U.S. presidents have classified information for political purposes.

It would be unwise to give the government additional power to control information. We believe, therefore, that the best policy is self-regulation by scientists. However, for self-regulation to be effective, scientists need to cooperate with government officials so that they will have sufficient information and advice to decide how to deal with papers that have national security implications. The government should provide journals with the information and resources they need to make these difficult decisions. It may be necessary, in some cases, to share classified information with journal editors. Since scientists have a vested interest in publishing research rather than restricting access to it, they should be mindful of this potential bias, and they should give a fair hearing to national security concerns (Resnik and Shamoo 2005, Miller and Selgelid 2007).

Electronic Publication

Many journals today, such as *Science, Nature*, the *New England Journal of Medicine*, and *JAMA*, publish simultaneously in paper and electronic form. Some journals are now entirely in electronic form, and almost all journals publish article abstracts in electronic databases. Many books are also now available online. It is likely that these trends will continue and that one day almost any article, book chapter, or book will be available in some form through the Internet and electronic databases and bibliographies. Internet information services, such as Google Scholar, Blackwell, Informaworld, and Springer, have already made tremendous strides in this direction. Given their high costs, printed journals may soon become obsolete (Butler 1999b).

Electronic publication offers many important benefits to researchers. It can allow more rapid review and publication. It can increase access to publication for those who do not have access to regular print journals. Researchers can more easily search for specific articles, topics, or authors with a search engine, which can locate in less than a second a document that may have taken a month to find using older methods. Electronic publication also increases the quantity of published material and reduces the cost.

Despite all of these advantages, electronic publication also has some disadvantages, as mentioned above. Chief among them is the problem of quality control: how to separate the wheat from the chaff, to tell the difference between good science, bad science, pseudoscience, and fraudulent science. Before electronic publication, researchers could rely on the reputation of the journal to help them make judgments about the quality of published material, a method still useful today. However, as more and more electronic articles are published outside of well-established journals, it will become much more difficult to rely on journal reputation. Also, many articles are published directly on Web pages, with no supervision by journal editors.

The controversy over the NIH's "E-Biomed" proposal illustrates the problem of quality control (Marshall 1999a,c). E-Biomed would link all electronic biomedical publications in one public Web site. Critics worried that E-Biomed would increase the quantity and speed of biomedical publication at the expense of quality control (Relman 1999). Although there are problems with the specific E-Biomed proposals, many researchers still endorse the idea of a central electronic publication index, such as PubMed. One way to make this proposal work would be to develop a Web site that links existing peer-reviewed journals but does not bypass the normal mechanisms of peer review (Butler 1999a).

Since 2001, an increasing number of journals have begun to publish articles via an open-access system. Under the traditional publishing system, authors transfer their copyrights to the publisher, and the publisher charges a fee for access to the articles. Publishers earn their revenue under this system by selling copies of the journal or access to articles. Under the open-access system, authors pay an up-front fee to publish their articles, which are then made available to readers free of charge (Smith 2004). Thus the publishers derive income not by selling products or by providing access to articles, but from fees paid by authors. PLoS, a private nonprofit organization, now publishes many Internet open-access scientific and medical journals; many high-impact journals, such as the *New England Journal of Medicine* and *JAMA*, also publish selected articles on an open-access basis. The chief advantages of an open-access system are that it promotes openness and the dissemination of knowledge, and it allows researchers who cannot afford to purchase articles, for example researchers from developing countries, to have access. A problem with open access is that it can be very costly to finance, with journals needing to charge authors several thousand dollars to publish an article. However, the costs of open-access publishing have declined since the system's inception: PLoS charges $1,500 to publish an article. In 2005, the NIH adopted an open-access policy (Smith 2004). It now requires that all of the articles that are funded by NIH are available to the public for free on its Web site. Publishers have objected that this policy will put them out of business by undercutting the fees they charge to access NIH-funded articles (Kaiser 2005b). It remains to be seen whether the traditional publishing model can survive in the electronic age.

Data Access

We discuss data access issues in depth in chapter 3, but we mention them briefly again here. Access to data has become an issue in publication in recent years as private companies have sought to protect their proprietary interests in research data while also publishing the data. Researchers are publishing DNA sequence data as they become available from the human, mouse, rice, fruit fly, and other genomes. For research funded through public funds, data are available free of charge. Private companies that publish these data, such as Celera Genomics Corporation, are making data available free of charge for academic scientists but for a subscription fee for all for-profit users. Many researchers cried "foul!" when *Science* magazine published Celera's human genome data because of Celera's policy of charging for-profit users a fee (Kennedy 2001). These scientists argued that all published genome data should be deposited in a public and free-access site, such as GenBank. Although we believe in openness in science, we recognize that data also have economic value, and the best compromise may involve some combination of publication with fees for access or use. Many private companies, such as Celera, are investing heavily in research and are producing important results that benefit science. Companies need some way to obtain a return on their investments, and a fee for data access to nonacademic researchers seems reasonable. Private companies may also charge fees for data services, such as indexing and analysis.

Science and the Media

Many scientific discoveries and results have a significant bearing on public health and safety, education, the economy, international relations, criminal justice, politics, and the environment. These research findings are newsworthy in two different ways: First, they are events that the public should know about; second, they are events that many people find intrinsically interesting. For many years, scientific discoveries and results have been reported in newspapers and magazines and on radio and television. Science is also discussed in popular fiction and movies. We believe that scientists have a social responsibility to report their findings to the media because these findings are often useful to the public. Scientists also have an obligation to educate and inform the public. Research findings can help prevent harm to individuals or to society and can promote general welfare. Reporting discoveries and results in the press can also enhance the public's understanding of science. However, some problems can arise when scientists interact with the media.

One problem concerns publication in the media prior to peer review, or "press conference" science. The controversy over cold fusion illustrates this problem. Stanley Pons, chairman of the chemistry department at the University of Utah, and Martin Fleischmann, a chemistry professor at Southampton University, announced their results at a press conference on March 23, 1989, in Salt Lake City, Utah. They claimed to have produced nuclear fusion at room temperatures using equipment available in most high school laboratories. Researchers around the world rushed to try to replicate these results, but the Utah press release did not provide adequate information to replicate these results. Pons and Fleischmann made their announcement to ensure that they would not be scooped by other research groups in the race for priority, and to protect their pending patent claims. Once other researchers failed to replicate their work, mainstream physicists came to regard their work as "careless," "irresponsible," or simply "loony." Although cold fusion research continues to this day, and Pons, Fleischmann, and others continue to publish results, the cold fusion community has become isolated from the traditional fusion community.

In retrospect, one wonders whether this press conference actually hindered the cause of cold fusion by calling too much attention to the subject before the work was adequately peer reviewed. If the work had been published in a low-profile physics journal and others with an interest in the subject had reviewed the work in an orderly fashion, then cold fusion research could still have been regarded as controversial but not "irresponsible" or "loony." Fleischmann (2000) considered this option and favored delayed publication in an obscure journal. Thus, this press conference may have hindered the progress of science. Another problem with this prior publication in the media is that it wasted valuable resources as researchers scrambled to replicate the results with insufficient information. This probably would not have happened if this research had been published in a peer-reviewed journal first.

The whole cold fusion episode also undermined the public's trust in science in that the researchers made extraordinary claims in their press conference that have not been verified. This makes scientists look foolish or incompetent. Other types of press conference science have the potential to do much more damage to science's public image as well as to the public. Premature announcements of research on diseases or new medical treatments can have adverse effects on public health and safety (Altman 1995). For example, if researchers announce in a press release that they have discovered that a blood pressure medication has some serious drug interactions, then patients may stop taking the medication, and some may have a hypertensive crisis as a result. If this claim has not been subjected to peer review and turns out to be false, then the press conference will have resulted in unnecessary harm to patients. However, if researchers have very good evidence, and they feel that it is important for the public to know their results as soon as possible, then they may choose not to wait for their work to go through the peer-review process, because a delay of several months could result in unnecessary loss of life or injury.

To deal with issues relating to prior publication in the media, many journals have adopted policies requiring authors to certify that they have not discussed their work with the media before submitting it for publication (LaFollette 1992). Many journals also forbid authors from prior publication of their works on Web pages as condition of acceptance. Top journals, such as *Science*, *Nature*, *JAMA*, and the *New England Journal of Medicine*, have adopted an embargo policy for science journalists (Marshall 1998). According to this policy, journals allow reporters to have a sneak preview of articles prior to publication on the understanding that the reporters will not disclose the information to the public until the article is published. This policy allows reporters to have early access to newsworthy stories, and it allows journals to ensure that communications with the media do not undermine peer review and that scientists get adequate publicity. Although embargo policies often work well, confusion can arise regarding the status of work presented at scientific meetings: If a reporter working on an embargoed story learns about the story at a scientific meeting, would it break the embargo to cover the story at the meeting? A similar confusion can arise with journals' own prior publication policies: Does presenting results at a scientific meeting count as prior publication?

Misunderstandings can also result when scientists communicate with the media. Consider a scientist discussing with a journalist her research on the beneficial effects of moderate alcohol consumption. Communication problems between the scientist and the journalist can arise in several ways. Even if the journalist writes a perfect science news story, communication problems can arise between journalist and the public. For example, a science journalist may title a story on the research "Moderate Alcohol Consumption Protects Against Heart Disease," and members of the public may read the headline (and possibly the first paragraph) and conclude that the best way to avoid heart disease is to drink more alcohol. Although this example may seem a bit contrived, the issues are real and important: How should scientists discuss their work with the media? How can they communicate openly and freely while helping people to make informed, rational choices? To avoid these communication problems, scientists may try to "prepackage" their results for public consumption. They may boil down their research to specific key points and attempt to simplify complex findings for a general audience. However, preparing research for public consumption has its own problems: When does simplification become oversimplification? When does useful communication become paternalistic communication? These problems occur in all areas of public scientific communication but are most acute in public health communication, because researchers are attempting to develop guidelines and educational materials that will be used by the public to make decisions about diet, exercise, lifestyle, sexual practices, child rearing, and so on. To explore these issues in depth, see Friedman et al. (1999) and Resnik (2001d).

Finally, it is important to mention that scientists and companies sometimes make disclosures in the media solely to enhance the value of intellectual property, stock, or equity related to their research. This type of situation is ethically problematic, since communications with the media should help to improve the public's understanding of science, not just a researcher's (or company's) financial interests. Journalists that write stories about science should be aware of the possible financial motivations for scientists' disclosures.

QUESTIONS FOR DISCUSSION

1. What is the premise of peer review—what are the reasons behind it? What is the goal? Can you change the premise? Why would you want to?

2. What suggestions would you make to improve the current manuscript review process for journal publication?

3. What is your personal experience or your colleagues' or mentor's experience with peer review?

4. How would you promote ethical practices among your colleagues regarding peer review?

5. Do you think researchers should disseminate their findings in the media? How and why?

6. Do you think it is ever a good idea to censor scientific research that poses a national security risk? Why or why not?

CASES FOR DISCUSSION

Case 1

A very busy researcher is also a member of an NIH study section. He received 15 proposals to review in four weeks. He is the primary or secondary reviewer for six proposals. The primary and secondary reviewers are supposed to prepare a report about each proposal. Also, all reviewers are supposed to read all of the proposals.

The researcher delayed reading the material until the night before the meeting. He read thoroughly two of the six proposals where he was the primary/secondary reviewer and prepared reports. Lacking in time, he skimmed the other five proposals and wrote reports laden with generalized statements of praise or criticism. He never read the other nine proposals.

- What should he do?
- What would you have done?

Case 2

The same busy researcher received the same 15 grant proposals to review in four weeks. He gave 10 of them (five for which he was primary/secondary reviewer and five others) to his most senior postdoctoral fellow to read and comment on. During the study section deliberation four weeks later, the chairman of the study section realized that the researcher had never read those five proposals, because he was not able to discuss them.

- What should the reviewer do?
- What should the reviewer have done?
- What should the chairman of the study section do?

Case 3

A member of an NIH study section hands over a proposal to her postdoctoral fellow to read, to learn how to write a good grant proposal. The postdoctoral fellow then copied the proposal so that he could read it at his leisure and underline the important points. He also found a number of good references that he could use. The researcher found out what her postdoctoral fellow did and admonished him.

- What should the researcher do?
- What should the postdoctoral fellow do?

Case 4

A senior scientist at a major eastern medical school was also a member of the grant review board for a large medical foundation. He was also a member of an NIH study section. During foundation review deliberations of a grant proposal from a relatively junior scientist, he stated, "We just turned down his NIH research proposal, which was basically on the same topic." However, the outside reviewers recommended this proposal be funded by the foundation. Prior to the senior scientist's comments, the discussions were moving in favor of funding. After his comments, the entire discussion became negative, and the proposal ultimately was not funded.

- What should the senior scientist have done?
- What should the junior scientist do if she hears what happened?
- What should the chairman of the review group at the foundation have done?

Case 5

An associate professor in a university is also on the editorial board of a major journal in her field. She receives a paper to review in her field. While reading the manuscript, she recognizes that the paper's researcher is on the path to discovering an important peptide that the professor's close friend is also pursuing. She calls her close friend and discusses her friend's project and how he is progressing with it. She never mentions that she is reviewing the submitted paper. But at the end of their conversation, she informs her friend, "If I were you, I would hurry up and submit the work quickly before someone else beats you to it." She repeats the same sentence twice and adds, "I would recommend that you send it as a short communication so that it can be published shortly." The associate professor then mails her review paper three months later. She recommends acceptance with major modifications and few additional experiments.

- What should the associate professor have done?
- What should her close friend have done?
- What should the author of the submitted paper do if he hears about it?

Case 6

An assistant professor receives from her chairman a manuscript to review informally. The chairman is the associate editor of the journal where the paper was submitted. The assistant professor prepares a thorough review and recommends publication. The chairman receives reviews from two editorial board members he selected. Both reviewers recommend rejection. The chairman tears up the informal review and rejects the paper.

- What should the chairman have done?
- What should the author do if he hears about it?
- What should the assistant professor do if she hears about what happened to her review?

Case 7

An editor of a journal receives a controversial paper in the field questioning the existing paradigm. The editor knows that if he sends the manuscript to two experts on his editorial board, they will most likely reject it. He sends them the manuscript for review anyway, and he also sends it to an outside third reviewer who is neutral on the subject. To his surprise, one of the two editorial board members recommended acceptance, while the other recommended rejection. The outside third reviewer also recommended acceptance. The editor decided to accept the paper but felt obligated to call the reviewer who recommended the paper to be rejected and inform him of his decision. The purpose of the call was to keep editorial board members happy and content. The editorial board member informed the editor that if he accepts this piece of garbage, she would resign from the board.

- What should the editor have done? And do?
- What should the two editorial board members have done?
- What should the author do if she hears about it?

Case 8

An editor of a major ethics journal receives a paper dealing with a controversial topic on the use of human subjects in research. The paper deals with surveying the open literature for questionable ethical practices toward a vulnerable population. The editor mails the article to several reviewers, among them two researchers whose own work was part of the survey. These two reviewers are not members of the editorial board, so they were ad hoc reviewers for this paper. The instructions to reviewers clearly state that this is a privileged communication.

The two ad hoc reviewers copy the manuscript and mail it to about a dozen researchers whose work is mentioned in the paper. After three months, the paper is accepted based on the other reviewers' comments, because the two ad hoc reviewers never mailed back their reviews. The author receives the page proofs, makes a few corrections, and mails it back to the production office. A few weeks later, the editor calls the author to inform him that the paper will not be published because of a threat of a libel lawsuit from "some researchers" cited in the paper. She tells the author that the publisher does not have the resources to fight a lawsuit.

- What should the editor do or have done?
- What should the ad hoc reviewers have done?
- What should the author do?

Case 9

A university's promotion and tenure committee is deliberating on the promotion of an assistant professor to an associate professor with tenure. This is a crucial step in a faculty member's career. The assistant professor's package before the committee contains six strong letters of recommendation (three from inside and three from outside the university), a strong curriculum vita that includes 30 papers in peer-reviewed journals, an R01 grant for $1 million for five years, and a good but not exemplary teaching record. The letter from the chair of the department is supportive but not glowing. The department chair's concerns are legitimate in that this faculty member is not a "good" citizen and is hard to manage. He resists any additional duties that are not related to research. The department chair wants the committee to turn him down so she will be off the hook. She took additional informal steps to stop his promotion, including talking to the chair of the committee and two of its members about her concerns.

- What should the department head do or have done?
- What should the chair and members of the promotion and tenure committee have done and do?
- What should the faculty member do if his promotion is turned down? If he hears about why?

Case 10

An associate professor at a large research university in the Midwest receives a package from a committee chair at an East Coast university, asking him to evaluate the promotion of an assistant professor within four weeks. The associate professor has a grant deadline coming up in a month and has a heavy teaching load. He does not want to say no to the chair, because he is an important figure in his field. However, he does not have first-hand knowledge of the candidate for promotion. He will probably have time to read one or two papers from the candidate's 40 publications, but no more.

Near the end of the four weeks, the associate professor writes a letter that neither recommends nor rejects the promotion, and in this way he protects himself from all sides. The chair is an experienced person and realizes that this associate professor did not do an adequate job. But he does not want to alienate a colleague by not using the evaluation.

- What should the chairman do?
- What should the evaluator have done?

Case 11

A faculty member is also a reviewer for a major journal. Her graduate student is having difficulty setting up a unique and novel method to test for low-temperature electron tunneling in a specific metal. Development of the method must be done before the student proceeds with work on his project. The faculty member receives a paper from the journal to review that lays out the method in detail. She gives the paper to the student to read.

- Should the student read the paper?
- If the student reads the paper, should he then use the method?
- Who are the parties involved and why?
- Is there a responsible way to deal with this issue?

Case 12

Three authors submit a paper on superconductivity to a prestigious journal. They become irritated when the journal does not make a decision after six months. Finally, after nine months, the journal decides to reject their manuscript. The journal does not provide them with any substantial comments and says that there was a delay in getting a review back from the second reviewer. The editor recommends that they try to publish their paper elsewhere.

The next month, a paper comes out in a different journal that is suspiciously similar to their paper. That research used the same methods and techniques, and its data appear to be almost identical to the data in their study. The paper uses several sentences in the discussion section that bare a strong resemblance to sentences from their paper. It also draws similar conclusions.

- What might have happened in this case?
- Can you find fault with the conduct of the editor or reviewer(s)?
- What should the authors do now?

8

Misconduct in Research

Since the 1980s, well-publicized examples of research misconduct increased public concerns and stimulated responses from government, universities, and other research institutions. Surveys indicate that the prevalence of misconduct may be larger than many researchers would like to acknowledge. As a result, policies and procedures have been designed to investigate, adjudicate, and prevent misconduct in research. There is now a functioning system in place designed to deal with misconduct allegations, and efforts to prevent misconduct are increasing. Some aspects of these policies are controversial and are still in progress. This chapter discusses the definition of scientific misconduct as well as policies and procedures for reporting, investigating, and adjudicating misconduct.

Although breaches of scientific integrity have been part of our culture for many years, a book by two science journalists, William Broad and Nicholas Wade, *Betrayers of the Truth: Fraud and Deceit in the Halls of Science* (1982 [1993]), played an important role in focusing public attention on research misconduct. The authors recounted both historical and current cases of scientific fraud and criticized the scientific community for its indifference to the problem.

The book challenged scientific icons. According to the evidence presented by the authors, Galileo made the data for falling objects better than they really were; Isaac Newton made his experimental results fit his theories better by fudging his predictions on the velocity of sound, the procession of equinoxes, and gravitational forces; John Dalton cleaned up his data on the ratios of chemical reactions, which remain hard to duplicate; Gregor Mendel manipulated the heredity ratios on his experiment with peas; Robert Millikan selectively reported oil drop data on his calculation of electronic charges; and even Louis Pasteur was guilty of announcing his anthrax vaccine before he completed his experiments (Geison 1978, 1995, Broad and Wade 1982 [1993], Shamoo and Annau 1989).

Among the most famous historical examples of misconduct is the story of the "Piltdown Man." In 1908, skull bones were found in Piltdown, a town not far from London. The bones were presented by a brilliant young curator as being thousands of years old and belonging to a person who had the characteristics of both monkey and man. It was sensational scientific news: Here was the "missing link" to prove that man had evolved directly from apes. Forty-five years later, however, some scholars concluded that the curator had pieced together contemporary skull bones from the two different species and had aged them chemically (Barbash 1996). At the time of the discovery, the curator's colleagues had accepted his findings without critical appraisal, largely because "the researchers [had] shaped reality to their heart's desire, protecting their theories, their careers, their reputations, all of which they lugged into the pit with them" (Blinderman 1986, p. 235).

A host of contemporary cases complement these historical examples. Some of them have been widely publicized by the media, while others have appeared only as footnotes (Broad and Wade 1982 [1993], Shamoo and Annau 1989). Below are selected cases from a much larger pool of cases:

- In the early 1970s, the fabrication of animal testing data by Industrial Biotech Corporation sparked the legislation for the Good Laboratory Practices regulations enforced by the U.S. Food and Drug Administration (Marshall 1983).
- In his 1950s studies of IQ and inheritance, Cyril Burt created nonexistent twin subjects and phony coauthors and fabricated data to support his theory that IQ is inherited (Shamoo and Annau 1989).
- In 1974, William Summerlin admitted to fabricating data in skin transplant experiments he was conducting at the Sloan Kettering Institute in New York. Summerlin joined the transplantation immunology laboratory of Robert Goode at Sloan Kettering in 1973. Goode was one of the country's top immunologists. Summerlin hypothesized that growing tissues in culture for several weeks prior to transplantation could prevent tissue rejection. In an experiment designed to answer critics of his work, Summerlin "transplanted" cultured skin patches from black-haired mice onto white-haired mice. While cleaning the mice, a laboratory assistant observed that alcohol could wash away the black hair color on the white mice. The assistant reported this to Goode, who suspended Summerlin and initiated an investigation. Summerlin soon confessed that he had drawn patches of black hair on the white mice with a black felt-tip pen. The committee conducting the investigation of Summerlin also found that he had fabricated data relating to several other "transplantation" experiments. The committee required Summerlin to take a medical leave of absence (for mental health problems), publish retractions, and correct irregularities in his work. The scandal ruined Goode's career even though he was innocent of any wrongdoing (Hixson 1976).
- In 1981, doctoral candidate Mark Spector published a series of papers on enzymes involved in cancer initiation. He fabricated data by intentionally placing radioactive phosphorus onto thin-layer chromatography sheets, thus falsely indicating that it was a reaction by-product (Fox 1981, Wade 1981, Lock 1993).
- Between 1980 and 1983, Steven Breuning of the University of Pittsburgh published 24 papers funded by a National Institute of Mental Health (NIMH) grant to study powerful neuroleptic antipsychotic drugs for the treatment of retarded patients. The overseer of the grant was Robert L. Sprague, director of the Institute of Child Behavior and Development. Breuning's results questioned the use of

neuroleptics in retarded children and led to a change in their clinical management nationally (Garfield 1990). In a renewal application for a four-year extension to the grant, Breuning submitted additional new data. Sprague questioned these data and informed NIMH of his concerns. According to Sprague, NIMH's first response was slow and accusatory. Ultimately, Sprague's frustration with NIMH spilled over to the media and led to congressional hearings. Meanwhile, Breuning remained active and funded. Finally, an NIMH panel in 1987 found that Breuning had committed scientific misconduct by reporting nonexistent patients, fabricating data, and including falsified results in a grant application. The panel recommended barring Breuning from receiving any grants from the U.S. Public Health Service (PHS), including NIMH, and referred him for criminal prosecution. Breuning was convicted in 1988—the first scientist to receive such action. He was sentenced to 60 days of imprisonment and five years of probation and was ordered to pay $11,352 in restitution to the University of Pittsburgh. During this process, Sprague lost all funding for his grants and was investigated by the NIMH. He was ultimately cleared of wrongdoing (Shamoo and Annau 1989, Monson 1991, Sprague 1991, 1993, Wilcox 1992).

- In a highly publicized case in 1981, John Darsee, a postdoctoral fellow at the Harvard laboratory of Eugene Braunwald, was accused of using fraudulent data in the assessment of drug therapy to protect against ischemic myocardium (Wade 1981, Lock 1993). Stewart and Feder (1991), dubbed "fraud busters" by the media, published a paper in *Nature* showing that Darsee's publications contained excessively favorable language, fabricated experimental data, and fudged control data. A committee investigating Darsee found that he fabricated or falsified data in five papers he published while at Harvard and eight papers while a graduate student at Emory University. Even though Darsee was found to have committed misconduct in research, he went on to practice medicine.

- In another well-known case, Robert Slutsky, a University of California, San Diego, cardiologist who was being considered for promotion in 1985, listed 137 publications on his curriculum vita, of which 12 were found to be clearly fraudulent and 48 were questionable (Engler et al. 1987).

- In 1986, postdoctoral fellow Thereza Imanishi-Kari was the lead author of an MIT team of five who published a now famous paper in the journal *Cell* (Weaver et al. 1986). Among the authors was the winner of the 1975 Nobel Prize in Medicine, David Baltimore. The paper claimed to show that foreign genes had stimulated the production of large amounts of antibody by the genes in normal mice, which was considered to be a remarkable achievement. However, Margot O'Toole, a postdoctoral colleague, was not able to reproduce a key part of the experiment. She also discovered that the published data did not fully match the data in Imanishi-Kari's laboratory notebook. A series of charges and countercharges once again attracted the "fraud busters" Stewart and Feder. A widely distributed 1987 paper detailing an analysis of their notes and a 1991 paper they published in *Nature* led to an investigation by the National Institutes of Health (NIH) (Stewart and Feder 1987, 1991). Congressional hearings were held that featured angry exchanges between the congressional committee chair and Baltimore, who spoke in defense of Imanishi-Kari's work. He described the hearings as a "witch hunt." The NIH investigations concluded that fraud had indeed been committed. However, after nearly nine years of controversy, the appeals board of the U.S. Department of Health and Human Services (DHHS) dropped the charges. Imanishi-Kari admitted to poor record-keeping practices, but she denied that she committed misconduct (Weaver et al. 1986, Baltimore 1991, Eisen 1991, Hamilton 1991, Imanishi-Kari 1991, Kuznik 1991, Stewart and Feder 1991, Friedly 1996a,b, Kevles 1996).

- In 1984, Robert Gallo and his colleague Mikulas Popovic published in *Science* the first paper that identified the HIV virus, based on research that was done at the National Cancer Institute. Luc Montagnier of the Pasteur Institute, with whom Gallo had previously collaborated, accused Gallo of stealing the HIV virus. The major question in the case was whether the virus given by Montagnier to Gallo in the early 1980s had contaminated Gallo's virus line or whether Gallo had perpetuated the same virus and claimed it to be his own discovery. Claims, counterclaims, and the loss of primary data made it difficult to prove the origin of the virus line (Culliton 1990, Cohen 1991). A significant factor in the difficulties related to the resolution of the case was the failure to keep good laboratory notebooks (Culliton 1990, p. 202). For more details, see case 1 in chapter 6.
- In 1997, a German panel of expert investigators discovered that two biomedical researchers published 37 papers containing falsifications and/or data manipulation (Koenig 1997).
- From 1996 to 1998, a series of reports appeared that exposed a decades-long abuse of human subjects with schizophrenia involving the introduction of chemicals into patients to induce symptoms of the illness. Additionally, patients were left in the community for months without medication, sometimes suffering psychosis and delusions, during washout/relapse experiments (Shamoo and Keay 1996, Shamoo 1997c,d, Shamoo et al. 1997, Shamoo and O'Sullivan 1998).
- In 1997, after a seven-year court battle, Carolyn Phinney, PhD, won a $1.67 million jury judgment against the University of Michigan. Phinney had accused her supervisor of stealing her data. The court's award was both for retaliation and for personal damages related to the use of her data without her permission (New York Times News Service 1997).

- In 1999, the claims made by physicists at Lawrence Livermore Berkeley National Laboratory that they discovered heavy elements 118 and 116 were the result of falsified data (Seife 2002).
- In 2002, a 32-year-old star physicist, Jan Hendrik Schön, at the world-renowned Bell Laboratories, was found by a group of independent investigators to have faked data in at least 17 publications (Service 2002). His publication rate was astounding: He published a paper every eight days. His publications appeared in most prestigious journals, such as *Science, Nature,* and *Physical Review Letters.* This large-scale fabrication and falsification of data is considered the most extensive scientific misconduct in the physical sciences in recent memory.
- In 2006, a group of graduate students at University of Wisconsin, after much heart-wrenching and soul-searching, alleged that their doctoral adviser, geneticist Elizabeth Goodwin, committed data falsification. The university's investigation did confirm the allegations and referred the matter to the Office of Research Integrity (ORI). This is the saddest case that we have encountered in terms of its negative effect on the careers of so many graduate students. Three of the students with many years in graduate school quit the university. One student moved to a different university, and two others started their graduate education anew (Couzin 2006b). Moreover, many faculty members were not receptive to their action. This case illustrates that the system in place to deal with misconduct is not yet accepted or handled well.
- In 2004–2005, Woo Suk Hwang, a South Korean professor at Seoul University, published ground-breaking results of cloning human embryonic stem cells in culture. Hwang admitted to fabricating data pertaining the cell lines and faced criminal charges. As of the editing of this book (July 2008), the legal case is still pending. For more details, see chapter 1.
- In 2006, Eric Poehlman admitted that he committed scientific misconduct in the fabrication and falsification of data that were submitted to journals and granting agencies while he was a professor of medicine at the University of Vermont (1987–1993, 1996–2001) (Office of Research Integrity 2005a,b). Poehlman was the first academic researcher sentenced to jail. For more details, see case 1 below.

One of the most celebrated cases of continuous claims and counterclaims of scientific misconduct occurred in the "cold fusion" debacle. The main question in this debate remains whether "cold fusion" is a real phenomenon or an artifact of poorly understood experimental procedures. The issue of whether there was scientific misconduct by the original authors and their supporters or by their detractors still remains unresolved. In this short description, we are not taking sides on the controversy, but the discussion is worthwhile in terms of how the norms of scientific research and discourse have been violated repeatedly by many on both sides of the argument. More important, this saga illustrates how not to conduct research. The best article for the general reader on the subject was published in 1994 by David Goodstein of the California Institute of Technology.

In 1989, Martin Fleischmann and Stanley Pons of the University of Utah reported that they observed a greater production of energy (in the form of excess heat) from an electrolytic cell at room temperature than would be produced through electrical or chemical reactions. The electrolytic cell consisted of a solution of lithium deuteroxide in heavy water (deuterium, or D_2O) and two palladium electrodes. Known physical laws could not explain this key observation, according to the authors. Fleischmann and Pons also claimed to have observed neutrons, the normal by-product of nuclear reaction (for details, see Chubb 2000, see also Beaudette 2000).

Unfortunately, the process of scientific research failed in many ways in this case. Many different parties on all sides of this dispute violated ethical norms. These transgressors included university administrators who attempted to lay claim to research findings and potential funds through newspapers and headlines; researchers who cut corners to prove their points and gain recognition and priority; researchers who were quick to denounce their peers and insinuate misconduct; journalists who assumed the worst; journalists who beamed with naive credulity; journal editors who became censors for one side or another; reviewers who divulged confidential information; federal grant administrators who decided funding based on public perceptions; researchers who failed to accurately cite the work of others; and universities that violated academic freedom. The parties demonstrated a lack of civility, respect for individuals, respect for society, and respect for science.

This saga continues to this day. A nuclear engineering professor at Purdue University, R. P. Taleyarkhan, in 2002 published in *Science* an article giving evidence for nuclear emissions during acoustic cavitations. Even though this approach differs from the use of electric current to produce nuclear emission (i.e., cold fusion), it is within the same general procedure for "cold fusion," and it was called "sonofusion." Opponents and detractors of this approach accused Taleyarkhan of scientific misconduct. In February 2007, the university cleared him from any wrongdoing. However, opponents claimed that the investigation was not in accordance with university procedures. The university has launched a second investigation (Service 2007). In 2004, opponents of "cold fusion" met with many respected physicists and requested and obtained a hearing from the U.S. Department of Energy of the new data on the subject (New York Times News Service 2004) and from the Office of Naval Research (ONR), the funding agency. The new committee issued a report claiming that Taleyarkhan committed two instances of scientific misconduct (Service 2008). Taleyarkhan is challenging the report. All of these cases raise many questions. What exactly is meant by misconduct? What is the range of its manifestations? Are some kinds of misconduct worse than others? What are the factors that lead someone to be involved in misconduct? How often does misconduct occur? What has been done thus far to prevent it? What is the process by which misconduct is being identified and dealt with? How is innocence or guilt determined? How effective are the current efforts to manage misconduct and to prevent it? How are whistle-blowers being protected, and is the process adequate? Are the accusations and thus the investigation of research misconduct used to smear those opponents of an existing paradigm or ideological enemies? What yet remains to be done?

In 1993, the secretary of the DHHS created the Commission on Research Integrity (CRI) in response to the continued failure to resolve issues related to misconduct in research. The function of the CRI was to provide advice to the secretary of the DHHS on such issues as a new definition of misconduct, improvements in monitoring and institutional compliance processes, and a possible regulation to protect whistle-blowers.

In 1995, the CRI issued its report, titled *Integrity and Misconduct in Research*. Many of the subsequent changes in the ORI policies and procedures were the result of this report (Office of Research Integrity 2008).

The CRI produced nine major recommendations designed to foster an environment that nurtures research integrity. They suggested that particular institutions and disciplines implement these recommendations with an eye toward their specific research methods, goals, and procedures. The recommendations were wide-ranging and advocated adoption of a new federal definition of research, education in research integrity for everyone supported by PHS funds, standards for protection of whistle-blowers, increased monitoring of intramural PHS programs, streamlined misconduct management mechanisms, and the promotion of organization codes of ethics in research (Commission on Research Integrity 1995).

Most of the recommendations were accepted. Of particular interest are the recommendations calling for more education in research integrity, improvements in the protection of whistle-blowers, and the redefinition of misconduct. With respect to education, the assumption is that an awareness of the foundational relationship between research and scientific integrity, an understanding of the current ethical issues, and knowledge of the accepted practices in research are likely to have a sanguine effect.

Before proceeding to these topics, we need to understand exactly how misconduct is defined. However, because the term has been defined by the agency whose responsibility it is to regulate it, we first turn to a discussion of the structure and function of the ORI.

CREATION OF THE OFFICE OF RESEARCH INTEGRITY

In the late 1970s and early 1980s, newspaper reports of high-profile cases of scientific misconduct and a series of congressional hearings spurred the NIH to deal seriously with the issue of scientific misconduct (Office of Research Integrity 1995, 2001, Pascal 1999, 2000). The pressure was intensified by the poor treatment of whistle-blowers and by the inadequate responses of research institutions to instances of scientific misconduct.

In 1989, the DHHS established the Office of Scientific Integrity (OSI) within the NIH Director's Office, and the Office of Scientific Integrity Review (OSIR) within the Office of the Assistant Secretary for Health (Office of Research Integrity 2008). Almost immediately, the presence of OSI within the NIH was criticized because of potential conflicts of interest: The NIH, the chief provider of funds for research, was now also responsible for the investigation and punishment of scientific misconduct. As a result of the criticism, reorganization took place in 1992, 1993, and again in 1995, resulting in the establishment of the ORI within the Office of the Assistant Secretary for Health to replace both OSI and OSIR (Office of Research Integrity 1996, 1999).

The responsibilities of the ORI are very broad, including everything from the promotion of responsible conduct and the formulation of policy, to the oversight and/or investigation of both the intramural programs of the PHS and also the larger group of extramural programs. These responsibilities in part are as follows (Office of Research Integrity 2007a):

- Developing policies, procedures and regulations related to the detection, investigation, and prevention of research misconduct and the responsible conduct of research
- Reviewing and monitoring research misconduct investigations conducted by applicant and awardees' institutions, intramural research programs, and the Office of Inspector General in the DHHS
- Recommending research misconduct findings and administrative actions to the Assistant Secretary for Health for decision, subject to appeal
- Providing technical assistance to institutions that respond to allegations of research misconduct
- Implementing activities and programs to teach the responsible conduct of research, promote research integrity, prevent research misconduct, and improve the handling of allegations of research misconduct
- Conducting policy analyses, evaluations and research to build the knowledge base in research misconduct, research integrity, and prevention and to improve HHS research integrity policies and procedures
- Administering programs for maintaining institutional assurances, responding to allegations of retaliation against whistle-blowers, approving intramural and extramural policies and procedures, and responding to Freedom of Information Act and Privacy Act requests

DEFINITION OF MISCONDUCT

As a beginning activity, the newly formed OSI defined misconduct in order to establish the range of activities that should be investigated and that deserve sanctions if guilt is determined. Since the 1990s, the office and the definition have undergone many changes. In 2005 (Office of Research Integrity 2008), the federal government finalized its definition of misconduct in research. The current definition found in the federal regulations is as follows:

> Research misconduct means fabrication, falsification, or plagiarism in proposing, performing, or reviewing research, or in reporting research results.
>
> (a) Fabrication is making up data or results and recording or reporting them.
> (b) Falsification is manipulating research materials, equipment, or processes, or changing or omitting data or results such that the research is not accurately represented in the research record.
> (c) Plagiarism is the appropriation of another person's ideas, processes, results, or words without giving appropriate credit.
> (d) Research misconduct does not include honest error or differences of opinion. (45 CFR 93.103)

In addition to defining misconduct, the new definition sets a standard of evidence for a finding of misconduct by a federal agency:

> A finding of research misconduct made under this part requires that
>
> (a) There be a significant departure from accepted practices of the relevant research community; and
> (b) The misconduct be committed intentionally, knowingly, or recklessly; and
> (c) The allegation be proven by a preponderance of the evidence. (45 CFR 93.104)

The federal policy separates the three phases of handling a potential case of misconduct: (a) an inquiry—the assessment of whether the allegation has substance and if an investigation is warranted; (b) an investigation—the formal development of a factual record, and the examination of that record leading to dismissal of the case or other appropriate remedies; (c) adjudication—during which recommendations are reviewed and appropriate corrective actions determined (45 CFR 93).

The policy also separates the investigative process from the adjudicative process. Institutions perform the investigations and then take actions that are within their prerogative, such as reprimand, demotion, or termination of employment. The federal agency then reviews the deliberations of the institution and can take additional action if necessary. The accused may then appeal the decision through the agency's appeal procedure. The policy basically conforms to recent past practices. Institutions performed the investigative step in more than 90% of all cases of misconduct (Pascal 2000). Allowing investigations to be conducted solely by the institutions has been criticized, however, because of the reluctance of institutions to reveal their own wrongdoing and the early lack of diligence by institutions in their pursuit of investigations.

Federal agencies are responsible for the implementation of these policies. However, federal agencies delegate that implementation to the research institutions receiving federal funding. In most cases, federal agencies will rely on research institutions to manage misconduct policies. The policy reserves the rights of the federal agency in question to proceed with inquiry and investigation under the following circumstances, among others: (a) the institution lacks the ability to handle the case; (b) the institution cannot protect the public interest, such as public health and safety; or (c) the institution is too small to conduct an investigation.

To conclude this section, we make several key points about the definition of research misconduct. First, the federal definition of misconduct focuses on only three types of inappropriate behaviors—fabrication, falsification, and plagiarism, or FFP. It is worth noting, however, that some organizations and governments have adopted definitions of misconduct that are broader than FFP (Buzzelli 1993, Resnik 2003b). The Wellcome Trust Fund's definition of misconduct includes FFP as well as "deliberate, dangerous, or negligent deviations from accepted practices in carrying out research" and "failure to follow established protocols if this failure results in unreasonable risk or harm to humans, other invertebrates, or the environment" (Koenig 2001, p. 2012). In 1989, the PHS and the NIH used a definition of misconduct that included FFP as well as other practices that seriously deviate from those accepted by the scientific community (Office of Research Integrity 2007a, 2008). The agencies decided to narrow their focus to FFP in response to concerns that the category of "serious deviations" was too broad and vague to be enforceable. While we agree that the category of "serious deviations" is too broad, we think that the federal definition of misconduct should be expanded to include interfering with or obstructing a misconduct inquiry or investigation, to ensure that people who investigate misconduct have the full cooperation of all the parties who are involved.

Second, although the definition of misconduct seems clear and easy to apply, this is sometimes not the case. In some situations it may be obvious that misconduct has occurred. For example, in the Summerlin case discussed above, it is obvious that he fabricated data. The Hwang case described in chapter 1 and the Poehlman discussed further below are also clear cases of misconduct. Some cases are more difficult to discern, such as the cold fusion case, the Gallo case, and the Baltimore case described above. We call attention to some types of questionable research practices that might be considered misconduct, depending on specific details of the situation:

- Enhancing digital images (discussed in chapter 3). This could be fabrication or falsification, depending on the facts of the situation.
- Excluding data from an article or presentation without a good statistical, technical, or methodological reason (also discussed in chapter 3). This could be falsification, depending on the facts of the case.
- Performing a statistical analysis in a deliberately dishonest way. This could be fabrication or falsification, again, depending on the facts.
- Not giving credit to someone who has assisted with a research project by providing ideas, helping to collect data, or developing processes or techniques. While this might seem to be an authorship matter, it

could fit the definition of plagiarism, depending on the facts of the situation (also discussed in chapter 6).

Third, research misconduct as defined by the U.S. government is a legal, not ethical, notion. It is part of a branch of law known as administrative law. A finding of misconduct is a decision made by a federal agency. The standard of legal proof for this finding is "preponderance of evidence" (i.e., 50% probability or greater). People who are found to have committed misconduct may have action taken against them by a federal agency, such as debarment from receiving federal funding for a period of time. An administrative finding of misconduct does not, by itself, imply any civil or criminal liability. However, someone who commits misconduct may be charged with the crime of fraud, and he or she may be sued in civil court for the damages caused by his or her fraud. As noted above, very few researchers have faced criminal charges for research misconduct, although some scientists, such as Hwang from Seoul University and Poehlman from the University of Vermont, recently have. Poehlman served more than a year in jail for his misdeeds (see discussion below). In some cases, the federal government sued research institutions for the damages caused by fraud committed by researchers working for those agencies. Civil liability for fraud can be very costly for institutions because they can be liable for "treble damages" or triple the amount of money the damages cost (Resnik 2003b). For example, if a researcher committed fraud on a $1 million grant, the university could be liable for $3 million.

Fourth, there is a use of the word "misconduct" in which misconduct is simply behavior that is an intentional violation of widely accepted ethical norms. When a community lacks agreement about ethical standards, an action may be ethically controversial or questionable but not a form of legal misconduct. We think that it makes sense to speak of misconduct in this general, ethical sense of the word, but we also understand why governments must focus their efforts on narrower, legalistic notions of misconduct, since federal agencies are in the business of enforcing rules, not teaching ethical norms.

WHAT IS THE INCIDENCE OF MISCONDUCT?

In their book, Broad and Wade (1982 [1993]) claimed that there was more misconduct in the research community than scientists want to admit. Furthermore, they said misconduct is probably greatly underreported. By the time misconduct is reported, it is probably the culmination of a spectrum of unethical activity of which misconduct is only the end point (LaFollette 1994b, 2000). Therefore, one may surmise that the reported cases of misconduct are but the "tip of the iceberg" (American Association for the Advancement of Science–American Bar Association 1988). Although scientists, politicians, and the public regard the problem of scientific misconduct as extremely important, the scientific community still lacks solid evidence concerning the incidence of misconduct. Estimates of the misconduct rate, based on surveys that asked researchers whether they know about misconduct that has occurred, range from 3% to 32% (Steneck 2000). One survey of scientists found that 32% suspected a colleague of plagiarism (Tagney 1987). Another study showed that the reported incidence of fraud affected 0.01–0.1% of all published research (Glick 1992). A 1993 survey of science students and faculty estimated that the percentage of questionable research studies, including fraud, may range from 6% to 12%. The results further showed that 44% of the responding students and 50% of the faculty reported being exposed to two or more types of misconduct or questionable research practices, and 6–9% reported having direct knowledge of plagiarism or data falsification (Swazey et al. 1993).

A survey of 3,247 scientists by Martinson et al. (2005) found that 0.3% (3 out of 1,000) admitted to falsifying or cooking research data in the last three years. The survey also found that scientists admitted to engaging in other unethical activities in the last three years, including ignoring major aspects of human subjects requirements (0.3%), using someone else's ideas without obtaining permission or granting proper credit (1.4%), and unauthorized use of confidential information in connection with one's own research (1.7%). Steneck (2000) estimated the rate of misconduct by extrapolating from confirmed cases: With 200 confirmed cases of misconduct involving NIH funds in 20 years, this works out to a misconduct rate of 1 out of 100,000 researchers per year. However, because this estimate relies on confirmed cases of misconduct, it is susceptible to a large underreporting bias.

Richard Smith was the editor of the *British Medical Journal* from 1997 to 2004. He claims that scientific misconduct is much more prevalent than the scientific community acknowledges or knows: "Most cases are probably not publicized. They are simply not recognized, covered up altogether; or the guilty researcher is urged to retrain, move to another institution, or retire from research" (Smith 2006, p. 234).

Most methods of measuring the misconduct rate from surveys or other reports of misconduct probably underestimate or overestimate the rate of misconduct. To see why this is so, consider how one might attempt to measure the rate of tax fraud. One could mail out surveys and ask people if they have ever committed fraud, but who would want to admit to this, even on a survey? One could ask people if they know of someone who has committed tax fraud, but why would they have access to this information? Most people probably will not learn about tax fraud because someone who commits fraud will try to keep it secret. Or, they could claim to know about something that never happened, because they have based their opinion on rumor or hearsay. One could also estimate the rate of tax fraud based on the rate of convictions for tax fraud, but many people get away with tax fraud without ever being indicted, and many people who are indicted are never convicted. Given difficulties like these, we believe that the only way to get a good estimate of the rate of misconduct is to randomly audit research procedures and research data—other methods would introduce underreporting or overreporting biases.

Regardless of what the exact rate of misconduct turns out to be, misconduct is still a very important concern for researchers, because it undermines trust, trustworthiness, integrity, and accountability in research (National Academy of Sciences 1992). For example, murder is still regarded as a serious crime worthy public attention even when the murder rate is low. Thus, those who maintain that misconduct is not an issue in science because it is so "rare" do not understand the seriousness of the crime. Moreover, given the growing influence of money in research, misconduct may be more common than people think.

WHAT ARE THE FACTORS THAT LEAD TO MISCONDUCT?

Before discussing how institutions respond to allegations of misconduct, it is worthwhile to consider why scientists sometimes engage in any of the various forms of misconduct. How important to the etiology of misconduct are the pressures to publish and to be funded? To what extent is misconduct due to problems in our institutions as opposed to the personal flaws of our scientists?

The etiology of scientific misconduct is probably as complex as any other form of deviant human behavior. The theories of nineteenth-century French philosopher Emile Durkheim and twentieth-century American sociologist Robert K. Merton (1973) provide a theoretical framework for its genesis (Zuckerman 1977a,b, Garfield 1987). According to these theories, the values and norms pertaining to the conduct of science ordinarily become internalized during the period of education and are reinforced later with a system of social control that includes both rewards and sanctions. Deviant behavior is likely to occur when values break down and an individual's aspirations and goals come into conflict with society's structure and controls.

Although it is tempting to blame misconduct on the occasional "bad apple" (American Association for the Advancement of Science–American Bar Association 1988, National Academy of Sciences 1992), it probably results from a variety of factors, including the ambitions, interests, biases, and personal flaws of individual scientists; economic and financial pressures to obtain and maintain funding; political and institutional influences; and inadequate education, training, supervision, and oversight at all levels of research (Shamoo 1989, National Academy of Sciences 1992). Misconduct may be the unintended result (or side effect) of a system of scientific research, education, and funding that overemphasizes career advancement, institutional prestige, money, and a "win at all costs" attitude.

Factors that have been identified as fostering misconduct include the pursuit of fame, money, and reputation; the quest for promotion; pressures to produce; poor training in research methodology; and an increased complexity of the research environment (Shamoo 1989, Shamoo and Annau 1989, National Academy of Sciences 1992). Inappropriate and large collaborative groups may diffuse accountability and thus also contribute to misconduct (Shamoo and Annau 1989, Davis 1990, National Academy of Sciences 1992).

Additionally, three common factors are shared between financial misconduct and scientific misconduct: conflict of interest (Krimsky 2007), the complexity of the research, and the remoteness of the laboratory manager from day-to-day operations (Shamoo and Annau 1989).

HOW HAVE SCIENTISTS RESPONDED TO THE ISSUE?

Scientists themselves (and their leadership) seem reluctant to confront the issue of misconduct. In the 1980s and 1990s, the scientific community failed to manage the problem of scientific misconduct quickly and adequately. In 1981, then Senator Al Gore (D-Tenn.) held hearings emphasizing that the research enterprise rests on the trust of the American people in the scientific process (LaFollette 1994b). In subsequent hearings, strong concerns were raised about resistance to change and the betrayal of public trust within the scientific community. In response, in 1981 the NIH began to formulate proposed guidelines for the management of misconduct that eventually resulted in the formation of the OSI (LaFollette 1994b). However, because of resistance from the scientific community and foot-dragging by the NIH, especially in the formulation of a definition of misconduct, the OSI was not created until 1989.

Other examples of the reluctance to accept the problem of scientific misconduct include the observation that corrections in the literature necessitated by misconduct have been slow in coming (Friedman et al. 1990). Additionally, the scientific community has suggested that the media has overblown the problem with its sensationalistic coverage of the few cases of misconduct (LaFollette 1992). Scientists also contend that the existence of fraud signifies only that science, like other disciplines, has its own share of pathological individuals and that this says nothing about the integrity of the system (National Academy of Sciences 1992). A survey of scientists showed that they would not report 50% of unethical scientific behaviors (by their own definitions) to the responsible person or agency either within the school or outside the school. Instead, they would opt for collegial solutions to the problems (Wenger et al. 1999). Finally, academic scientists have resisted calls for random data audits to determine the prevalence of scientific misconduct (Shamoo and Annau 1987, Shamoo 1988, 1989, Loeb and Shamoo 1989, Rennie 1989a,b, Glick and Shamoo 1991, Glick 1992). Data auditing is common in industrial research, and financial auditing is a normal part of sound business practices.

THE PROCESS OF DEALING WITH MISCONDUCT

The ORI has formulated general guidelines for the process of dealing with misconduct and requires that each institution receiving PHS funds have specific procedures that ensure compliance with the guidelines. Consequently, there are many similarities and some differences among these documents around the country (Office of Research Integrity 1999). The ORI requires that institutions be the primary sites for handling misconduct allegations, and the process described below deals with that issue. We use the procedures currently in use at the University of Maryland, Baltimore (UMB) to illustrate the process (University of Maryland, Baltimore 1998). (These procedures conform to the ORI's model policy for dealing with scientific misconduct.)

A complaint can be made by anyone connected with the research. The complainant may direct the allegation either to the dean of the appropriate college or to the campus vice president for academic affairs, who is UMB's responsible official. The vice president consults with the complainant, reviews the allegations, gathers all necessary documentation, and determines whether to initiate the inquiry process within 10 days after receiving the complaint. If so, the vice president appoints an inquiry committee within 30 days after the decision to proceed. In circumstances where additional information is required, the UMB process provides for a preliminary review that takes place prior to the inquiry decision.

The inquiry committee consists of three tenured faculty members without conflict of interest with respect to either the complainant or the accused. It has the authority to meet with individuals, including the accused, who may have information about the case. It may also request information, records, expert opinions, and so forth. The accused may consult legal counsel. The inquiry committee should, if possible, complete its proceedings within 60 days. The purpose of its deliberations is to determine whether misconduct has occurred and whether there is sufficient justification to warrant the formal stage of investigation. If the committee so decides, there is no assumption of guilt. In this way, it is analogous to an indictment by a grand jury. The committee's findings are communicated to the vice president, who consults with legal counsel and decides within 15 days whether to proceed to a formal investigation, terminate the process, or undertake alternative actions, such as a correction of the literature, that do not require a full investigation. If an investigation is warranted, the vice president appoints an investigation committee within 30 days after the decision to proceed.

The investigation committee at UMB consists of three to five tenured full professors or professors emeriti without conflict of interest in the case. At least one member must not be connected with the entire University of Maryland system. While due process and confidentiality are integral components of all phases of the process, the formal assurances during the investigation are significantly more rigorous. For example, meeting and hearing dates are provided in writing 72 hours in advance, and the proceedings are tape-recorded, with transcripts made available to the accused if so desired.

The investigation should be completed, if possible, within 120 days. The accused again may choose to consult legal counsel. During the course of the investigation, the committee may expand the documentation provided during the inquiry process and may also expand the scope of the investigation beyond the scope of the original allegations. During its final deliberations, the investigation committee applies the standard of "preponderance of evidence." The committee must write a report that details the findings and its recommendations. The committee, if it finds that academic misconduct has been committed, may recommend sanctions. Note that the committee is also obliged to consider the restoration of damaged reputations, retractions, or disclaimers as necessary.

The report of the investigation committee is sent to the vice president. A copy is then sent to the accused individual, who has 15 days to offer a written comment. Appropriate portions of the report are also sent to the complainant, who has the same opportunity to respond. The final report, along with any additional required documentation, is sent to the director of the ORI.

Within 30 days after receiving the report of the investigation committee, the vice president, after consulting with legal counsel and other academic personnel, should make a determination as to the finding of misconduct. The vice president informs the president and then submits a report to the ORI that includes the decision and the description of the case and reason for the decision. The process is then complete, except for the possibility of appeal, which is provided for in UMB's procedures.

Simply counting the number of days required to navigate the process just described begins to illustrate its cost in time, dollars, and emotions. Assuming that the timetable is met precisely, and that no hitches occur—never a safe assumption—the process takes almost 10 months from allegation to final institutional action. Add to that an appeal and the deliberations at the ORI, and more than a year has elapsed. The gravity of the process is apparent from beginning to the end. Lives can be affected significantly, especially those of the accused and the complainant. Tens of thousands of dollars can be spent on legal counsel. Hundreds of difficult hours of university personnel time are spent in activities related to judging colleagues. The process is significant from every aspect—including the necessity for its existence.

What have been the actual experiences of research institutions around the country and their interactions with the ORI in the management of scientific misconduct? The ORI compiled and published statistics on their experiences with the management of scientific misconduct (Rhoades 2004) over a ten-year period from 1994 through 2003. Data compiled from these experiences are presented in tables 8.1–8.3.

These tables show that medical schools haves experienced the lion's share (71%) of investigations, with the remaining 29% spread among hospitals, research institutes, and other organizations (table 8.1). In most instances, 96% of the allegations (206 of 214) were handled by institutions instead of the ORI.

Table 8.2 shows that during 1994–2003, the ORI received 1,777 allegations of misconduct from institutions around the country, and that most allegations did not meet the definitions of misconduct. Whistle-blowers were the main source of the allegations: In these ten years, there were 289 whistle-blowers (out of 259 investigations), most of whom (57%) came from the ranks of faculty members (Rhoades 2004, p. 29). Some of cases involved such issues as non-PHS funded research, sexual harassment, or problems with human subjects that were handled in other agencies or divisions of the institutions. But in

Table 8.1. Types of Institutional Settings in 259
Investigations (1994–2003)

Institutional setting	Total	Percent
Medical schools	184	71
Colleges/universities	32	12
Hospitals	7	3
Research organizations	26	10
Public Health Service agencies	9	4
Other	1	0
Total	259	100

Adapted from Rhoades (2004).

Table 8.2. Disposition of 1,777 Allegations to the Office
of Research Integrity (1994–2003)

Disposition	Total	Percent
Inquiry or investigation	329	19
Referred to other agencies	15	12
No action possible	1,230	69
Total	1,777	100

Adapted from Rhoades (2004).

Table 8.3. Types of Misconduct in 133 Cases

Type of misconduct	Total	Percent
Fabrication	29	22
Falsification	53	40
Plagiarism	8	6
Fabrication/falsification	36	27
Falsification/fabrication	5	4
Other	2	1
Total	133	100

Adapted from Rhoades (2004).

most cases (69%), no actions were possible; only 19% of all allegations went to
the inquiry or investigation stage or beyond (table 8.2). Of those that reached
the inquiry stage, the great majority (62%) were sent on to the investigation
stage, and approximately half of those (51%) concluded with findings of misconduct (Rhoades 2004, p. 6). Table 8.3 shows that the great preponderance

of misconduct (99%) involved dishonesty in the presentation of data, with falsification observed more often than fabrication. Plagiarism and the "other" category together represented only 1% of the total (Rhoades 2004, p. 7).

What do these data mean? Is the system working well, or should the definition of misconduct be broader? Are other components of institutions handling too many cases of true scientific misconduct? Do these findings tell us anything about the existence of unacceptable scientific behavior? Are those who submit the allegations excessively zealous? It seems that many questions remain unanswered.

OTHER APPROACHES TO MISCONDUCT PREVENTION

At the beginning, the ORI emphasized defining, reporting, investigating, and adjudicating misconduct in its approach to the problem of scientific misconduct. However, the ORI has moved more in the direction of empirical research and professional education. Although it is important to formulate and enforce specific rules and policies on misconduct, such a legalistic approach has limitations. It is much more effective to encourage good conduct than to attempt to catch and punish criminals. One of the most important ways to prevent misconduct, according to many scholars (notwithstanding two recent reports cited later) who have studied research ethics, is to promote education and instruction in ethics and integrity in research (Hollander et al. 1995). The ORI Web site has extensive information and resources in education of responsible conduct of research (RCR). Moreover, the ORI has funded numerous research projects in RCR education. Current initiatives emphasize education in research ethics during graduate school, but undergraduate and even high school students can benefit from ethics education. Also, postdoctoral fellows and even established researchers can benefit from education and training. Education can take place through formal courses on research ethics or during informal discussions in the laboratory, small group exercises, brown bag lunches, workshops, or other settings. Role modeling also plays a key role in ethics education: Beginning scientists learn about ethics and integrity in research by following the examples set by their teachers, advisers, mentors, and peers. The current generation of scientists has an ethical responsibility to help transmit standards of conduct to the next generation.

In 2002, the Institute of Medicine on the behest of the ORI issued a report on promoting integrity in science (Institute of Medicine 2002a). The report encouraged research institutions to go beyond following proscribed policies and procedures. The impact of these recommendations is yet to be felt. An ORI survey of 2,910 researchers regarding research integrity measures used in biomedical research laboratories indicates that there is a need for more educational material, written laboratory research guidelines, assessment of outcome measures, and electronic data recording with an audit trail (Office of Research Integrity 2003). A pilot study based on interviews of 23 senior researchers and research administrators found that they favor education and training being imparted by researchers to students, with emphasis on professional development and practical wisdom (Deming et al. 2007).

Two different surveys cast some doubts on the effectiveness of the current process of RCR education. Anderson et al. (2007) found that formal courses in RCR were ineffective in reducing unethical behavior among research scientists funded by the NIH, but that mentoring did have a positive effect on behavior. Another, more limited survey (Funk et al. 2007) of those who received an NIH training grant fellowship found that ethics training had no effect on their ethical behavior and knowledge regarding authorship and publication. It is clear from these studies that further research is needed to understand the impact of education, training, and mentoring on ethical behavior in science. In this regard, Steiner et al. (2004) found in a pilot study that mentoring alone was effective in promoting ethical behavior. Koppelman-White (2006) argues that mentoring is more effective than classroom education when it is discipline specific and evolving.

The National Science Foundation (NSF), under a 2007 law called the America Competes Act, introduced sweeping ethical reforms for the conduct of research. NSF requires that "each institution that applies for financial assistance from the Foundation for science and engineering research or education describes in its grant proposal a plan to provide appropriate training and oversight in the responsible and ethical conduct of research to undergraduate students, graduate students, and postdoctoral researchers participating in the proposed research project" (National Science Foundation 2007). The law also requires open communication of research results as well as sharing research data with responsible individuals when requested. It is hoped that other federal agencies will follow suit.

In addition to education, the idea of data audit was first introduced in the late 1980s as an approach to the prevention of misconduct (Shamoo and Annau 1987, Shamoo 1988, Loeb and Shamoo 1989, Rennie 1989a,b, Glick and Shamoo 1991, Glick 1992). It was patterned after experience in the financial field, where cooperative experience between auditor and corporate executives is the norm. Auditors have total access to records, and personnel answer questions forthrightly. In the field of science, however, data audit is looked upon with suspicion.

In 1991, J. Leslie Glick and Adil E. Shamoo (one of the authors of this book) published a prototypical example of how data audits can work in the scientific world. Shamoo, an experienced biophysicist, agreed to have his work audited by Glick, a much-audited biotechnology corporate executive as well as an experienced scientist with a scholarly interest in data audit. The audit was to be a retrospective evaluation of a research project that culminated in the publication of a paper in 1983. The ground rules were that Glick had total access to any records, could ask any question, and could arrive at any conclusion. The audit process involved tracking the data from the published paper back to the original raw data in the laboratory notebooks. At the conclusion of the audit, a report was written providing the details of the audit process and its results (Glick and Shamoo 1991).

Data audits could be funded through the regular funding process. Every grant proposal would contain funds for data audit (Shamoo 1988, Shamoo and Dunigan 2000). Glick (1992, 1993) has argued that a scientifically conducted data audit can become an effective management tool in saving the public not only from poor data but also from unnecessary expenses. He estimated that 10–20% of all research and development data funds result in questionable data because of inappropriate research activities ranging from sloppy work to outright misconduct. If data audit were to result in a 50% reduction of questionable data production, there would be a savings of $5–10 per audit dollar spent. Twenty years into the data audit program, the public would save $40–90 of research expenditure per audit dollar (Glick 1992).

Subsequent to the introduction of the proposal for data audits, Grimlund and Doucet (1992) developed detailed scientific statistical auditing techniques for research data based in part on financial auditing methods. The key new element in this technique is the development of stratified sampling for rare events. Two years later, Doucet and Grimlund collaborated with others, including one of the authors of this book (Shamoo; Doucet et al. 1994), to apply their statistical technique to existing data in toxicology that had been audited using a 100% random auditing method. They found the stratified method to be faster, more reliable, and more economical than 100% random auditing, because fewer actual data were audited.

Additional activity pertaining to data audits includes a 1992 report by the ORI (Price and Hallum 1992) on their use of several data analysis methods to detect scientific misconduct, including statistical auditing and sampling techniques. Similarly, a statistical approach that identified lack of randomness in detecting fabricated data was published by Mosimann et al. (1995). Biotechnology and pharmaceutical companies audit data as a quality assurance method (Glick 1993).

Finally, it may also be important to provide an educational and policy framework pertaining to research practices and arrangements that can tempt, challenge, or corrupt even those researchers who have a strong sense of integrity and ethics. While these initiatives would not deal specifically with misconduct, they would complement efforts to prevent misconduct by helping to create and maintain a research environment that emphasizes honesty, openness, fairness, integrity, responsibility, and other ethical norms (Kopelman 1999). Some of these complementary policies and educational efforts could address the wide variety of ethical issues in science that go beyond mere misconduct, such as issues relating to data storage, data sharing, authorship, publication, conflicts of interests, contractual arrangements, peer review, whistle-blowing, human subjects research, animal research, and intellectual property, to name but a few. To be sure, the research community already has policies and procedure relating to most of these issues. We discuss these as well as other ethical issues in the remainder of this book.

QUESTIONS FOR DISCUSSION

1. Based on the current information available about the incidence and significance of misconduct, do you think the press has exaggerated its implications? Does the process currently mandated by the federal government adequately manage the issue? How could it be improved? Do you think more or fewer regulations are needed to manage scientific misconduct?

2. What comments do you have on the causes of scientific misconduct? Which factors do you think are the most important? Do you think that the etiology of scientific misconduct resides primarily within the individual or within the system?

3. Can you give an example of possible scientific misconduct from your own experience or one that you have heard about? How was it dealt with? How should it have been dealt with?

4. How would you evaluate the process (inquiry, investigation, etc.) for the management of misconduct described in the text? Is it fair to the accused? To the accuser? To the university? To the public? Can you think of any changes that would improve the procedures?

5. If you were charged with scientific misconduct, would you hire legal counsel? If so, at what stage in the process?

6. Do you think that it could be possible to fully protect a whistle-blower such as Robert L. Sprague (see description in text)? What would be the essential safeguards of an effective policy to protect whistle-blowers?

7. Is data audit contrary to the personal liberty of scientists? Is it destructive to the principles of the scientific method? Could data audit stifle creativity and engender distrust? What system you would design to ensure scientific integrity?

CASES FOR DISCUSSION

Case 1: Eric T. Poehlman

Eric Poehlman was a tenured professor at the University of Vermont (1987–1993, 1996–2001) and a professor at the University of Maryland (1993–1996) and held an endowed chair at the University of Montreal (2001–2005). He resigned his position in Canada in 2005 after a lengthy investigation by the University of Vermont and the ORI (Office of Research Integrity 2005a,b). Poehlman conducted research with human subjects to study aging, menopause, and hormonal replacement therapy. He received funding from the NIH, Department of Agriculture, and Department of Defense. Between 1992 and 2000, Poehlman submitted grant proposals to federal agencies totaling $11.6 million. Poehlman admitted to falsifying and fabricating "preliminary research data" in these applications to impress reviewers and bolster his case. In this period, he received $2.9 million in federal funding. He also falsified and fabricated research data in his published research reports.

The University of Vermont investigation lasted from 2000 to 2002. The investigation was prompted by allegations by his research assistant, Walter F. Denino. In the course of the investigation, Poehlman destroyed evidence in his electronic data that showed his falsification and fabrication, he falsified documents, and he presented false testimonies. The university's investigative report was referred to the ORI, which in turn submitted it to the U.S. Attorney's Office for the District of Vermont.

The areas of Poehlman's scientific misconduct were as follows:

1. The Longitudinal Menopause Study (1994–2000): In a paper published in *Annals of Internal Medicine* in 1995, he falsified and fabricated results for all but three of the 35 women in the study.
2. The Longitudinal Study of Aging (1996–2000): In grant submissions, Poehlman falsified and fabricated data for subjects on physical and metabolic data.
3. The Prospective Hormone Replacement Therapy Study (1999–2000): Poehlman fabricated preliminary test results from a double-blind study on women for a grant proposal.

In March 17, 2005, Poehlman pleaded guilty, accepting a comprehensive criminal, civil, and administrative settlement. He agreed that he committed scientific misconduct, that he would pay $180,000 in restitution and $16,000 to the lawyer of the whistle-blower, that he would be barred for life from receiving federal grants, and that he would retract and correct the literature (10 papers were retracted). On June 29, 2006, he was sentenced to jail for one year plus one day.

- What you think drove Poehlman to misconduct?
- Should there been a system in place to help discover, or at least discourage, the misconduct sooner?
- Should the University of Montreal have investigated the research conducted at its location? Why?
- Should the University of Maryland have investigated the research conducted at its location? Why?
- Do you think Poehlman would have been caught if the whistle-blower had not come forward?

Case 2

A graduate student was conducting a behavioral study on drug addicts as part of a larger NIH grant that had been obtained by his mentor. The study was to be the core of his PhD thesis. However, the intended sample size of 50 was proving even more difficult to recruit than he had expected. Despite his most diligent efforts and the help of a social worker who had recently joined the grant team, he was able to recruit only 44 subjects. The prospects for getting more seemed dim. He was more successful, however, in his rapport with the social worker, and they began living together. Still, he was faced with the practical academic problem of completing his thesis. The conclusions that he could make from the statistical data from his 44 subjects supported his hypotheses. Nevertheless, he was still short on subjects. His solution to the problem was to make up fictional data for the four remaining, nonexisting subjects. The conclusions from the statistical data were basically the same whether he used 50 or 44 subjects, but he presented all 50 for his thesis.

Within a year he had defended his thesis successfully and received his PhD. He submitted two papers based on his thesis in a refereed journal and they were accepted for publication. He took a job at another college in the area so that his living arrangements could remain the same. A year later, after a nasty fight, he broke off his relationship with the social worker. Unbeknownst to him, she had been aware of the entire fraudulent episode. In anger, she wrote a letter to the ORI. The ORI informed the university and requested a report.

- What responsibility does the social worker have for the actions of the graduate student? Should she be considered a candidate for charges of misconduct?
- Should the graduate student's mentor have known what was going on in the activities of her grant?
- Does the mentor bear any responsibility for her student's actions?
- Can the university and the ORI take action even though the PhD degree has already have been given?
- Assuming a guilty verdict, what do you consider appropriate sanctions in this case?

Case 3

A new assistant professor was accused of misconduct by a technician at the university where she had trained. The accusation was that she published a questionable abstract for a national meeting based on research that was supported by an NIH grant. The problem was that the abstract contained data that the technician could not find in any of the laboratory notebooks. The technician also determined that these data had never been used in any other full-length published paper but had been part of the professor's PhD dissertation. The technician discussed his concerns with professor's former mentor, who in turn called the professor. Subsequently, the professor said that she was not able to find the data and claimed that it must have been misplaced. The technician made an official complaint both to the ORI and to the university.

- If the grant had been from a non-NIH source, would the university still have jurisdiction over this issue?
- What, if any, is the mentor's responsibility for the action of his former graduate student, either while she was a student or in the preparation of this abstract? Does the mentor have any obligations beyond his call to his former student?

- How do you assess the technician's response, including both the scope of his fact-finding activities and his decision making regarding his complaint?
- Now that the complaint has been made, what do you think should be the response of the assistant professor? The mentor? The ORI? The university?
- How could the situation in this case have been prevented?

Case 4

An assistant professor presented his department chair with a Howard Hughes Foundation grant proposal for her signature prior to sending it off. The dean's signature and those of other university officials were already on the routing form. As usual, like most overworked administrators, the chair checked the budget but did not read the grant. However, unlike many administrators, this chair, after the grant was submitted, took the time to read it. She became suspicious that several figures and tables in the grant application had not been based on valid data. A discussion with the assistant professor did not alleviate her concerns, and she filed a complaint to the university. After the university completed its investigation, the chair's suspicion was borne out: Several tables and figures had been altered to fit the overall hypothesis of the grant proposal. The assistant professor finally admitted the falsification.

- In this case, misconduct has already been determined. The question here is whether the misconduct simply involved the actions of a "bad apple" and was unavoidable, or whether the necessity of the university's investigative process could have been prevented, "bad apple" or not. If it was avoidable, how could prevention have occurred? By individuals doing their jobs better within a system that works, or by changing the system? For each alternative, how could misconduct have been prevented?
- Should the university have ever investigated this problem, given that the grant proposal is submitted to a private foundation?

Case 5

An associate professor in a large university submitted a major grant proposal to the NIH. During the grant review at the study section, a reviewer noticed that the p-value calculated for the table was based on eight experiments. However, the methods section had stated that the experiments had been conducted four times in duplicate. A preliminary inquiry by the study section staff raised further questions about the authenticity of the data. The executive secretary of the study section felt compelled to write to the university's grant office for clarification, with a copy to the ORI. The university responded by conducting an inquiry. The faculty member admitted that she always uses the number of experiments times the number of replicates to obtain the total number of experiments. The university chose not to proceed with the investigation. Instead, it warned her to stop using such methods. The university informed the ORI of its decision.

- What failed, and what preventive remedies you would recommend?
- Do you think the university's action was appropriate? Why or why not?
- Now that the case is in the ORI's hands, should it accept the university's decision?

Case 6

A newly appointed assistant professor applied for his first NIH grant. A member of the study section noticed that parts of the proposal looked unusually familiar. The reason was that entire paragraphs in the methods and the perspective/significance sections had been lifted from her own grant proposal that she had submitted last year. Although it was obvious that the new assistant professor had not been a member of last year's review process, the current reviewer knew that the submitter's department chair and former mentor had been a member of that study section. The reviewer called the department chair to find out what had happened. The chair reluctantly told her that he had showed the grant proposal to his then postdoctoral fellow for an opinion because he was too busy to read it thoroughly.

- Did the assistant professor do anything wrong when he agreed to read the grant proposal?
- Besides the assistant professor, does anyone else share responsibility for what happened?
- Was it appropriate for the member of the study section to first call the chair? Did the study section member have other alternatives?
- If the university hears about these events, what should it do?
- Was this is a personal or a systemic failure? Why?

Case 7

A new graduate student working in a senior professor's laboratory conducted experiments testing the modification of telomerase activity. The student was careful to do triplicate experiments for every variable. She found that the compound had no effect on telomerase activity. The professor himself took over and conducted the same experiment with slight modifications, but did only duplicate experiments for each variable. The professor found an important inhibition of telomerase.

By this time, the student had moved on to other areas of interest, but she was suspicious that the professor may have falsified the data, especially given that the experiments had not been done in triplicate. She made a complaint to the university. The university conducted an inquiry, during which the professor claimed that he had not altered any data. He explained that he had not used the student's data because of lack of confidence in her ability to carry out the procedure. The inquiry concluded that there was insufficient evidence to recommend moving to an investigation.

- Discuss the appropriateness of the student's action. Were there other alternatives she should have considered?
- Should the professor have accepted the student's data without further direct experimentation on his part?
- Do you agree that the case should have proceeded to the stage of inquiry? Should it have gone further?
- At this point, does the university have any obligations to the accused? To the complainant?
- Now that the university has effectively closed the case, could the ORI reopen it?

Case 8

A paper was written and submitted for publication as a result of the collaboration of three investigators. One of the authors noticed that the data in a table in the paper that was finally submitted was not the same as those in the penultimate draft. This author made several inquiries to his coauthors about what had happened but received unsatisfactory answers. He then accused the other two authors of scientific misconduct and informed the university. The university conducted an inquiry, which was not conclusive because no one could trace who actually modified the content. The data in the laboratory notebooks justified the data in the table up to the final draft but not in the submitted paper. The university's inquiry found that it could not place blame on any individual and closed the case.

- Where was the failure among the three authors?
- Was there a need for an inquiry because the paper was not yet published?
- Was the university's response appropriate? Did the university make a sufficient attempt to find out who changed the data? Are there other responses that should have been considered? Should the paper have been withdrawn?

Case 9

An administrative assistant, after leaving her current job to take other employment in the same institution, accused her former boss of falsifying a letter of support from a collaborator. The letter was appended to an NIH grant proposal. The assistant wrote directly to the ORI, which requested that the university initiate the misconduct process. An inquiry found that the professor had actually falsified the letter. However, the collaborator had verbally agreed to write such a letter of support but was unable to do so because of travel commitments. If the researcher had waited until the collaborator had come back, he would have missed the grant submission deadline. Based on the collaborator's confirmation that the agreement of collaboration had already been made, the university found that there was no scientific misconduct.

- Although the professor falsified the letter, the university found that scientific misconduct had not occurred. Do you agree with this action? What are the implications of this action? In what way does the professor's falsification of the letter meet or not meet the definition of misconduct?

- Could the professor have handled his dilemma differently?
- What other options did the assistant have when she discovered the falsification of the letter? Did she make a good choice?

Case 10

A postdoctoral fellow made a complaint to the university that his mentor, an associate professor, had falsified the data in a newly published report. The fellow alleged that patients' reports of ovarian cancer did not represent the actual data. The university convened an inquiry committee, which determined that among the 120 subjects in the study, only 80 reflected actual data. The records of the other 40 subjects contained various types of errors as well as information that did not support the investigator's hypothesis. The inquiry committee concluded that although the investigator showed considerable carelessness, there was no proof of intentional falsification of the data.

- Do you agree with the university's decision to not find that scientific misconduct had been committed? Do you think there are circumstances where sloppy work should be considered misconduct, regardless of the ability to prove intent?
- Do you think that the university has any responsibility to comment further on the investigator's management of data or to require assurances of appropriate data management in future research?
- The postdoctoral fellow went directly to the university without first discussing the issue with his mentor. Did the fellow have any obligation to contact his mentor first?
- Should the investigator's report be rescinded even though she was not found guilty of misconduct?

9

Intellectual Property

This chapter discusses the history of the intellectual property and its ethical and legal foundations. It provides an overview of the U.S. intellectual property system and discusses patents, copyrights, trademarks, trade secrets, and ownership of research data. The chapter also examines some key pieces of intellectual property legislation and intellectual property cases and discusses some ethical controversies, such as patents on biological materials.

HISTORY AND OVERVIEW OF INTELLECTUAL PROPERTY

When most people think of their property, they imagine their house, their land, their car, their book collection—something that they can touch, see, feel, hear, smell, or taste. Many of the property rights that people have pertain to tangible objects located in time and space. But people also claim to own things that are not located in any particular time or space, such as a song, a poem, computer software, a play, a formula, or any invention. These kinds of intangible things that we claim to own are known as intellectual property (Foster and Shook 1993). In general, property rights are collections of rights to control some thing, such as a house. Someone who owns a house has a right to sell, rent, modify, paint, use, or tear down the house. People who have intellectual property have rights to control intangible objects that are products of human intellect (Garner 1999). For instance, if you have a copyright on a play, you are granted the right to prevent other people from performing the play without your permission. You also have the right to sell your copyright on the play.

Modern property right laws have their basis in Roman laws, which influenced the development of legal systems in Europe and the United States. Nations recognized property before the advent of the Roman Empire—Jewish laws dating to the time of Moses address property issues, for example—but the Romans developed what was at that time the world's most comprehensive and precise legal system. The U.S. Constitution draws heavily on the property rights theories of the eighteenth-century English philosopher John Locke.

Although the Western world has recognized property for thousands of years, intellectual property is a more recent development. While ancient Greek and Roman authors and inventors were concerned about receiving proper credit for their discoveries, the Greeks and Romans did not have

Note: Nothing in this chapter should be taken as a legal advice. Engaging an intellectual property attorney is recommended in the event of contemplating issues related to a patents, copyrights, or trademarks.

intellectual property laws per se. Although the origins of patents are obscure, some of the world's first patents were granted in England in the 1400s when the monarchy granted privileges, known as letters patent, to manufacturers and traders. King Henry VI granted the first known English patent in 1449 to John of Utynam for a method of making stained glass (U.K. Patent Office 2001). During the next 200 years, patents became a routine part of commerce and industry in England, although disputes arose concerning the length of the patent period and the conditions for patenting.

The steam engine (1769) was probably the single most important patent awarded by the British government. This invention helped to provide additional justification for patents and served as a model of science–industry collaboration. James Watt (1736–1819) developed a more efficient version of the steam engine, which had been developed by Thomas Newcomen and Thomas Savery. He was awarded a patent in 1769 titled "A New Method of Lessening the Consumption of Steam and Fuel in Fire Engines." Watt collaborated with the entrepreneurs John Roebuck and Matthew Boulton. Roebuck made two thirds of the initial investment required to develop the steam engine but went bankrupt. Boulton bought Roebuck's share of the patent and helped to market this new product. Watt's steam engine was initially most useful in draining mines but was later used for machinery in factories. Watt and Boulton made a considerable sum from the steam engine, which was the product of scientific ingenuity and private investment and marketing (Burke 1995).

The need for copyright coincides with the development of the printing press in the 1500s. Before the printing press, copying of books and author's writings was rare, and most people were illiterate. Books and other documents were copied laboriously by hand in Europe. Because it took so much effort to copy a book, the problem of unauthorized copies did not arise often. After the printing press was invented, it was possible to make thousands of copies with relative ease, and literacy increased. The question naturally arose as to who would control the making and selling of these copies, and whether "unauthorized" copies would be allowed. Thus, the idea of a "copyright" was developed in eighteenth-century England as a way of giving authors and publishers some control over printed works. In 1710, the English Parliament passed a statute granting copyright protection to books and other writings. Prior to this statute, copyrights were protected by common law (U.K. Patent Office 2001).

The first U.S. patent was awarded in 1641 to the Massachusetts Bay Colony for the production of salt. The framers of the U.S. Constitution were aware of the scientific and technical developments that were occurring before their eyes and the need to grant intellectual property rights to authors and inventors to encourage the advancement of science, technology, industry, and the practical arts. The primary author of the Constitution, Thomas Jefferson, was himself an author and inventor. Benjamin Franklin, who helped draft the Constitution and the Declaration of Independence, was both a statesman and a prolific inventor whose inventions included the harmonica and the lightning rod. Given their familiarity with science and technology and their appreciation of the importance of free enterprise and commerce, it should come as no surprise that the founding fathers included a provision about intellectual property rights in the U.S. Constitution. Article 1, Section 8, provides the basis for intellectual property laws in the United States when it states that Congress shall have the power "to promote the progress of science and useful arts, by securing for limited times to authors and inventors the exclusive right to their respective writings and discoveries." In 1790, Congress enacted the first patent and copyright laws, long before the U.S. Patent and Trademark Office was officially established in 1836. The patent laws have been amended numerous times, including significant revisions in 1952 and 1984, and most recently in 1995 (U.S. Patent and Trademark Office 2001). Congress also enacted laws establishing the U.S. Copyright Office. These laws have also been revised several times, with the most significant revision occurring in 1976 (U.S. Copyright Office 2001).

During the 1800s, science–industry and government–industry collaborations continued to bear fruit, led by the German universities and the German dye industry. The modern chemical industry began when German companies developed private laboratories for making synthetic dyes. William Perkin, a student of the organic chemist August Hoffman at the Royal College of Chemistry at London University, discovered a purple dye during his experiments on coal-tar compounds. Perkin realized that this invention would have great commercial value, because purple was a very rare and expensive dye at this time. He opened his own laboratory and began commercial production of the dye (Ronan 1982). By the end of the century, many companies had their own laboratories and employed scientists, engineers, and technicians. The great master of invention, Thomas Edison, obtained over a thousand patents from his private laboratory in Menlo Park, New Jersey, including the electric light bulb (1879), the phonograph (1877), the stock ticker, and the duplex repeating telegraph (Burke 1995).

In the twentieth century, many new science–industry and government–industry collaborations produced more inventions and discoveries, such as the automobile, the airplane, plastics, synthetic fabrics, and computers. In most of these cases, governments funded the basic research that laid the foundations for practical applications and commercial products (Dickson 1995). For example, computers were developed in government laboratories to solve difficult problems in rocket telemetry. Private companies, such as IBM, Apple, Dell, and Microsoft, have invested a great deal of money in computers and information technology, but the basic ideas were developed by academic scientists. The biotechnology industry follows a similar trend. Most of the important basic research, such as James Watson and Francis Crick's discovery of the structure of DNA in 1953 and the development of recombinant DNA techniques, used government funds. For the most part, private biotechnology and pharmaceutical companies, such as Millennium Pharmaceuticals, Genentech, Amgen, and Monsanto, have capitalized on the basic research conducted by academic scientists. However, in recent years private companies are beginning to invest heavily in basic research. For example, Celera Genomics has sequenced the human and mouse genomes. (We consider genome sequence data to be basic rather than applied research.) We return to this issue further below.

To help ensure a smooth transfer of technology from the public to the private sector and to private university–industry collaborations, Congress passed an amendment to the patent laws known as the Bayh-Dole Act of 1980 (35 USC 200), which was amended by the Technology Transfer Act of 1986 (Public Law 99–502). These laws allow individuals and companies to commercialize research developed with the aid of government funds (Dreyfuss 2000). Companies or universities that plan to develop inventions from government-funded research can sign cooperative research and development agreements (CRADAs) with the government. CRADAs specify how data and materials will be used and delineate intellectual property rights.

Issues relating to intellectual property and university–industry collaboration were not clearly defined before these laws were passed. Prior to Bayh-Dole, more than two dozen different policies pertained to intellectual property claims resulting from federal grants (LaFuze and Mims 1993, Lentz 1993). Many different institutions and individuals could make some claim to "own" the data generated, in part, by a government grant, including different researchers, technicians, different universities (if the research was conducted at different sites), private companies (if they funded any of the research), and private endowments (if they funded any research), as well as the government itself. The laws passed in the 1980s have helped those involved in research to define their roles in research and intellectual property rights. Although many university researchers believe that they own the data generated through their work, data ownership is determined by agreements faculty members have with institutions (Shamoo and Teaf 1990, Fields and Price 1993). Today, most universities have policies pertaining to intellectual property and require researchers to disclose potential inventions. In a typical arrangement, the university and the researcher may split the royalties from a patent on an invention developed at the university. For example, the University of Maryland, Baltimore offers researchers a 50% share of royalties. Researchers who work for private industry usually must transfer part or all of their intellectual property rights to the company as a condition of employment.

One of the most significant effects of the Bayh-Dole Act is that universities have become increasingly interested in patents and licenses (Bowie 1994, Resnik 2007b). The rate of university patenting has increased dramatically, and universities have opened technology transfer offices designed to encourage patenting and to protect intellectual property rights (Dickson 1995, Washburn 2006). In the United States, research conducted in university laboratories constitutes the major output of basic research, with close to $20 billion annual investment mostly by the federal government (Shamoo 1989, Walsh et al. 1997). The Bayh-Dole Act was therefore a natural outgrowth of government policy to protect its investment and expedite its use for public good in promoting progress. To compensate for decreases in government or other sources of funding, many universities view intellectual property as a potential source of revenue. However, income generated from intellectual property has not yet become a significant component of the budget of most universities (Guston 2000). Income from royalties to universities is estimated at $350 million annually, representing about less than 2% of their total annual research budget (Waugaman and Porter 1992, Walsh et al. 1997). However, it is likely that universities will begin to realize increased revenues from intellectual property due to innovations in biotechnology, medicine, and information technology.

In recent years, universities have established technology transfer offices to boost their income from royalties related to patents and copyrights. Technology transfer offices usually consist of an officer who is familiar with intellectual property and technology transfer laws and guidelines and has access to an intellectual property attorney. These technology transfer offices attempt to educate and train faculty members on measures designed to protect intellectual property, such as keeping dated and witnessed laboratory notebooks, rules on data sharing, and early filing of patents within the legal limit. Technology transfer officers also instruct researchers on the universities intellectual property policies and how to protect the university's interest in collaborative research with other investigators at other universities or in industry. Many of these financial interests of universities and the close ties between universities and industry raise ethical concerns about conflicts of interest (Rule and Shamoo 1997), which we discuss in chapter 10. Moreover, the interest of the university may collide with the interest of the faculty/inventor. The Bayh-Dole Act calls on the university to develop the invention and, if not, to turn the invention over to the inventor.

There are also some international treaties pertaining to intellectual property. The first international intellectual property treaty was the Paris Convention of 1883, which was adopted by 20 countries initially and has been adopted by many others since then. Other important intellectual property treaties include the Agreement on Trade-Related Aspects of Intellectual Property Rights (TRIPS), which was signed by 120 countries in 1994. Although a patent or a copyright grants legal protection only in the country in which it is issued, nations that abide by international intellectual property treaties agree to honor each other's intellectual property laws. For example, a nation that abides by TRIPS does not allow the importation of pirated software or unauthorized generic drugs. However, TRIPS allows for some compulsory licensing to address public safety or public health crises (Resnik 2001c, 2003c,d). For example, a nation facing a devastating epidemic such as HIV/AIDS could use the compulsory licensing provisions of TRIPS to require a pharmaceutical company to license another company to manufacture HIV/AIDS medications.

As one can see from this brief history, the main rationale the people have offered for intellectual property protection is utilitarian: Intellectual property laws promote social welfare by encouraging ingenuity and progress in science, technology, and the arts (Kuflik 1989). They encourage ingenuity and progress because they provide authors and inventors with economic incentives to produce original works and inventions and to share the products of their labor with the public. Without such protections, authors and inventors may decide to not pursue their original works or inventions or to keep them a secret (Foster and Shook 1993). When an inventor is granted a patent, the patent application becomes a public record, which enables other scientists and inventors to learn from the invention. This allows researchers to share information while also granting them intellectual property rights. The intellectual property laws also protect the financial interests of businesses and therefore encourage business to invest in research and development (R&D). Businesses view R&D funding as risks that can be justified only if there is some expectation of a reasonable return on investment. Intellectual property laws enable businesses to take these risks by allowing them to control the products of their R&D investments (Kuflik 1989, Resnik 2001c, 2007b).

Another type of justification for intellectual property comes directly from the work of John Locke (1764 [1980]), who was a strong defender of individual rights (i.e., a libertarian). According to Locke, all human beings have some inalienable rights relating to life, liberty, and property. The main function of government is to protect these rights and prevent citizens from violating each other's rights. Thus, on the Lockean view, laws against theft can be justified on the grounds that they protect rights to private property. We can acquire property, according to Locke, through original acquisition or transfer, such as through commercial transactions or gifts. Original acquisition of property occurs when one mixes or adds one's labor to a thing or common resource. For example, if we view the forest as a common resource, if I remove a piece of wood from the forest and carve it into a flute, then the flute becomes my property because I have added my labor to the wood. This libertarian approach implies that laws can be crafted to protect intellectual property rights and that people can acquire intellectual property through original acquisition or transfer. For example, one might acquire a new invention (e.g., a better mousetrap) by adding one's labor to previous ideas and inventions (e.g., the old mousetraps). One could acquire property rights to a song by using or putting together melodies, words, and harmonies from previous songs to make a new one (Kuflik 1989, Resnik 2001c).

Regardless of whether one adopts a utilitarian or a libertarian approach to intellectual property, the most basic theoretical issue with respect to intellectual property rights is finding the proper balance between public and private control of intellectual property (Resnik 2007b). Most theorists agree that some form of private ownership is necessary in order to provide rewards and incentives to individuals and corporations, and most theorists agree that a public domain of information is needed to ensure that people have freely available resources to create new inventions and make new discoveries. But finding the proper balance between public and private control is not always easy, and that balance may change as new technologies, such as computers and recombinant DNA, emerge and social institutions, such corporations and universities, evolve. This is one reason that it is necessary to reevaluate and revise intellectual property laws as the situation warrants.

TYPES OF INTELLECTUAL PROPERTY

In this section, we discuss four basic types of intellectual property recognized by U.S. law—patents, copyrights, trademarks, and trade secrets—and their relevance to sharing data in a research setting. We also discuss the status of data as a form of intellectual property.

Patents

Under U.S. law, a patent is a type of intellectual property granted by the U.S. Patent and Trademark Office to an inventor. A patent gives inventors exclusive rights to prevent anyone else from using, making, or commercializing their inventions without their permission. Inventions may include machines, products of manufacture, methods or techniques, compositions of matter, or improvements on any of these. Individuals as well as corporations or the government can own patents. The length of a patent is 20 years from the filing date of the patent application (Kayton 1995). Patents are not renewable. Inventors can sell their patent rights, or they can grant others a license to use, make, or commercialize their inventions. In exchange for the licenses, the licensee may provide the licensor with royalties in the form of a one-time payment or a percentage of profits. An exclusive license is a license between the licensor and only one licensee. A nonexclusive license allows the licensor to license the invention to more than one licensee. Different companies often reach agreements allowing them to use, make, or commercialize each others' inventions, known as cross-licensing agreements and reach-through licensing agreements. These agreements enable companies to avoid costly and time-consuming patent infringement litigation.

Patent infringement occurs when someone makes, uses, or commercializes a patented invention without permission of the patent holder. If someone infringes a patent, the inventor can file an infringement claim with any U.S. federal court (Foster and Shook 1993, U.S. Patent and Trademark Office 2001). The patent holder can sue the infringing party for damages and obtain an injunction requiring the infringing party to cease infringement. The party accused of infringement may challenge the patent, and the court will determine whether the patent is valid. Because all appeals in patent cases go to the same federal court, the Court of Appeals for the Federal Circuit in Washington, D.C., patent laws are uniform throughout the United States. Appeals of cases heard at this court, such as *Diamond v. Chakrabarty* (1980) (discussed below), are heard by the Supreme Court.

Research (or experimental) use is a narrow provision in patent law that allows a researcher to make or use an invention for research purposes, but not for commercial purposes. For many years, academic researchers believed that they were protected from patent infringement under the research use exemption in patent law. However, an important decision made in a legal case, *Madey v. Duke University* (2002), has limited the research use. In this case, the Court of Appeals for the Federal Circuit ruled that Duke University had infringed John Madey's patent on a laser by using the device without his permission. Duke University argued that its researchers could use the laser under the research use exemption, but the court ruled that the university could not claim this exemption because it was using the laser to further its business interests. The court treated the university as similar to a private company. Because most scientists work for either universities or private companies, *Madey* effectively ended the research exemption in the United States.

To obtain a U.S. patent, the inventor must file an application with the U.S. Patent and Trademark Office. In deciding whether to award a patent, the office considers the following criteria (Foster and Shook 1993):

1. *Product of human ingenuity:* Is the invention a product of human ingenuity or a product of nature? Only products of human ingenuity can be patented. The courts have ruled that laws of nature, natural phenomena, mathematical formulas, and naturally occurring species are products of nature and cannot be patented. In a landmark case for the biotechnology industry, *Diamond v. Chakrabarty* (1980), the Supreme Court ruled that living things could be patented. The Court referred to an earlier patent case, *Funk Brothers Seed Co. v. Kalo Inoculant Co.* (1948), and stated that "anything under the sun made by man" is a human invention. Ananda M. Chakrabarty had filed a patent claim on a genetically modified bacterium useful in cleaning up oil spills. Prior to this case, the only living things that could be patented were hybrid species of plants protected by special plant patenting laws, such as the Plant Variety Protection Act of 1930 (9 USC 2321). A lower court had rejected the patent on the grounds that the genetically modified bacteria were not human inventions, but the Supreme Court negated this decision. This Court's decision paved the way for patenting many products of biotechnology and bioengineering, including genes, proteins, and transgenic animals (Eisenberg 1995, Resnik 2001c).

2. *Novelty:* The invention must be new or innovative. It must not be previously patented or disclosed in the prior art, which includes patents, publications, or public uses. Because publication or public disclosure of an invention can jeopardize the patent, most inventors keep their patentable work secret until they are ready to file an application. In the United States, the first person to conceive of an invention is awarded the patent, provided that she exhibits due diligence in filing the patent application. In the United States, the inventor has a one-year grace period to file a patent application after publication. If a person fails to exhibit due diligence, another inventor may be awarded the patent, if he files the application first (Miller and Davis 2000). In Europe, the first person to file a patent application

is awarded the patent. To prove that one is the first inventor of an invention, one must keep good research records (see chapter 3).

3. *Nonobviousness:* The invention must not be obvious to a person trained in the relevant discipline or technical field. Whether an invention is or is not "obvious" is subject to a great deal of debate (Duft 1993).

4. *Usefulness:* The invention must serve some worthwhile practical use. "Trivial" uses, such as filling a landfill or a subject of meditation, do not count as practical uses, nor do uses as research tools. For instance, the U.S. Patent and Trademark Office has recently ruled that basic genome sequence data is not in itself patentable; DNA patents must specify specific uses for DNA in drug development, diagnostics, or bioengineering (Resnik 2001c). If the invention has a military use or implications for national security, the U.S. government has the right to co-opt the invention and compensate the inventor, who may also sign a contract with the government.

5. *Enabling description:* Inventors must reduce their inventions to practice; that is, they must describe the invention in enough detail that someone trained in the relevant discipline or field could make and use the invention. This description of the patent becomes a public document and is part of the patenting "bargain."

Although the patent system encourages public disclosure in exchange for intellectual property rights, some corporations use the system not to develop new inventions but to prevent competing companies from developing new inventions (Resnik 2001c). For example, to secure the market for their trademarked drugs, large pharmaceutical companies have purchased patents on competing generic drugs owned by smaller companies. Other companies have developed "blocking" patents designed to prevent competitors from developing new products. For example, if a company is developing a new internal combustion engine, a competing company could block production of this engine by acquiring a patent on a part needed to make the engine.

Copyrights

Copyrights are exclusive rights granted by the U.S. legal system that allow the authors of original works to make copies of the work, make other works derived from the original work, perform or display the work, and distribute, sell, or rent copies of the work. People who perform any of these actions without the permission of copyright holders violate copyrights. Original works include written works, such as books, papers, software, databases, and poems; performances, such as plays or dances; audiovisual recordings, such as movies, music, photographs, and televisions shows; and artistic works, such as paintings and sculpture. A work can be original without being new or novel because the author is the first person to put the work into tangible form. A copyright extends for the lifetime of the author(s) plus 70 years, and may be renewed. To register copyright, one may file for a copyright with the U.S. Copyright Office at the Library of Congress. However, authors of original works have copyright protections even if they do not take this step. To ensure that others are aware of their claims to a copyright, many copyright holders write "copyright" on their original works, such as "Copyright © 2008 Shamoo and Resnik, all rights reserved." Copyrights protect original works but not the ideas expressed by those works (Chickering and Hartman 1980, Office of Technology Assessment 1990). Although it is illegal to sell copies of the book *Jurassic Park* without permission of the copyright holder, it is perfectly legal to discuss (or profit from) the ideas expressed in the book without the owner's permission.

Many copyright holders sell their rights to publishers or other distributors for a one-time fee or a percentage of profits. A work produced by an employee of a business is considered a "work for hire," so the copyright belongs to the business, unless the business grants the employee some portion of the copyright as part of contract negotiations. For example, most academic institutions allow faculty members to retain copyrights over their works. In some instances, universities or colleges may seek copyrights to special, commissioned works, such as Internet courses or educational software. Works created by U.S. government employees as part of their official duties are considered to be in the public domain, and the employees have no copyrights pertaining to such works.

One important exception to copyright law is the doctrine of fair use. According to this doctrine, it is permissible to copy portions of the author's work or even the whole work without his or her permission if the copying is for personal, educational, or research purposes and does not jeopardize the commercial value of the work (Foster and Shook 1993). For example, the doctrine of fair use allows a person to use a VCR to record a television show to watch later, but it does not allow a person to use a VCR to make or sell copies of the television show for a large audience. During the 1980s, the copying company Kinko's compiled, copied, and sold course packets—selections of readings from journals or books—for professors teaching university or college courses. Publishing companies sued Kinko's for copyright violations, and the courts ruled in their favor in 1991 (*Basic Books, Inc. v. Kinko's Graphics Corp.*1991). Even though the use was for an educational purpose, it was not a fair use because it defrayed the commercial value of the published works. The dispute over the legality of the Napster Web site illustrates the continuing evolution of the doctrine of fair use. Many companies from the recording industry sued Napster for copyright violation because the company was distributing copyrighted music over the Internet without permission. Also note that many types of knowledge, such as government documents, public records, weight conversion tables, temperature measures, calendars, known titles, phrases, and lists of ingredients, are considered to be in the public domain and are not copyrighted (Chickering and Hartman 1980).

Trademarks

A trademark is a distinctive symbol or mark, such as a name, phrase, device, stamp, logo, or figure, that businesses use to distinguish themselves. Some examples of trademarks include the name "Coca-Cola," the phrase "have it your way," the McDonald's arches, the Hot Wheels® flame logo, and the Planters peanut man. Trademarks are useful to businesses for marketing their goods and services, because they provide consumers with a way to easily recognize the business and its products. Trademarks are protected by state and federal laws and are important in commerce, but they play only a minimal role in research. To obtain federal trademark protection, a business may submit an application to the U.S. Patent and Trademark Office. Trademarks are renewable indefinitely for 10-year periods (Office of Technology Assessment 1990, Adler 1993, U.S. Patent and Trademark Office 2001).

Trade Secrets

The forms of intellectual property discussed above—patents, copyrights, and trademarks—are rights granted by the government designed to promote the dissemination of information while protecting proprietary interests. The key policy issue in these forms of intellectual property is finding the proper balance of public and private control of information. Trade secrets, on the other hand, are designed to prevent information from becoming publicly available. The law recognizes trade secrets because they promote the interests of commerce and industry and can be more useful than other forms of intellectual property for some businesses (Foster and Shook 1993). For example, consider the formula for Coca-Cola, one of the best-guarded trade secrets. If the company had patented its formula when it invented Coca-Cola, then the patent would have expired decades ago and the company would have lost its share of the market for this product. As far as the company is concerned, it can make more money by keeping the formula a secret instead of patenting it. The company can do this because the secret is well guarded and difficult to discover. Although many companies have manufactured products that taste similar to Coca-Cola, no company has manufactured a product that tastes just like "the real thing" because it is difficult to master all the subtle variations in ingredients and manufacturing processes that the company employs. Other types of trade secrets may include other formulas, instruments, business plans, customer lists, and company policies. State laws protect trade secrets provided that the company makes an attempt to keep the secret and the secret has commercial value. Someone who discloses a trade secret to a competing business can be prosecuted under civil liability or for theft and can be fined up to and exceeding the monetary value of the secret.

The main problem with trade secrets is that they are difficult to protect. First, employees may disclose trade secrets either intentionally or inadvertently. Second, there are legal methods that competitors can use to discover trade secrets, such as reverse engineering. It is perfectly legal for another company to "reverse engineer" Coca-Cola by purchasing some of the product and analyzing it to determine how it is produced. Trade secrets are not protected if they are derived from independent research, open meetings, and a host of other methods. In the biotechnology industry, it is virtually impossible to keep trade secrets due to the open nature of biotechnology R&D.

Most of the materials used in biotechnology are available to the public, such as organic compounds, organisms, common tools, and techniques. Thus, in biotechnology and pharmaceuticals, patents are generally a better form of intellectual property protection than are trade secrets (Office of Technology Assessment 1990, Adler 1993, Resnik 2001c).

The biotechnology and pharmaceutical industries have invested heavily in R&D in the last two decades with the aim of developing patentable products, such as new drugs or biologics. Private investment in biomedical R&D in the United States rose from $2 billion per year in 1980 to $16 billion per year in 1990 and is probably now more than $70 billion per year (Resnik 2007b). Most of the funding comes from pharmaceutical companies, which spent more than $44 billion on R&D in the United States in the year 2007 (Pharmaceutical Research and Manufacturers Association 2008). In deciding whether to develop a new product, a company must determine whether it will be able to obtain an adequate return on its investments. It takes, on average, more than $500 million of R&D and 10 years of testing to bring a drug to the market (Resnik 2007b). Many biotechnology products, such as proteins, genome sequence data, and transgenic organisms, also require investments on the scale of hundreds of millions of dollars (Resnik 2007b).

DATA OWNERSHIP

Individuals, corporations, universities, and government agencies often assert ownership claims over data. For example, private companies claim to own all the data produced in their laboratories or using their funds. Universities also claim to own data generated by means of university funds or resources. The following policies are typical of those found at many other universities:

> The primary owner of research records is the University. The University has the right of access to the supporting records for all research carried out through the University with the understanding that information or data that would violate the confidentiality of sources or subjects involved in the research should not be disclosed. In addition, extramural sponsors providing support for research at Duke University may have the right to review any data and records resulting from that extramural support. (Duke University 2007, p. 25)

> The Wake Forest University School of Medicine (WFUSM) is the owner of research data developed by the faculty, fellows, students and employees of WFUSM in the course and scope of their employment at WFUSM unless specifically modified by contract with WFUSM. Research data collected under usual and customary laboratory practices include recorded information, regardless of the form or the media on which it may be recorded, as well as unique research resources developed at WFUSM. This right of ownership applies to all research activities of WFUSM regardless of the sources of funding that supported those activities. (Wake Forest University School of Medicine 2007)

Individuals, institutions, and organizations claim to own research data mainly to protect their financial interests in the data. Suppose a researcher does animal toxicology studies on a drug that a pharmaceutical company is developing. She has a contract with the company and is employed by a university. Suppose a dispute arises over publishing the data: The researcher wants to publish but the company objects to publication. Who owns the data, the researcher, the company, or the university? Or suppose a researcher obtains data at a university through a government grant and then leaves the university for a job at another institution—can he take his data with him, or must it stay at the university? If questions such as these are not resolved up front by means of contracts or agreements, such as CRADAs or material transfer agreement (MTAs), then costly legal battles may arise (Shamoo 1989, Shamoo and Teaf 1990).

There is certainly a sense in which one might view data as property, but there are no laws designed specifically to protect data. To treat data as property, one must therefore apply existing copyright, patent, trade secrecy, or property, contract, criminal, or civil laws to research data or enact specific institutional policies. For example, an employee's contract with a university or private company can settle data ownership issues. If the employee discloses or uses data without permission, the university or private company can sue the employee for breach of contract. Similarly, trade secrecy laws can permit companies to prevent employees from disclosing data. Research records, such as lab notebooks or computer disks, are physical objects that can be protected by property laws. Someone who takes a lab notebook from an institution without permission can be charged with theft. Trade secrecy laws can provide companies with some ownership of data, provided that the data are properly protected trade secrets. For example, tobacco companies sought for many years to protect their research on nicotine's addictive properties under the cloak of trade secrecy (Hurt and Robertson 1998, Resnik 1998a).

In the private sector, data are treated as "propriety information." The data submitted by private companies in support of a drug application to the U.S. Food and Drug Administration (FDA) are not all made public, even after the drug has been approved. Although U.S. law protects this type of information, researchers may still face an ethical dilemma in situations where promoting the good of society requires them to break the law. For example, a scientist conducting secret research for a drug company who discovers a problem with the medication, which the company does not want to report to the FDA, must choose between abiding by the company's policies and serving the public good (Resnik 1998a).

Some laws and regulations require researchers receiving government funds to share data. Under the 2002 Freedom of Information Act amendments, data generated by the use of government funds are public property. Interested parties can request research data once a researcher has completed a study by contacting the appropriate agency of the U.S. government. The agency will gather the information and may charge a reasonable fee for its services. However, the public cannot gain access to some types of data, such as confidential information pertaining to human subjects or patients, proprietary business information, classified information, information used in law enforcement, or personnel records (Fields and Price 1993, Cohen and Hahn 1999, Freedom of Information Act 2002). According to the National Institutes of Health (NIH) policy on data sharing, NIH-funded researchers are required to share published research data with qualified scientists. Researchers may charge a reasonable fee for the costs of sharing data with interested parties (National Institutes of Health 2003). There is no evidence that this policy is widely complied with. In fact, there is evidence to the contrary (see chapter 3 for a discussion of data sharing).

OTHER TYPES OF PROPERTY IN SCIENCE

In addition to intellectual property, many other different types of property are related to scientific research, including:

- *Research materials*, such as pharmaceuticals, reagents, cell lines, antibodies, genetically engineered organisms, fossils, and blood samples
- *Research tools*, such as microscopes, telescopes, mass spectrometers, satellites, assays, computers, satellites, and test tubes
- *Research venues*, such as laboratories, greenhouses, archeological sites, jungles, reefs, and lecture halls

Ethical and legal issues can arise concerning the stewardship of these resources. One might argue that scientists have an ethical obligation to share these resources so that other researchers may benefit from them. However, some reasons that scientists may refuse to share research resources include the following:

- *Financial interests:* Researchers and research sponsors may have financial interests related to research materials, tools, or venues. Sharing these resources adversely affects those interests.
- *Scientific priority:* A researcher who shares research materials, tools, or venues may give an edge to his competitors and lose the race for priority.
- *Cost and inconvenience:* Sharing research materials and tools can be costly and inconvenient. Researchers may not want to maintain stockpiles of materials for others to use or ship them out upon request.
- *Scarcity:* Some research materials and tools are scarce resources. Researchers may want to limit their sharing to preserve resources for their own work.

The sharing of research resources, therefore, raises many of the same issues that arise in data sharing. Deciding whether, and how, to share resources used in research requires a careful examination of the facts, circumstances, and options pertaining to the decision. Given the importance of sharing information and resources in science, the burden of proof falls on those who refuse to honor legitimate requests to share. Many government agencies require researchers to share not only data but also materials that have been developed or purchased with public funds. Grant applications may include plans for sharing research materials. Researchers may charge a reasonable fee for sharing, or they may license private companies to help them answer request to share materials. Obligations to share materials need not undermine researchers' plans to publish or patent (National Institutes of Health 1999).

ADDITIONAL ETHICAL CONCERNS

Before concluding this chapter, we return to some more fundamental issues in intellectual property. This first issue is the question of *who* has an ethically defensible claim to ownership of intellectual property; the second is the question of *what* ethically can be treated as intellectual property. Here we are not concerned with what the law says about intellectual property, but with what the law *should* say.

Who Has Intellectual Property Rights?

Consider an invention or an original work. Many different individuals and institutions may assert ownership claims over an invention, such as a transgenic mouse, or an original work, such as computer software. Who should be granted patent rights on the mouse or copyrights on the software? In many ways, this question parallels questions about authorship discussed in chapter 6, where we argue that authorship and accountability should go hand in hand: An author is someone who makes an important contribution and can be held publicly accountable for the research. If all authors have copyrights, then the question of who should have copyrights is the same as the question of who should be an author. One may argue that the same point applies to patent rights: An inventor, like an author, is someone who makes an important contribution to the invention and can be held accountable for the invention. That is, the inventor could describe the invention, explain how it works and what it does, and show how it is useful and original (Dreyfuss 2000). Interestingly, an invention often has more authors listed in the paper describing it than it has inventors listed on the patent application (Ducor 2000). One might speculate that in these cases people are being listed as authors who do not deserve to be or people are not listed as inventors who deserve to be.

Although relying on some principle of accountability may settle many concerns about intellectual property rights, it does not address the role of "contributors" and any morally legitimate claims they may make to intellectual property. Because contributors do not, by definition, play a significant role in research, most of the intellectual property claims made by contributors relate to concerns about fairness, not about accountability. Consider the following examples:

- A lab technician carries out a great deal of the work in developing a patented mouse and is listed as an author on the paper but not as an inventor on the patent. Should the technician have any patent rights? If she has no patent rights, then should she not be listed as an author on the paper?

- A medicine man in the Amazon jungle teaches a team of botanists and pharmacologists about some of the healing powers of a native plant, and they develop a new drug by isolating and purifying a compound in the plant. Should the medicine man (or perhaps his community) be granted some share of royalties from the patent?
- An oncologist develops a valuable cancer cell line from cancerous tissue extracted from a patient's tumor. Should the cancer patient have intellectual property rights over commercial products from his tissue? (This example is based on the famous *Moore v. Regents of the University of California* [1990]; see case 1 below.)
- A graphic artist develops images for a textbook and is listed as a contributor but not as an author. Should she be granted a share of copyright on the book?

Questions about "fairness" raise fundamental issues about how to allocate benefits and burdens. According to the libertarian approach exemplified by Locke, fairness is strictly a matter of contribution or merit: If you contribute something to a project, than your fair share (i.e., your benefits) should be in proportion to your contribution. According to the utilitarian approach, what is fair is what best promotes society's good, and intellectual property principles and laws should promote the social good. Thus, it may follow, on this view, that it is fair not to allocate benefits, such as royalties, on the basis of contribution. The best way to maximize utility may be a system that rewards authors and inventors who are the first to create an original work or invention.

What Can Be Treated as Intellectual Property?

The final issue we consider in this chapter concerns the ethical or moral limitations on intellectual property. Are there some things that should not be treated as intellectual property? In recent years, many biological materials have been treated as intellectual property that were previously viewed as belonging in the public domain, such as organisms, cell lines, genes, and proteins, all of which biotechnology or pharmaceutical companies have now patented. Many people find the idea of "owning" products of nature to be morally or even religiously offensive or at least not in the best interests of science, medicine, and technology. For example, James Thomson of the University of Wisconsin and the University of Wisconsin Research Foundation came under attack from many prominent stem cell researchers for patenting human embryonic stem cells (Holden 2007). Other controversial types of patents include patents on computer software, business plans, and medical diagnostic tests.

In reflecting on these issues, we stress again the importance of the *Diamond v. Chakrabarty* (1980) decision and the U.S. Supreme Court's ruling that a human invention is "anything under the sun made by man." (Courts in other countries have followed the Supreme Court's reasoning in this case.) The fundamental issue decided by the Court was where to draw a line between *products of nature*, which are not patentable, and *products of human ingenuity*, which are patentable. One way of making this distinction is to say that a product of human invention is something that does not exist in a natural state; it would not exist without human manipulation, control, refinement, analysis, and so on (Shamoo 1995). For example, Benjamin Franklin discovered that lightning is static electricity, but he invented the lightning rod. Galileo discovered that Jupiter has moons, but he invented a type of telescope. This seems clear enough. However, consider some less clear-cut cases:

- Digitalis occurs naturally in the foxglove plant, but human beings can make an isolated and purified form of the compound, digoxin.
- Mice exist in nature, but human beings can create genetically altered mice with genes for specific human diseases, such as cancer or diabetes.
- Neural stem cells occur in the human body, but human beings can culture purified cell lines.
- The gene that codes for the protein erythropoietin occurs in the human body, but human beings can use recombinant DNA and molecular cloning techniques to produce mass quantities of an isolated and purified form of the protein.

While people many have a general sense of the difference between discovery and invention, the question we face as a society is how to draw the line between invention and discovery in difficult cases. In the examples above, the items in question are structurally and functionally similar, yet we may decide to call some discoveries and others inventions. To settle these issues, one might appeal to some metaphysical or scientific theory that uniquely divides the world into "products of nature" and "products of human invention." For instance, one could argue that inventions are produced by human beings and cannot exist without human intervention. The trouble with this suggestion is that if one takes a close look at the many different products of research and technology, one finds that all inventions and discoveries have human and natural causes. Moreover, these causes are interdependent and interacting. For example, many scientific laws and theories, such as the natural gas laws and general relativity, would not be known unless some human being had developed and confirmed them. These laws and theories may indeed reflect regularities in the world that are independent of human beings, but their discovery would not occur without human intervention. Furthermore, although artifacts and inventions result from human labor, they make use of natural resources and raw materials. For example, without steam there can be no steam engine, without electricity there can be no electric light bulb, and so forth. Our general point is simply this: Questions about what can or cannot be treated as patentable often cannot be settled by the "facts" relating to humanity's interaction with the natural world. Therefore, we must appeal to ethical, political, or economic considerations to decide what is or is not patentable (Resnik 2001c, 2004e).

As noted above, the main argument for patenting is that it stimulates invention and investment in R&D. Inventors and investors are lured by the economic rewards of patenting. However, scholars, politicians, advocacy groups, and others have raised a variety of objections to patenting related to biotechnology and biomedicine. We briefly mention a few of the important ones here:

1. Patenting human body parts, such as genes, proteins, or cell lines, violates human dignity by treating people are property.
2. Patenting human DNA appropriates the human genome for private use, but the genome should be in the public domain, because it is mankind's common heritage.
3. Patenting any form of life demonstrates a lack of respect for nature.
4. Patenting drugs, medical devices, diagnostic tests, and other inventions used in diagnosis, treatment, or prevention increases the costs of medical care and therefore undermines access to care.
5. Patenting often involves the exploitation of research subjects, patients, communities, or indigenous peoples for commercial purposes.
6. Patenting plant crops increases the costs of agricultural production and therefore harms economically disadvantaged farmers living in the developing world.
7. Patenting can slow the progress of science by encouraging secrecy and creating numerous financial and legal problems with licensing of technologies used in research.

These objections address social, cultural, economic and ethical complex issues that we do not explore in depth here. However, we encourage readers to consider arguments for and against various types of patenting as well as other forms of intellectual property in science and technology. (For further discussion, see Resnik 2004e, 2007b, as well as Chapman 1999, Nuffield Council 2002.)

QUESTIONS FOR DISCUSSION

1. Do you think that people have a right to intellectual property?
2. Is it unethical to copy software without the copyright owner's permission?
3. If you are familiar with the Napster case, do you think the company acted unethically?
4. Should pharmaceutical companies be allowed to charge whatever price the market will bear for patented drugs? Should there be moral or political limits on drug prices?
5. Should developing nations violate international intellectual property agreements in order to make drugs available to patients with HIV/AIDS, dysentery, or malaria?

6. Do you have any objections to patenting DNA, cell lines, proteins, tissues, animals, or other human biological materials?

7. Would you have any objections to a patent on a genetically modified human?

8. Should scientists have free access to patented DNA for use in basic research?

9. Do you think that researchers, research institutions, and companies put too much emphasis on intellectual property?

10. Do you think that intellectual property undermines free inquiry, openness, and the academic ethos?

CASES FOR DISCUSSION

Case 1: John Moore's Cell Line

John Moore went to University of California, Los Angeles (UCLA) Medical Center in 1976 to receive treatment for a rare type of cancer known as hairy-cell leukemia. Moore's physician, David Golde, recommended that Moore have his spleen removed. After the surgery, Golde asked Moore to provide samples of blood, skin, bone marrow, and sperm, which required him to travel a considerable distance to make several visits to the UCLA Medical Center. Moore was led to believe that Golde needed the samples to monitor his health, but, in fact, the purpose of gathering the samples was to develop a cell line from Moore's cancerous tissue (Resnik 2004e). Golde was interested in growing this cell line because of its scientific and commercial value. The cell line had a mutation that caused it to overproduce lymphokines, which are proteins that help to regulate the immune system. The estimated market for the lymphokines was $3 billion. Golde signed an agreement with the University of California and several private companies to develop the cell line. Golde and his research assistant Shirley Quan applied for patents on the cell line. When the patents were awarded, they assigned them to the University of California (Resnik 2004e).

Moore eventually discovered that the true purpose of his visits to the UCLA Medical Center and became very upset. He sued Golde, Quan, the private companies, and the university for two separate harms: (a) medical malpractice and (b) conversion, that is, substantially interfering with another person's personal property. The case wound its way to the California Supreme Court. The court ruled that Moore could not prove his conversion claim because he did not have a property interest in the cell line (*Moore v. Regents of the University of California* 1990). According to the majority opinion of the court, the researchers who isolated and cultured the cell line had property rights on the cell line because they had invested their time and labor in developing it. According to the majority, granting patients or research subjects property rights in their biological samples would interfere with biomedical research and innovation. Only inventors and companies should have property rights over biological materials left over from medical procedures or donated research. In separate dissenting opinions, two judges from the court argued that Moore should have property rights to the cell line because his cells are no different from other materials that are exchanged on the market, such as sperm or hair, and a person should be able to control his own body and its parts (*Moore v. Regents of the University of California* 1990).

Although the court did not recognize Moore's property rights pertaining to the cell line, it did rule that Golde had committed medical malpractice by violating his fiduciary obligation to Moore to disclose his financial interests in Moore's tissue as part of the informed consent process. A fiduciary has an obligation to promote the best interests of specific parties. Physicians and other health care providers have obligations to promote their patients' best interests. One interesting legal (and ethical) question related to the Moore case is whether researchers who are not acting as health care providers have fiduciary obligations to research subjects. Another important question is whether researchers have an obligation to disclose financial interests to research subjects (Resnik 2004b, Morreim 2005). In a more recent case regarding who owns the patient's tissue sample, the federal court in Missouri similarly ruled that the tissue belonged to Washington University in St. Louis and not to the patient (Kaiser 2006).

- Do you think that Moore was treated fairly?
- Should Moore have property right pertaining to patented cell lines developed from his tissue?
- Should Moore's doctors have told him about their financial interests?

Case 2

A faculty member at a U.S. university met a very bright young postdoctoral fellow at a meeting in Germany. The postdoctoral fellow was working for a company in Germany and had a new chemical entity that could be developed into a product to prevent one type of urinary tract infection. Animal testing had already been conducted, with very promising results. The postdoc then came to the United States to work for the university faculty member, with support from the German company. The German company, through its subsidiary in the United States, submitted an investigational new drug application to the FDA in order to start phase I clinical trials on humans. Later that year, a French company acquired the German company. In the meantime, the postdoctoral fellow met a colleague at a meeting at a different U.S. university and collaborated to test the new chemical entity. They discovered a new modality to treat another disease with the same chemical entity.

At this time, the faculty member's U.S. university was negotiating with the French/German company to have the research conducted at the university's facilities. From the start, the French/German company demanded sole proprietorship of the drug and wanted to control all aspects of its R&D. The university then asked the postdoctoral fellow to sign a visiting fellowship agreement assigning all intellectual property claims to the university. The postdoctoral fellow refused, on the advice of the French/German company. Meanwhile, the German/French company filed for a patent alone without any mention of the part of the work conducted at the university.

- What should the university do? What should the university faculty member do?
- Who owns the patent? Is the patent application valid?
- What should each party have done in the first place?

Case 3

A university investigator is in hot pursuit of a project that received a medium-sized grant from the NIH. Her grant depends on the use of reagents from an independent company. She has had a long and fruitful relationship with the company—she has given several seminars at the company, and her counterpart in the company has given several seminars at the university. Also, the investigator has earned a few thousand dollars annually from the company through consulting. She has signed a consultancy agreement with the company without clearance from the university. Her agreement relinquishes all her intellectual property rights regarding the subject of the consultancy to the company.

The investigator asked the university to sign a material transfer agreement (MTA) so that she could start using the reagents. The university refused because the MTA gave too many rights to intellectual property to the company in exchange for the reagents. The university investigator was anxious to start her project and make progress so that she would be able to renew her grant. One day during her frequent seminars at the company, she was asked to sign the MTA agreement, and she did.

- What should the university do?
- What should the faculty member do?
- What should the company do?

Case 4

In early 1990s, a university faculty member was collaborating with a Danish company in a joint venture on a compound with a view toward clinical trials within a year. The company has already submitted a patent application on a portion of the project. In written correspondence between the university faculty member and her counterpart at the Danish company, both pledge full cooperation without mentioning anything about intellectual property.

A year later, the Danish company enlisted a U.S. company to conduct certain experiments. The university and the Danish and U.S. companies all entered into negotiation regarding intellectual property. The negotiations failed, and the university ordered the faculty member to stop any further collaboration.

- What should the faculty member do?
- What should the university do?
- What lessons are learned in this scenario?

Case 5

A member of an NIH study section, while reviewing a grant proposal, realized she could do part of the proposed research faster and better with a method already available in her laboratory. Under normal conditions, she would not be conducting such research. After getting back to her laboratory, she gave the project to her most reliable postdoctoral fellow.

One year later, they submitted a paper to a prestigious journal. One of the reviewers was the investigator who wrote the original grant proposal. The original investigator has not yet published his paper on the subject because he has applied for a patent, which delayed writing of the paper. The original investigator complained to the U.S. Office of Research Integrity.

- What should Office of Research Integrity do?
- What should the reviewer do?
- Who should get the patent?

Case 6

An anthropologist and his team from a university have discovered a human skeleton in a national forest in Montana. Carbon dating shows that the skeleton is more than 10,000 years old. Further study of the skeleton will prove to be extremely useful in determining human migration to the North American continent. However, a group representing Native Americans is taking legal action to have the skeleton returned to their custody. They say that the skeleton is one of their ancestors and should not be the object of scientific study.

- Who owns this skeleton?
- How should this dispute be settled?

Case 7

A tissue collection and storage company has signed a contract with a medical school to collect human tissue for research purposes. The company plans to collect tissues from human donors at the medical school. Tissue donors will sign an informed consent form giving the company their tissue. They will not receive any money for their tissue but will be informed that their tissue may benefit other people. Once the tissue is donated, it will be placed in a tissue bank. All personal identifiers linking the tissue to the donor will be removed. The company expects to profit by charging access to its tissue database. It also plans to patent valuable cell lines. The medical school will receive a portion of the profits. Do you see any ethical problems with this proposal?

10

Conflicts of Interest and Scientific Objectivity

> Researchers and research institutions have a variety of financial, personal, and political interests that sometimes conflict with their professional, ethical, or legal obligations. These situations can create conflicts of interest or the appearance of conflicts of interest. This chapter discusses how conflicts of interest affect research, how they are defined, and how they should be managed. It also describes how government agencies and research institutions have responded to conflicts of interest in research and discusses some cases from science.

Individual scientists and research organizations daily encounter situations where personal, financial, political, and other interests conflict with professional, ethical, or legal obligations or duties. Although conflicts of all types are a normal part of human existence, some are called "conflicts of interest" (COIs) because they involve conflicts between interests and duties. Most of the concern with COIs in research arises because personal interests can undermine duties relating to scientific objectivity (Shamoo 1992, 1993, Resnik 2001e, 2007b). Consider some examples of situations that might be considered COIs in research:

- A clinical investigator with stock in a pharmaceutical company that sponsors his research overestimates the clinical significance of his research on one of the company's drugs, which drives up the price of his stock.
- An academic researcher in a National Institutes of Health (NIH) study section reviews a grant from a competitor in the same field (Shamoo 1993).
- A scientist receives orders not to publish results that are contrary to the interests of the company that funds her research.
- A medical journal runs an advertisement for a new drug.
- A clinical investigator receives $3,000 in patient care costs for each patient she recruits into a clinical trial.
- A university's institutional review board (IRB) reviews a research proposal sponsored by a company that has recently given $10 million to the university.

In each of these situations, researchers have financial or other interests that may interfere with the ability or tendency to fulfill ethical or legal duties to colleagues, patients, sponsors, or the public. Another way of putting this is to say that these interests compromise the researchers' integrity. There are two ways that COIs can undermine integrity. First, a COI can affect a person's thought processes (i.e., judgment, perception, decision making, or reasoning) (Davis 1982, Shamoo 1993, Thompson 1993, Resnik 1998a, 2007b, Bradley 2000, Angell 2001, 2004, Krimsky 2003). A person with a COI is like a thermometer that does not work properly. In some of these cases we might say that the person's thought processes are biased; that is, they tend to be skewed in a particular pattern or direction that can be attributed to the particular interest. Second, a COI can affect motivation and behavior (Resnik 1998a, Porter 1993). A person with a COI may be perfectly capable of sound thinking but may fail to implement it due to temptations that affect his or her motivation and behavior. All people are tempted from time to time by a number of interests and desires, but the problem with COIs, one might argue, is that they place people in situations where only people with extraordinary characters will not succumb to temptation (Resnik 2007b).

There is a substantial body of literature demonstrating a relationship between the source of funding for scientific research and the outcome of research (Krimsky 2003, Angell 2004, Resnik 2007b). Many studies of articles published in biomedical journals have shown that a study sponsored by a pharmaceutical company is more likely to favor the company's products than a study funded by the government. For example, Ridker and Torres (2006) studied 349 randomized clinical trials related to cardiovascular disease and found that 67% of trials funded by for-profit organizations, compared with 49% of trials funded by not-for-profit organizations, favored a new treatment over the standard of care. The authors also found that in medical device studies, 50% of trials with not-for-profit sponsors favored the new treatment compared with 82% of trials with for-profit sponsors. A study by Friedberg et al. (1999) found that 95% of industry-sponsored articles on drugs used in cancer treatment reported positive results, as opposed to 62% of non-industry-sponsored articles. Cho and Bero (1996) found that 98% of drug studies published in symposium proceedings favored the sponsoring company. Bekelman et al. (2003) reviewed 37 papers on financial relationships and their influence on research and found that there is a statistically significant relationship between industry sponsorship and pro-industry results. A plausible explanation for these associations between sources of funding and research outcomes is that financial interests have biased the judgments and decisions made by investigators (Agnew 2000, Resnik 2007b).

Although COIs can affect thought processes and behavior, they do not necessarily operate at a conscious level. People may be influenced by financial, personal or other interests even though they are not aware of their effects. People often claim that they would never deliberately allow personal, professional, political, or financial interests to affect their thought processes or behavior. While this may be often true, studies have shown that interests can exert subconscious influences on thought processes (Katz et al. 2003). Additionally, even small interests can have an effect. People can be influenced by small gifts and small sums of money. Pharmaceutical companies give small gifts, such as pens, notepads, and free lunches, because they are trying to influence physicians' behaviors. Physicians often deny that they could be influenced by something as small as a pen, but studies indicate otherwise (Katz et al. 2003).

Because COIs can undermine integrity, they can also have a negative impact on trustworthiness. Throughout our book, we have emphasized the importance of trust and trustworthiness in research: Research subjects must trust investigators, students must trust mentors or supervisors, principal investigators must trust subordinates, and colleagues must trust each other. Researchers must trust journals, granting agencies, and research institutions. Society must also trust investigators to honor their ethical and legal obligations in conducting research. When investigators have COIs, people may suspect that their research is biased, unreliable, or morally tainted. Because personal, professional, political or financial interests can threaten or undermine trustworthiness, it is incumbent upon the research community to address such concerns.

Putting all of this together, we offer the following definition of a COI for an individual:

> An individual has a conflict of interest when he or she has personal, financial, professional, or political interests that are likely to undermine his or her ability to meet or fulfill his or her primary professional, ethical, or legal obligations.

In theory, it is fairly easy to define COIs, but in practice it is often difficult to determine whether people have COIs, because we may not know how their personal, financial, professional, or political interests are affecting (have affected or will affect) their thought processes or behavior. Many people with personal, financial, professional, or political interests will acknowledge them but maintain that they have a great deal of integrity and will not allow their interests to undermine their ethical or legal responsibilities. How can we know which conflicts are likely to undermine a person's ability to meet his or her primary obligations? How much money does it take to affect reasoning, judgment, motivation, or behavior—$10,000? $1,000? $100? We cannot address these difficult practical issues adequately here, but we adopt a standard that we call the average person standard: A person has a COI when the average person under similar circumstances would be affected. Although some extraordinary people may be able to maintain their objectivity and integrity when faced with such circumstances, most people will succumb to biases or temptations.

Situations may arise where it is clear that the person does not have a COI, but it might appear to a reasonable outside observer that they do. The situation might create an issue related to trustworthiness, even it has little or no impact on integrity. Because trustworthiness is very important in research, situations that affect trustworthiness should be addressed, even when they are not likely to undermine integrity. Thus, it is important to distinguish between true and apparent COIs for individuals:

> An individual has an apparent conflict of interest when he or she has personal, financial, professional, or political interests that create the perception of a conflict to a reasonable outside observer.

These two definitions of COIs apply to many different types of conduct, including scientific research. As discussed elsewhere in this text, scientists frequently have ethical and legal responsibilities to other members of their profession, clients, patients, students, and society. Scientists also have personal, financial, or political interests that could interfere with their ability or tendency to fulfill these responsibilities. Many of these personal interests relate to career ambitions, such as the quest for reputation, fame, power, priority, funding, tenure, and promotion. These interests can affect scientific thinking or behavior. For example, a scientist who has stock options in a pharmaceutical company and is conducting research on that company's drugs has a responsibility to conduct objective (or unbiased) research. However, her financial interests may prevent her (or appear to prevent her) from fulfilling that responsibility; she may conduct, or appear to conduct, research that is biased toward the company. Money or ambition can also interfere with a researcher's ability or tendency to promote the rights and welfare of human subjects. A researcher with strong financial or personal interests in recruiting human subjects for a clinical trial may oversell the trial to his patients or take other steps to compromise the informed consent process. When students are supported by funds from industrial sponsors, conflicts may affect duties to the research profession, patients, and students (Blumenthal 1995).

Virtually any part of the research process can be affected by COIs. Some of the parts most susceptible to the effects of COIs include selection of the research problem, statement of objectives, research design, choice of methods, and data analysis and interpretation, as well as publication, peer review, clinical trials, selection and recruitment of subjects, and funding (Blumenthal et al. 1996a,b, 2006, Spece et al. 1996, Bradley 2000, Resnik 2001e, 2007b, Krimsky 2003, Campbell et al. 2004).

Because COIs can have such a great impact on scientific objectivity and integrity, they have been linked to fraud and misconduct. For example, due to pressures to obtain funding, a scientist might fabricate or falsify data in a grant application or paper (see, e.g., case 1 in chapter 8 regarding the scientific misconduct of Eric Poehlman). Or perhaps a researcher working for a pharmaceutical company manipulates the data to make his results appear to be more statistically or clinically significant than they are. Indeed, money has played an important role in most of the cases of research misconduct (or alleged misconduct) discussed in chapter 8. However, having a COI is not, in itself, research misconduct: A COI is a risk factor for misconduct (U.S. Congress 1990).

To summarize this section, COIs are an important ethical concern in science because they can prevent researchers from fulfilling their professional, ethical, or legal obligations. COIs can affect scientists' thought processes or behavior and cause them to make biased decisions and judgments in protocol development, research design, data analysis and interpretation, peer review, and other important research activities. COIs not only can affect the integrity of the research process, but also can undermine the public's trust in the science by creating the perception that research is biased, unreliable, or morally tainted.

CONFLICTS OF COMMITMENT AND DUTY

It is important to distinguish between COIs and other types of conflicts that occur in research. Consider this example: A molecular biologist who is hired by a public university also consults with several private and public clients, including a biotechnology firm, a hospital, and the local health department. As a result, she often has a difficult time balancing her commitments to the university and to these other organizations. This type of situation is known as a conflict of commitment (Resnik 1998a). A researcher who frets over whether to prepare his lecture or finish a paper would also have a conflict of commitment. Conflicts of commitment, unlike COIs, do not normally involve situations that compromise a person's thought processes or behavior. They primarily have to do with the prudent and responsible management of time and effort. People with conflicts of commitment usually can manage their different commitments. However, if one cannot manage one's many different commitments, the responsible thing to do is to prioritize these commitments and eliminate some of them. Sometimes conflicts of commitments can lead to COIs. For example, a university scientist may become so involved with time-consuming consultations that she can no longer meet her professional, contractual, or other obligations (Davis 1995b).

A conflict of duties is also not a COI (Davis 1982). For example, a doctor may have to decide between respecting a patient's autonomy and promoting the patient's best interests when an otherwise healthy patient chooses to refuse life-saving medical care, such as refusal of dialysis. Or someone may have to choose between honoring a prior commitment to attend a meeting and taking care of some important business that just came up. A researcher may face a conflict between the duty to share data and the duty to respect intellectual property. Although conflicts of duties can lead to COIs when the person in question also has an interest in meeting a specific duty, COIs are not the same as conflicts of duty. COIs involve the collision between personal interests and duties, not conflicts among duties.

INDIVIDUAL CONFLICTS OF INTEREST

There are several ways to protect the integrity of the research process and promote trust in response to COIs. One strategy that has been adopted by many institutions is known as disclosure (Bradley 2000, McCrary et al. 2000). Researchers should disclose COIs and apparent COIs to the relevant parties. A relevant party would be someone in a supervisory role that does not have a vested interest in the particular conflict situation, such as a department chair, section head, dean, or director. The person should have a high degree of independence or objectivity. Disclosure is a useful strategy because it puts conflicts out in the open and allows people to assess them. For example, if a researcher discloses a COI in a paper, scientists may pay more careful attention to the paper's methods, results, and interpretations. The paper may demand some additional scrutiny, but it is not automatically invalidated (Resnik 1998a). Moreover, disclosure can promote trust because people are likely to be more suspicious and untrusting when they discover previously hidden conflicts than when they have access to them up front (Cho 1997). In the last decade, many scientific journals and other organizations have adopted financial disclosure policies.

Disclosure alone may not always be a sufficient response to a COI, however (Cech and Leonard 2001, Elliott 2008), and something more than this may be needed in some situations. A strategy that goes beyond mere disclosure, which has been adopted by many organizations, is known as management of conflicts (McCrary et al. 2000). According to this strategy, people with COIs should disclose them to the relevant parties, who consult with other people to develop rules or policies for controlling the interests that are at stake. They may also decide to review these COIs on a regular basis and take additional steps to promote trust, if necessary. A committee could be formed to review the COIs and ensure that there are no ethical or methodological problems with the research.

Most research institutions have policies that require some form of disclosure of COIs, and many have methods for managing COIs (Cho et al. 2000). Fewer institutions make much use of a third strategy, known as avoidance/prohibition. In this strategy, which is frequently used in government, law, and medicine, organizations make rules that prohibit specific situations that create COIs. For example, government ethics rules place limits on the amount of money a legislator may receive as a gift or honoraria. Medical ethics codes prohibit kickbacks and self-referrals. The NIH has COI rules for its study sections that prohibit scientists from reviewing grant proposals from colleagues at the same institution. The NIH has adopted rules that forbid NIH intramural researchers from consulting with industry for money and prohibits senior staff members from owning any stock in pharmaceutical or biotechnology companies (Tanne 2005). (See further discussion below.) COI avoidance may also occur when a person chooses to remove (or "recuse") himself or herself from the situation that creates the conflict, for example, when a reviewer for a journal declines to review a paper because of a personal bias against the authors.

Like disclosure and management of conflicts, conflict avoidance can promote integrity and trust. Elliott (2008) argues in favor of more conflict avoidance in research because current means of managing COIs with disclosure, management, and divestiture are insufficient to promote research integrity. However, conflict avoidance can sometimes lead to outcomes that are worse (from a utilitarian point of view) than the outcomes that result from nonavoidance. For example, suppose that three-fourths of the members of a zoning board own land that will be affected by their decision. If they all recuse themselves, either the board will lack a quorum and be unable to make a decision, or it will lack informed debate on the issue. In either case, decision making would be worse off. Situations like this can occur in research when it is difficult to find qualified people to make decisions who have no COIs.

When dealing with a COI requires more than mere disclosure, how should one decide whether the COI should be avoided/prohibited or managed? Three factors help guide these decisions:

1. *The strength of the COI:* How strong are the interests involved? How likely are they to affect thought processes or behavior?
2. *The difficulty of managing the COI:* How difficult is the COI to manage? Is there an appropriate mechanism or process for managing it? Could one be created? Are there sufficient financial and human resources to manage the COI?
3. *The consequences of prohibiting or not prohibiting the COI:* What will happen if the COI is prohibited? Will prohibition have a negative impact on science or society? What will happen if the COI is not prohibited? Will this have a negative impact on science or society? Will this create a public perception problem?

When the COI is strong and difficult to manage, and not prohibiting it has large negative consequences, then it should be prohibited. For example, a researcher should not be allowed to review a grant proposal from a colleague at his institution because this situation is likely to have a significant impact on his thought processes and behavior, it is difficult to manage, and allowing it could have a very negative impact on grant reviewing.

INSTITUTIONAL CONFLICTS OF INTEREST

It is important to realize that organizations or institutions can also have COIs. Consider the following situations:

- A television network decides to not broadcast a story that is unfavorable to a company because that company threatens to stop advertising on the network.
- A medical society endorses a medical product made by a company that donates a great deal of money to the society.
- A university's IRB approves a human research proposal over the objections of several members because a failure to approve the proposal could threaten a $10 million grant.
- A university disciplines a faculty member for speaking out on an environmental issue because several state legislators want him punished.
- A university is trying to silence a faculty member because he spoke against research conducted on his campus that in his opinion is unethical.
- A university holds stock in a chemical plant but also hires employees who monitor its emissions.
- A university president owns stock in a company that sponsors research on campus and has made large gifts to the campus.
- A wealthy CEO for a media company wants to donate $50 million to the journalism school. In return, he wants the school named after him and to have some input into the school's curriculum.

These examples illustrate how institutions, including research institutions, government agencies, professional associations, and journals, can also have COIs and apparent COIs. These organizations, like individuals, often have collective duties to professionals, clients, students, patients, and the public. Their duties usually stem from their primary missions (or goals), such as research, education, and public service. Institutions, like individuals, may also have financial, political, or other interests that can adversely affect (or appear to affect) their ability to carry out these responsibilities. Although institutions do not "think," they do make collective decisions that result from the judgments, decisions, and actions of their members, and they do perform actions. These decisions and actions occur through the various components of the organization's management structure, such as faculty members, committees, advisory boards, deans, and vice presidents. Institutional COIs, like individual COIs, can also threaten the integrity or trustworthiness of research, except the damage may be more extensive because it may affect more researchers and members of the public (Association of Academic Health Centers 1994, 2001, 2002, Bradley 2000, Moses and Martin 2001). The Association of American Medical Colleges recognized, though belatedly, that the institutional COI is important and issued new recommendations (Association of American Medical Colleges 2002). However, these recommendations do not prohibit any relationships, and they leave loopholes for institutions. This is illustrated by the case of the Cleveland Clinic and its venture/investment in a device company called AtriCure involved in the treatment of atrial fibrillation (Armstrong 2005, 2006). Case 1 below gives the details of the case.

We define an institutional COI as follows:

> An institution has a conflict of interest when its financial, political, or other interests are likely to undermine its ability to fulfill its professional, legal, ethical, or social responsibilities.

As with individual COIs, the same three strategies to protect research integrity may apply: disclosure, conflict management, and avoidance. For some organizations, it may be enough to simply disclose the conflict; for others, avoidance may be the best option.

The research community has just begun to think about how to deal with institutional COIs. In some ways, institutional COIs can pose more significant problems for research than individual COIs, because institutional COIs can exert systematic, widespread effects. For example, if a university has a COI resulting from its relationship with a private company, then this COI could affect many different faculty or research projects at the university. Another concern with institutional COIs is that they may be more difficult to manage or control than individual COIs, because often there is no independent body with the authority and willingness to control the institution. At a university, a faculty member can disclose her interests to her department chair, who is independent from the faculty member and has the authority to make decisions pertaining to the faculty member. But who can oversee a university's or federal agency's COIs? While there may be some organizations with the oversight authority, such as the board of trustees for a university or Congress for a federal agency, these outside authorities many be unwilling or unable to do anything to deal with these COIs.

Some authors have suggested some specific strategies for dealing with institutional COIs (Moses and Martin 2001, Association of American Medical Colleges 2002):

- *Building firewalls:* An organization builds firewalls to prevent some parts of the organization from having an adverse impact on other parts. For example, a university could try to shield committees that oversee research, such as IRBs and institutional animal care and use committees (IACUCs), from parts of the university that deal with money, such as the contracts and grants office, the technology transfer office, or the development office. While this strategy sounds good in theory, it may not work in practice, because members of a university often know each other and know about the university's business. Subtle or not so subtle influences can still occur even when there is no direct line of communication between different parts (Resnik and Shamoo 2002). Moreover, it may not be a good idea to shield these different parts, since they often need to communicate. For example, an IRB needs to communicate with the contracts and grants office when it is reviewing a research project sponsored by a federal agency or private company.
- *Manage investments and properties:* An organization could take steps to manage its investments and properties to prevent inappropriate influences. For example, many universities have established separate institutes (or foundations) to hold stock and equity, invest in startup companies, and hold intellectual property. This is another approach that sounds good in theory, but probably has little effect in practice because there are close connections between universities and these institutes. The fact that an institute associated with a university holds stock in a company, instead of the university, probably makes little difference to how decisions are made related to the company or the public's perception of those decisions (Resnik and Shamoo 2002).
- *Require COI committees to also review institutional COIs:* Many institutions have committees to review and deal with COIs disclosed by employees. Some authors have suggested that these committees could also deal with institutional COIs (Association of American Medical Colleges 2002, Moses and Martin 2001). While we think this is also a good idea, there may be some problems with implementing it, since committees that review individual COIs may lack the independence to review institutional COIs. COI committees are usually composed of members of the institution (i.e., faculty) and, at most, a few outside members. They are also supported by the institution and report to a leader at the institution (e.g., a vice president or president). We

question whether these committees can give an adequate review of institutional COIs, given their composition, funding, and structure (Resnik and Shamoo 2002).

- *Establish policies and rules for institutional officials and other important employees or associates:* An institution could establish different policies to prevent some types of COIs or minimize the effect of some COIs. For example, IRB or IACUC members with COIs could be prohibited from voting on or discussing submissions where they know an investigator on the study or have some other interest. Leaders of the institution, such as presidents, deans, or directors, could be prohibited from having some types of stock or equity. While it is a good idea to establish rules and policies to deal with institutional COIs, we wonder whether there will be sufficient interest in developing these, given the lack of independent oversight of institutions (Shamoo 1999). A recent study (Ehringhause et al. 2008) indicates that only 38% of all medical schools have institutional COI policies. This survey indicates that most institutions have little interest in dealing with institutional COIs while they continue to push for individual COI policies.

LESSONS FROM PUBLISHED CONFLICT OF INTEREST CASES

We now discuss some COI cases and situations that offer some important lessons for research ethics and policy. We present the following five examples of COIs in research primarily to illustrate the intertwining financial interests that either hinder scientific objectivity or are perceived to do so by society. Since these cases came to light, the NIH and National Science Foundation (NSF), along with universities, have promulgated better policies on COI disclosures. Additionally, many journals, such as the *Journal of the American Medical Association* (*JAMA*) and *New England Journal of Medicine*, and the NIH now require their authors and researchers to disclose COIs.

Pharmatec and the University of Florida

In early 1980s, Nicholas S. Boder, a professor at the University of Florida, invented a chemical delivery system (CDS) to carry drugs directly to the brain by temporarily linking them to another molecule that crosses the blood–brain barrier. Boder entered into an exclusive license agreement with a corporation to form a new startup company, Pharmatec, with an initial investment of $6 million. Boder was to receive $1 million of the $6 million to develop and further test the CDS. Boder's time was then divided in a complicated fashion among the university and Pharmatec. Boder was also the vice president and director of Pharmatec, as well as the chairman of the Department of Medicinal Chemistry at the Florida College of Pharmacy. The dean of the College of Pharmacy was given 1,000 shares of stock. Boder hired three other College of Pharmacy faculty members to work for Pharmatec, who also received stock from the company, as did some additional key administrators of the college. The dean, Boder, and other faculty members of the college became members of Pharmatec's advisory board (U.S. Congress 1990).

One of the advisory board members, Dr. Sloan, an associate professor in the college, raised concerns regarding the safety of the CDS because of its structural similarity to the toxin MPTP. He suggested that, to ensure safety to students and faculty researchers at the college, the compound should be tested on primates. Sloan approached the dean and others and made presentations of his data at conferences indicating the possibility of a toxicological effect of the compound. There was considerable delay before any toxicological tests were conducted, with the result that Sloan resigned from the advisory board. It is alleged that his career suffered from the incident (U.S. Congress 1990).

The U.S. congressional report states: "The story of Pharmatec demonstrates how the financial interests of universities and their faculty can conflict with such important scientific questions as the safe conduct of research as well as the fair treatment of faculty" (U.S. Congress 1990, p. 14).

tPA: A Multicenter Trial

Each year, 700,000 patients suffer from myocardial infarction. During the 1980s, the NIH funded research at medical centers across the country to evaluate thrombolytic agents such as tissue plasminogen activator (tPA) that have the potential for dissolving blood clots. The clinical trials were to compare tPA, which had not yet been approved by the U.S. Food and Drug Administration (FDA), with streptokinase (SK), which had been FDA approved for intracoronary administration. tPA is a product of Genentech Corporation. In 1985, the *New England Journal of Medicine* published a paper showing that tPA was twice as effective as SK in dissolving blood clots. No data were presented indicating the lack of difference between tPA and SK in ventricular function, an important indicator with bearing on mortality. Also not reported was the fact that five of the coauthors owned stocks in Genentech. Two years later, the data on ventricular function were publicly revealed. Members of the NIH Safety and Data Monitoring Committee expressed concerns about the tPA results. The NIH dissolved the committee and formed a new one that included only one member from the former committee. Genentech stocks rose and fell precipitously, depending on the latest news about tPA. Later, it became known that 13 of the researchers involved in the nationwide tPA studies owned stocks in Genentech (U.S. Congress 1990).

At roughly the same time as the U.S. studies, two European groups had been conducting clinical trials comparing SK and tPA based on large populations of 12,000 and 20,000 patients. One of their reports, published in the journal *Lancet*, showed that SK alone reduced mortality by 12% and that SK plus aspirin reduced mortality by 42%. In 1987 an FDA advisory committee decided to recommend approval of SK for use in dissolving blood clots, but not tPA. This action resulted in an uproar from the business community. Subsequently, the FDA reversed its decision on tPA and approved it (U.S. Congress 1990).

This case illustrates the impact that COIs can have on the general public and on public institutions when university scientists enter the private arena, becoming stockholders in companies whose products they are evaluating.

The Retin A Case

Retin A is a cream product of Johnson and Johnson Corporation. In 1988, *JAMA* published an article and an editorial claiming that Retin A could reverse the effects of aging on the skin, such as wrinkles. After the article was published, about 1.2 million tubes of Retin A were sold in one month, compared with less than 250,000 for the same month a year earlier. Later a reporter disclosed that the author of the *JAMA* editorial and one of the authors of the research article received financial compensations from Johnson and Johnson. Furthermore, it was subsequently determined that the safety of Retin A had been exaggerated and that the study had not been well controlled for bias because of the absence of a double-blind design. Also, an NIH consensus conference has concluded that the safety and effectiveness of Retin A have not yet been established (U.S. Congress 1990).

This case illustrates the impact COIs can have on the objectivity of scientists. Moreover, it illustrates that public confidence in our scientists and scientific institutions such as scientific journals can be undermined by a COI or the appearance of a COI.

Clinical Trials of Therapeutic Drugs

In 1983, Thomas Chalmers evaluated 145 clinical trials of therapeutic drugs for myocardial infarction. He placed each study in one of three levels of bias control. Highly controlled studies included those that were randomized and blinded. Partly controlled studies included those that were randomized but not blinded. The category of no controls included those that were neither randomized nor blinded. He found that the effectiveness of the new drugs was inversely related to the degree of bias control. When the bias control was categorized as highly, partly, or not controlled, the drugs were effective 9%, 24%, and 58% of the time, respectively (Porter 1992). This example illustrates that even though there was no direct financial interest for the investigator, investigators' bias for their sponsor or their own success as a researcher can seriously corrupt the results.

Concerns about such practices prompted the American Medical Association's Council on Scientific Affairs and Council on Ethical and Judicial Affairs to make the following statement: "For the clinical investigator who has an economic interest in the outcome of his or her research, objectivity is especially difficult. Economic incentives may introduce subtle biases into the way research is conducted, analyzed, or reported, and these biases can escape detection by even careful peer review" (quoted in Porter 1992, p. 159).

Familial Hypercholesterolemia and the Nobel Prize

Michael Brown and Joseph Goldstein were awarded the Nobel Prize in Physiology or Medicine in 1985 for discoveries related to the elucidation of the mechanism of familial hypercholesterolemia (FH), which in part enhanced the understanding of the larger problem of atherosclerosis. The mechanism could not have been determined without three crucial and separate discoveries related to issues of genetics, receptor dysfunction, and blood–cell relationships. Brown and Goldstein made the discovery that the disorder is caused by a receptor on the cell membrane, but Avedis Khachadurian made the other two discoveries, before the discovery by Brown and Goldstein. As early as 1964, Khachadurian had shown that FH was due to a single Mendelian gene disorder. In addition, at least four years before the Brown and Goldstein publications, Khachadurian discovered that the disorder was not located within the cell and would therefore be revealed in some interaction between the blood and the cell. Thus, although Khachadurian made two of the three key discoveries, the Nobel Prize was awarded to Brown and Goldstein (Shamoo 1992).

An investigation of the literature showed that there was a consistent pattern by Brown and Goldstein in not quoting Khachadurian's work in the 1970s and early 1980s (Shamoo 1992). However, after receiving the Nobel Prize, Brown and Goldstein as well as others gave clear recognition to the importance of Khachadurian's discoveries. For example, N.B. Myant's 1990 book *Cholesterol Metabolism, LDL, and the LDL Receptors* referred to Khachadurian's 1964 key findings as follows: "This work, a landmark in the development of our understanding of FH, is worth placing on record" (quoted in Shamoo 1992, p. 403). Brown and Goldstein wrote the foreword for Myant's book.

The importance of this case can be seen in the remarks of Philip Siekevitz, a leading U.S. scientist at Rockefeller University: "The intense competition for recognition by peers and by the general public, for prizes, for commercial gain, is slowly eroding the scientific ethic.... And if this ethos is disappearing, then the citation indices no longer reflect worth but a lack of the scientific communitas. The future of the scientific endeavor depends on regaining the scientific soul" (quoted in Shamoo 1992, p. 72).

DISCUSSION OF SOME CONFLICT OF INTEREST POLICIES

Association of Academic Health Centers

Despite increasing concerns about scientific COIs in the 1980s from both Congress and the general public, due in part to cases such as those described above, universities were generally reluctant to take a leadership role in tackling the issue and being accountable for the actions of their employees (Bradley 1995). In 1994, following a series of congressional hearings and media revelations related to the erosion of scientific objectivity and the potential loss of public confidence in the scientific process, the Association of Academic Health Centers (AAHC; see 1990) issued a document titled "Conflict of Interest in Institutional Decision-Making."

The AAHC document called for the preservation of the traditional missions of teaching, research, and service during this period of increasing funding from industry. Oversight procedures would play a key role in this effort. It further called for full and accurate disclosure of COIs, on the part of both university employees and the institutions themselves. Although it suggested no criteria for minimum disclosure, the AAHC pointed out that some states and universities required disclosure of leadership positions in a company, investments of more than $25,000 in any work-related company, corporate ownership of at least 10%, annual consultation agreements paying more than $5,000, and any research involvement with industry at all. Additionally, the disclosure should hold not only for the university personnel themselves, but also for those related by blood and for close relationships outside the family. At the institutional level, disclosure should pertain to those who have decision-making authority and extend to the making of investment decisions involving companies with faculty interests. The AAHC guidelines also called for the prompt publication of the results of research, fairness in the allocation of the financial benefits of intellectual property, and the administration of patent issues in the public interest. They also addressed the free utilization of knowledge by students and the conduct of sponsored research under the terms of university policies and in the same context as federal funds. Finally, COIs should be avoided in the pursuit of the financial benefits of technology transfer.

As we noted above under institutional COIs, the Association of Academic Health Centers (2001, 2002) has issued two documents regarding the institutional and individual COIs. Although these two documents recognize the importance of the impact of COIs on scientific integrity and the public perception of science, they do not go far enough in dealing with serious COIs. The inadequacies of these two documents prompted 11 leading scholars from across the country to advocate for eliminating or drastically reducing the amount of funding from drug industry to academic institutions (Brennan et al. 2007). The authors recommend the elimination or modification of ghostwriting and consulting and research contracts. Prior to this recommendation, a few medical centers (e.g., University of California in Los Angeles and Stanford University) had already taken some steps in this direction.

The effects of these guidelines are uncertain. The AAHC has no authority to control the behavior of its membership. There is little evidence of voluntary compliance by the members of the AAHC. On the contrary, our impression is that academic health centers continue to compete for private funding with a vengeance. Nevertheless, these guidelines have helped stimulate the discussion of approaches to COIs. Their breadth of coverage can serve as a framework by which to compare the management of COIs by organizations that do have the ability to control the behavior of their membership.

Federal Conflict of Interest Policies

In July 1995, after a prolonged period of revisions and redrafting (Warsh 1989, Pramik 1990, Zurer 1990), the NIH and NSF both issued final rulings under the heading "Objectivity in Research; Investigatory Financial Disclosure Policy; Final Rule and Notice" (U.S. Department of Health and Human Services 1995, 2005b). The essence of the NIH ruling is the minimal requirement for the disclosure of any financial COI in the form of a $10,000 equity interest by funded extramural investigators or their immediate family members in any organization whose products are related to the research of the investigators. The scope and wording of the disclosure statement and the monitoring of financial COIs are left to the individual research institutions that receive federal funding. All universities and other research institutions that receive federal funding have subsequently issued their own guidelines fulfilling these requirements.

The U.S. Public Health Service issued some draft interim guidelines concerning relationships with the pharmaceutical industry for clinical investigators, institutions, and IRBs (Public Health Service 2000a). Institutions should

- Collect information on COIs and to relay that information to concerned parties, such as IRBs
- Develop COI policies and form committees to review COIs
- Develop educational programs on COIs
- Become aware of their own COIs and take steps to manage them
- Establish independent oversight committees to manage institutional COIs

The guidelines also recommend that investigators consider carefully the impact of COIs and make relevant disclosures, and take part in educational programs on COIs. IRB members are encouraged to disclose COIs, to have policies and procedures for managing its own COIs, to pay greater attention to COIs in clinical trials, and to develop policies or mechanisms for minimizing the impact of COIs in clinical trials (Public Health Service 2000a). No one knows yet their effect on research.

Recently, the NIH issued a new ruling (National Institutes of Health 2005) for NIH employees, enacted by NIH director Elias Zerhouni under pressure from Congress after an investigation by the *Los Angeles Times* found many violations of the existing rules (Willman 2005). The investigation found that numerous intramural researchers at the NIH had violated NIH COI rules and tried to hide it. An internal review by the NIH found that 44 NIH researchers violated the ethics rules on consulting fees with industry. The new ruling prohibits all NIH staff from consulting with pharmaceutical companies or nonprofit health care organizations. Moreover, researchers cannot own more than a $15,000 financial interest in a biotechnology, drug, or medical device company, with staff being exempted from this ruling (Kaiser 2005). Researchers are also banned from giving paid lectures at organizations that are affected by the NIH, such as universities, with a few exceptions. These are some of the strictest COI rules for researchers anywhere.

The FDA has also issued its own COI policies. Its policies cover the same issues as the NIH and the NSF policies, but its requirements are directed toward participating institutions rather than their individual employees. The FDA has dozens of advisory committees composed of researchers all across the country. These advisory committees are formed by the FDA to advise them regarding issues related to drug approval and labeling. However, many advocacy groups and newspapers have expressed concerns regarding many members of advisory committees having a relationship with drug industry. Using the FDA's COI disclosure forms of advisory committee members (Food and Drug Administration 2007b), Lurie et al. (2007) found some frequent and serious monetary financial relationships between committee members and the drug industry. However, they found that there was only a weak relationship between advisory committee members' interests and their voting behavior. Nevertheless, there was the appearance of COI. Therefore, the FDA issued a draft guidance regarding COIs of advisory committee members (Food and Drug Administration 2007). The FDA's new guidance sets a $50,000 relationship with industry as a goalpost to decide whether these researchers should be advisory members. In short, if an individual has a financial relationship with industry of more than $50,000, he or she cannot be a member. If the relationship is less than $50,000, then the individual must be a nonvoting member. The FDA commissioner can waive these requirements in certain narrow circumstances. Advocacy groups and the editors of the *New York Times* objected to the $50,000 limit as too high (New York Times 2007).

It is often taken as an article of faith that government advisory groups, such as advisory councils, study sections, and review panels, render objective opinions regarding important public policy issues, such as grant funding, approval of drugs, regulation of toxic chemicals, and public health. The reliance on such groups is primarily based on two premises: Group members have expertise in the subject area, and group members are objective and impartial. Shamoo (1993) analyzed 10 years (1979–1989) of funding decisions regarding grant applications submitted by members of the NIH advisory councils and found that they received nearly twice the percentage of grants funded despite the fact that their merit ratings by the study sections were the same as those for the rest of the grant applicants. Science journalists Mooney (2005) and Shulman (2007) have also described in detail how political interests can undermine objectivity of other types of government advisory groups.

The Conflict of Interest Policy at the University of Maryland, Baltimore

The policies of the University of Maryland, Baltimore (UMB) are an example of how institutions have met their responsibilities under NIH and NSF COI rules. The policies, available at UMB's Web site, were in response to a 1996 legislative act that encourages institutions of higher education to increase their financial resources through various arrangements with industry, including collaborative research and development. Such relationships must avoid violations of the law, as determined by the Maryland State Ethics Commission. Like the NIH policy, UMB's guiding principles revolve around the fundamental principle of disclosure. However, the scope of UMB's frame of reference, like that of many other institutions, is broader than that of the NIH. Its principles include an open academic environment, research that is appropriate for an academic environment, the freedom to publish, the avoidance of misuse of students and employees, the routine use of university resources for purposes of teaching and research (not commercial purposes), and the monitoring and management of the outside interests of employees in ways that are consistent with their primary commitment to the mission of UMB.

UMB has chosen to monitor these relationships through an appointed COI officer. Each faculty member is required to complete a form that identifies any relationships with "an entity engaged in research or development, or an entity having a direct interest in the outcome of research or development." The forms are reviewed by the appropriate chair or division head and by the dean for their approval. The COI officer reviews those relationships that are identified as potential COIs. The vice president for academic affairs makes the decision about whether a given relationship represents a "harmful interest." When a relationship is confirmed as a harmful interest, under two defined circumstances the board of regents allows the relationship to continue. One circumstance is that the success of the research or development activity depends not only upon the employee's participation, but also upon the financial interest or employment of the employee. For example, a waiver could be granted for a project that would fail without a particular university scientist's investment in or simultaneous employment by, a biotech company. The other circumstance is when a given COI can be managed in ways that are consistent with the purposes of the relevant provisions of the law. These legal provisions deal with constraints related to the acceptance of gifts and honoraria, advocacy representation by an employee to the board of regents on behalf of a corporation, and the misuse of one's position for personal gain. University employees who seek waivers must make a formal request that is acted upon by the vice president for academic affairs. Individuals whose requests are disapproved must discontinue their relationships within a reasonable time or risk losing employment by the university.

CONCLUDING REMARKS

Although we believe that it is important to respond appropriately to individual and institutional COIs, it is not possible (or even desirable) to avoid or prohibit all situations that can create COIs or apparent COIs. Individuals and institutions have many different duties and interests. Only a hermit could avoid all COIs and apparent COIs. Since most institutions need private funding to continue to exist, it is also doubtful that they can avoid COIs or apparent COIs. Given the ubiquity COIs and apparent COIs, the best response is to take appropriate steps to disclose or manage these situations and to sometimes avoid or prohibit these situations. Many situations in research that create COIs also have important benefits. For instance, intellectual property rights can create COIs, but they can also encourage innovation and investment in research. From a utilitarian perspective, one might argue that society is best served by allowing researchers and research institutions to own intellectual property but to require them to make appropriate disclosures.

COI policies should strike a reasonable balance between accountability and freedom in the conduct of research. On the one hand, it is important to have rules that promote accountability, because these rules help ensure the objectivity, integrity, and trustworthiness of research. On the other hand, excessive and highly strict rules can impede research by interfering with scientific creativity and innovation, by undercutting financial rewards, or by impeding the funding of research. Clearly, we need rules and policies that address COIs in research. But how should we craft them and implement them? These are not easy questions to answer. Deciding how to respond to COIs raises ethical dilemmas for researchers, research institutions, government agencies, journals, and businesses. Almost all parties agree that any policy must involve an educational component. Researchers should help students and colleagues understand what COIs are, how they can occur, why they create serious problems relating to objectivity and integrity in research, and how best to respond to them.

As noted above, COIs vary a great deal in their degree of seriousness. The spectrum of COIs ranges from the benign and trivial to unethical and criminal misconduct. Policies such as those discussed above attempt to provide a balance between regulatory accountability and the freedom necessary to conduct research. Considering the complexity of our society, the increased financial underpinning of COIs, and the heightened demands for greater public accountability, how good are these policies? How close do the guidelines of the NIH and others come to ideal guidelines? Should accountability be enforced?

In today's society, it is difficult to imagine the scientific process without some sort of mechanisms that deal with the issues of misconduct and COIs. One view of how to proceed is to cultivate an improved moral culture that attracts idealistic individuals with high ethical standards to the scientific enterprise. Another view is that scientists, like everyone else, are subject to frustrated dreams, resentments, and hopes for fulfillment, and that the best we can do in these circumstances is to work hard to foster, especially for scientists in training, an understanding of the crucial importance of scientific integrity and its relationship to the day-to-day processes of science. And we also need a reasonable system of accountability.

In conclusion, it is important to remember that a nation's ability to deal with problems, such as COIs in research, is a sign of a stable democratic society with the rule of law, reasonable public confidence in the system and the willingness of its population to abide by its rules. From this perspective, the national will to deal with such issues is an apparent luxury compared with the constraints presented by the poverty, limited health care, genocide, war, and corruption that exist in many other countries. Nonetheless, if society does not deal effectively with COIs in research, the erosion of public confidence in science that may occur can be extremely damaging and corrosive to democracy. Thus, addressing COIs is important not only to protect the integrity of science but also to protect the integrity of society.

QUESTIONS FOR DISCUSSION

1. The approaches to the management of COIs in science appear to rely primarily on disclosure, without the severe penalties that can be levied for violations of COI policies in finance or government. Considering the goals of science and other dissimilarities between science and the other disciplines, do you believe the differences in management are warranted?
2. Can you think of situations where scientists should remove themselves from a COI situation?
3. Should some COIs be prohibited? Which ones?
4. What is the difference between an actual and an apparent COI? Does it matter?
5. How much money (or stock, etc.) has to be at stake before a person has a COI?
6. How are COIs are risk factor for misconduct?
7. How important is COI to science? Should it be viewed as just another issue of life, no more or less important than others? Or, given the importance of integrity in science, should it be viewed as a crucial issue for scientists?
8. What is the best approach to sensitize young scientists to the predicament of COI? To approach established scientists?
9. Can you give an example of scientific COIs from your own experience or that you have heard about? How was it dealt with it? How should it have been dealt with?
10. How do you evaluate the process for managing COIs? Do you view it as primarily preventive or primarily remedial? Is it fair to the individual scientists? To the research institutions? To the public? Do you think that existing policies favor one group over others? Can you think of any changes that would improve the procedures?
11. In the world of finance, confidentiality is a key aspect in managing COIs. How important is confidentiality to integrity in research?
12. Journalists believe that the public's right to know the truth is all-encompassing. Should the same be true for scientists? Journalistic codes of ethics state that fidelity to the public interest is a primary value. Should the same be true for scientists? What bearing do these questions have on COIs in science?
13. How would you analyze the "Pharmatec and the University of Florida" example discussed in this chapter, considering the primary and secondary interests, the types of COIs that occurred, and how they would have been managed under the several COI policies presented above?

14. Consider the "tPA: A Multicenter Trial" example described above: How would it have been managed under the several COI policies presented above? What are the most important differences between this example and the example in question 13?

15. In the examples "Clinical Trials of Therapeutic Drugs" and "Familial Hypercholesterolemia and the Nobel Prize," what do you think are the likely primary and secondary interests?

CASES FOR DISCUSSION

Case 1: Cleveland Clinic CEO and AtriCure

The Cleveland Clinic Foundation, with its eight hospitals, is one of the most prominent organizations in health care. It has a prestigious cardiac center and is a national leader in oncology research. It has nearly $4 billion in revenue and 30,000 employees. Delos "Toby" Cosgrove, a world-renowned cardiologist, was the chief executive officer until 2005. The clinic created the Foundation Medical Partners (FMP) in 2001 to raise money. Cosgrove was an investor in the FMP (Armstrong 2005). The clinic contributed $25 million of the total limited partnership in FMP and was entitled to 38% of its profits.

FMP invested in several startup companies involved in treating atrial fibrillation, including a $3 million investment in AtriCure. Atrial fibrillation affects 2.2 million Americans; 15% of them will have a stroke due to the illness. The purpose of AtriCure's device was to control heart rhythm using radiofrequency to scarred tissue near the pulmonary veins that contribute to irregular heart beat, which would block the impulses.

Cosgrove and the Cleveland Clinic promoted AtriCure's device in articles, conferences, and brochures. The Cleveland Clinic claimed that Cosgrove mistakenly declared that he had no COIs on forms he was required to fill out by the clinic's board of directors.

AtriCure's device was approved by the FDA for soft-tissue surgery, but it has not been approved to treat atrial fibrillation by the FDA, which has rejected its device application three times. But the clinic went ahead with its use for other surgeries under "off-label" use. (An "off label" use of a medical product occurs when the product is used for a purpose other than the indication [i.e., specific disease symptoms] that has been approved for.)

However, the device was also used in patients on an "off label" basis at other hospitals. Four patients died as a result of using the device, but AtriCure did not notify the FDA. (The FDA requires manufacturers to report adverse events.) The Cleveland Clinic has used the device on 1,247 patients/subjects, which includes 16 subjects who are part of an ongoing clinical trial.

Cardiologist Eric Topol and blood specialist Alan Lichtin, both at the Cleveland Clinic, reported AtriCure to the COI committee at the Cleveland Clinic. Cosgrove announced that Topol lost his post as a provost of the medical school and thus automatically lost his membership in the COI committee and the board of governors of the clinic. Topol is an outspoken critic of those involved in a relationship with industry that is entangled with a COI. For example, Topol was critical of the now known case of Vioxx, manufactured by Merck.

The COI committee began to investigate the case after it learned about it and after the IRB had informed them of the conflict. The IRB directed the researchers to voluntarily stop their work. The FMP partners will receive 2.75% of money raised plus other fees totaling $1.7 million a year. Cosgrove stepped down from the AtriCure board and promised to resign from the FMP.

The board of trustees, based on a request from Cosgrove, appointed a special committee to oversee the investigation, and the committee hired a law firm to assist them. The board decided that it should tighten the COI policies and that there should be better communications between the COI committee and the IRB. Cosgrove admitted his shortcomings in this case. The chairman of the board of trustees and its members have admitted having continued investments in medical equipment companies (Armstrong 2006).

- Do you think Cosgrove had a COI? Why or why not?
- Should the clinic have such a relationship with a medical device company?
- Was the IRB doing its job?
- Was the COI committee doing its job? Why or why not?

Case 2

A faculty member of a midwestern university has a postdoctoral fellow who had been working for three years on one aspect of a new drug that has just been approved by the FDA. The drug company pays the mentor $5,000 per year in consulting fees. This year, the company is sponsoring a three-day conference in San Francisco, all expenses paid. The conference brings together approximately 15 scientists from around the country who are leaders in research on this drug. Their job is to speak to 300 invited guests, physicians who are potential prescribers of the drug. All the physicians' expenses are paid. The speakers will each receive a $3,000 honorarium. The faculty member accepts the invitation, and the postdoctoral fellow hears about it by accident. He informs his mentor that he has new data indicating that the drug may have serious side effects. The mentor points out that the work is not yet published and that she will deal only with the published data for now.

- Under the terms of the COI policies described above, is this a COI?
- If the answer is yes, what is the COI? What type is it?
- Are there COIs in this case that are not covered by the policies?
- Is this a case of misconduct?
- Should the mentor have done anything differently up to the point that the postdoctoral fellow told her about the new data? What should the mentor and postdoctoral fellow do now?

Case 3

A senior researcher in a major research university has two grants from the NIH and research contracts from two different large pharmaceutical companies. She also serves on the National Heart, Lung, and Blood Institute Advisory Board and one of the FDA's advisory panels. Her work pertains to drugs that lower cholesterol. The FDA panel that she is on was convened to decide whether to recommend the final approval of a drug for consumer use. The drug is manufactured by one of the two companies with which she has a contract. However, the contract has nothing to do with this particular drug. Prior to the deliberations, she disclosed her research contract as a COI and proceeded to participate in the panel. She was one of those who voted to recommend approval of the drug, but the recommendation would have passed anyway, without her vote.

- Is this really a COI case? Was it necessary for the senior researcher to have disclosed her contract with the company as a COI?
- Should the FDA have allowed her to deliberate on this panel?
- Despite the fact that she was allowed to participate on the panel, do you think she can deliberate without being unduly influenced by the contract she holds?
- Should universities encourage or discourage their researchers from placing themselves in such positions?

Case 4

A junior scientist at a university made a major discovery for a new anti-breast cancer drug. With the help of the university's Office of Research Development, she applied for a patent and established an agreement that splits the profits between her and the university. The new drug had the potential for bringing in millions of dollars for both parties. Using venture capital, the university formed a startup company. The junior scientist and her department chair were on the board of directors. Other board members included several university administrators and a few employees of the venture capital company. The patent was awarded few years later. The startup company still has not made a profit, but it is very close to marketing the drug.

The junior scientist, while at a conference, heard that a researcher at another university has come up with a drug very close in structure to hers, but one that is much more effective and has fewer side effects. The inventor of the new drug plans not to patent his drug but instead to place the information in the public domain. A different group of venture capitalists formed a new company to manufacture the drug as quickly as possible. The junior scientist, with prodding from her venture capitalists, is now thinking about challenging the new company with a lawsuit on the basis of patent infringement. The patent battle could delay and may eventually prevent the new company from ever producing the new drug.

- What are the COIs in this case? Who has COIs?
- Because NIH funds have not been used for this research, should the NIH and university COI policies be applied to this case? If so, what impact would these policies have?
- At this point in the case, what should the interested parties do?

Case 5

A member of an NIH study section was the primary reviewer of a grant proposal submitted by a very close personal friend and colleague. The extent of their relationship was well known among the close-knit research community, and the primary reviewer chose not to recuse himself. He proceeded to review the proposal and gave it his highest ranking. During the deliberation, another member of the study section raised the issue of the perception of COI. The chair of the study section and the reviewer in question rejected the implications of the comments, pointing out that the member's signature on the COI declaration was sufficient indication of his ability to make an objective judgment. Additionally, the NIH scientific review administrator said that study section members were obligated and expected to put personal relationships aside as part of their commitment to an objective evaluation. The study section proceeded with the overall approval of the grant proposal. However, the administrator informed the chair of the study section that the person who raised the issue concerning the COI has the right to go through the formal appeal process.

- Was the scientific review administrator justified in making the statement about expected objectivity? What should the policy be in such circumstances? Should personal friendships ever be considered as a disqualifier for an objective review?
- Should the member of the study section in question have done anything differently?
- Was the other member of the study section justified in bringing the subject of COI?

Case 6

A member of an NIH study section has been asked to review a grant proposal submitted by a direct competitor to her own research. They are both working on practically the same project and are very familiar with each other's strengths and weaknesses. Their mutual animosity is common knowledge; during scientific conferences, their exchanges almost amount to shouting matches. The reviewer prepared a very thoughtful and lengthy evaluation. She highly criticized the proposal and gave it a poor rating. The proposal received an overall average rating, which removed it from consideration for funding.

- Under the terms of existing federal or university COI policies, does this represent a COI?
- Given the general awareness of their competitive positions and their mutual animosity, was the scientific review administrator justified in not intervening? In selecting the reviewer in the first place?
- Should the member of the study section have done anything differently?

Case 7

A newly appointed bright and industrious assistant professor became a member of an NIH study section. His research was in a new area of X-ray diffraction of membrane proteins. He was the primary reviewer of a grant proposal from a senior department chair at a different university. He recommended the rejection of the proposal, and the committee concurred. A few months later, the chair whose grant was turned down got the bad news. The chair reviewed the membership of the study section and noticed that one member—the reviewer in question—was someone he had rejected for a job in his department. The chair wrote to the scientific review administrator and expressed his concern, particularly if the assistant professor was one of the two primary reviewers. The administrator confirmed with the reviewer the truth of the chair's statement about the job application. After assessing the notes of the deliberations, the administrator dismissed the complaint.

- Is there a COI in this case? If so, is it within the scope of any of the COI policies discussed in this chapter?
- Was the chair's concern reasonable? Should the assistant professor have acted differently?
- Was the administrator justified in his actions? Were there other options that the administrator might have taken?

Case 8

A well-known researcher was ready to publish some important findings made after three years of work with his two postdoctoral fellows. The findings clearly add a new and crucial dimension to the activity of a receptor discovered six years ago by someone in India. However, the original discovery is not well known in the United States. The two postdoctoral fellows wrote the paper with a clear citation to the earlier, foundational discovery made by the Indian scientist. The senior researcher admonished the two postdocs for not recognizing the originality and importance of their discovery. His viewpoint was, "Why dilute the impact of your paper? Why publicize someone else's work unnecessarily?"

- What are the COIs in this case? Do they fall within the scope of COI policies?
- Is this a case of COI, scientific misconduct, or both?
- At this point in the development of the case, what options should the two postdoctoral fellows consider? What should be their guidelines for decision making?

Case 9

A senior staff scientist and section chief at the NIH outsourced a portion of her research with a subcontract to a junior researcher at a nearby university. The subcontract was $150,000 per year for five years. Although the research was a component of the NIH researcher's work, it was to be carried out independently. After two years, the university researcher wrote a paper without using the name of the NIH scientist as a coauthor. The NIH scientist received a draft of the paper and asked for her name to be added, because the research had been her conception, even though she had not contributed to the work.

- Does this case represent a COI, or is it merely a dispute about authorship?
- If you think it involves a COI, what type of COI is it? Does it fall under the category of any of the COI policies mentioned in this chapter?

Case 10

A university scientist/ethicist who is often cited by the national media on issues related to research on human subjects is called by a journalist from a major newspaper in a distant city. The journalist wants to know what he thought of the letters just received from the FDA and Office for Human Research Protections warning his university about its continued and frequent failure to comply with federal regulations. The letters in question had been addressed to a university vice president, who happens to chair a search committee formed three months ago to find a top-level assistant on research affairs. Two weeks ago the scientist/ethicist had applied, with many others, for the job.

- Is the COI fairly obvious here? Why or why not? Does it fall under the umbrella of any COI policies?
- What should the scientist/ethicist do? Should he tell the journalist about his application? Should he inform the vice president of the journalist's request for his opinion? Should he give the interview off the record or find a reason not to give the interview at all?
- Is there any reason for the scientist/ethicist to consider the university's interests as more important than his own? The public's right to be informed as more important than his own interests?

Case 11

A clinical researcher receives $3,200 per patient from a drug company to enroll patients in a clinical trial of a new hypertension medication. The money covers patient care costs and administrative costs for the duration of the study and includes a $200 finder's fee. After the initial screening and enrollment, patients will make a total of eleven 15-minute office visits during the study. At each visit, nurses will take blood, record vital signs, and ask them questions about their hypertension. The clinician will do a physical exam.

- Is this financial arrangement a COI or apparent COI?
- Should it be prohibited?
- During the informed consent process, should patients be told about the clinician's financial interests?

Case 12

A clinical researcher has partial ownership of a patent on a new gene therapy technique. Her work is being funded by a biotech company, which owns the other part of the patent. She also has stock in the company. She plans to use her patented technique in clinical trials for gene therapy to treat breast cancer. If her cure works, she will make millions of dollars, and so will the company. Even if the cure doesn't work, she still may make a great deal of money if the value of her stock rises.

- Does the researcher have any COIs or apparent COIs?
- Should any of these financial relationships be prohibited or avoided?

11

The Use of Animals in Research

This chapter provides a brief history of animal research and examines the ethical arguments for and against animal experimentation. It discusses the animal rights views of Peter Singer and Tom Regan and considers some morally significant differences between animals and humans. The chapter also discusses some principles for the ethical treatment of animal in research, such as the "three Rs"—reduction, replacement, and refinement—as well as animal research regulations.

Experimentation on (nonhuman) animals is one of the most controversial issues in research ethics. Like the abortion debate, the issue has been hotly contested and often violent. Animal rights activists have freed laboratory animals and destroyed research records, materials, equipment, and buildings to protest what they consider to be immoral uses of animals (Koenig 1999). Animal welfare organizations, such as People for the Ethical Treatment of Animals (PETA), have enlisted volunteers to penetrate institutes, spy on researchers, and uncover abuses of animals. In response to threats from individuals and organizations, universities have tightened security measures. Researchers have rallied around the cause of animal research and have formed professional organizations, such as the National Association for Biomedical Research and the Foundation for Biomedical Research, to promote the humane use of animals in research. Given the highly polarized nature of the debate about the use of animals in research, one wonders whether there can be any hope of some consensus (DeGrazia 1991).

Estimates of the number of animals used in research vary from 17 to 70 million animals per year (LaFollette and Shanks 1996). Advances in technology have made it possible to eliminate some uses of animals in research and replace animal models with other testing procedures, such as tissue cultures and computer simulations (Barnard and Kaufman 1997). Researchers are also finding ways to obtain valid results using fewer animals. Additionally, universities are using fewer live animals in educating graduate and undergraduate students. The number of animal experimental procedures conducted in the United Kingdom declined from 5.2 million in 1978 to 3 million in 1998. However, this trend may reverse as researchers increase the number of animals used in transgenic research (Stokstad 1999). For example, researchers participating in the National Institutes of Health (NIH) Knockout Mouse Project plan to make approximately 25,000 strains of genetic knockout mice, one for each gene in the mouse genome (Austin et al. 2004). The project will use about 10 million mice (Williams et al. 2003).

The percentages of animals used in different research activities are as follows: 40% in basic or applied research, 26% in drug development, 20%

in safety testing, and 14% for other scientific purposes (Pence 1995). Much of what is known about human and animal physiology, anatomy, biochemistry, embryology, development, genetics, cytology, neurology, immunology, cardiology, and endocrinology has been gained through experiments on animals. Animals are commonly used in applied research to test new medical therapies, such as drugs, vaccines, medical procedures, or medical devices. Indeed, U.S. Food and Drug Administration (FDA) regulations require that new drugs and medical devices be tested in animal populations before they are tested in human populations (Bennett 1994). Animals are used in environmental studies to determine the toxic or carcinogenic effects of compounds that are released into the environment, such as pesticides, herbicides, or pollutants. Animals are also used in agricultural research in the development of hybrid breeds, clones, or transgenic species and in cosmetic research to test the toxicity of mascara, shampoo, hair dye, lipstick, and other products. Transgenic animals are playing an increasingly important role in research: Researchers have developed varieties of transgenic mice that contain genes for specific diseases, such as diabetes, obesity, cancer, and Parkinson's disease. Transgenic animals are playing an increasingly important role in agriculture, as well: Researchers have developed sheep that produce human hormones in their milk, and they are attempting to develop pigs that will produce organs suitable for transplantation into human beings (Wilmut 1997, Marshall 2000c).

Although people have spoken on behalf of animals for many years, a book by the philosopher Peter Singer, titled *Animal Liberation* (1975 [1990]), spurred and buttressed the modern animal rights movement. But before examining Singer's views on animals, we provide a brief history of the animal rights movement.

HISTORICAL PERSPECTIVE

For many years, scientists who used animals in research were influenced by the views of the seventeenth-century French philosopher René Descartes (1596–1650), who argued that animals are like machines and that their behaviors result from instincts, reflexes, and other internal mechanisms that do not involve consciousness or rationality (Descartes [1970]). Descartes's view probably had a significant influence in the early vivisection practices: Seventeenth- and eighteenth-century vivisectionists nailed dogs to boards and cut them open without anesthesia or analgesia. They interpreted their howls and cries as mere noises produced by machines (LaFollette and Shanks 1996). Pain-relieving measures were not used in animals before the discovery of anesthesia in 1846. In the twentieth century, the behaviorists, such as B.F. Skinner and J.B. Watson, wielded considerable influence over researchers' attitudes toward animals. The behaviorists held that science can only study animal and human behavior: It is not possible to have scientific knowledge of the inner workings of the mind, such as thoughts, feelings, and emotions. People who were influenced by this doctrine did not pay sufficient attention to animal pain or suffering, because they believed that it is impossible to know whether animals think or feel (Rollin 1989).

Two intellectual forerunners of the modern animal rights movement were British philosophers Jeremy Bentham (1748–1832) and John Stuart Mill (1806–1873), who advanced the notion that animals can suffer and thus deserve moral consideration. The views of Mill and Bentham provided a philosophical and moral basis for the nineteenth-century's animal welfare movement. From the mid-1800s until the early twentieth century, there was a strong anti-vivisection movement in England and the United States. Leaders of this movement opposed the use of animals in experiments and opposed all forms of cruelty to animals.

The American Society for the Prevention of Cruelty to Animals (ASPCA) was formed in 1866, and local societies for the prevention of cruelty to animals soon followed. The American Humane Organization, founded in 1874, opposed the use of animals in experiments as well as inhumane experiments on human beings. Many of the leaders of this movement, such as Caroline White and Mary Lovell, were women with strong religious and moral convictions. The anti-vivisection movement also received support from the Women's Christian Temperance Movement, the Department of Mercy, and *Life* magazine. The anti-vivisectionists made two main arguments against animal experiments. First, they argued that these experiments cause unnecessary and unjustifiable suffering to animals. Second, they argued that our attitudes toward animals could influence how we treat human beings: Cruelty to animals can lead to cruelty to human beings. In this light, the anti-vivisection movement also drew connections between exploiting animals in research and exploiting human beings and helped draw attention to unethical experiments on children, mentally ill people, poor people, African Americans, and prisoners (Lederer 1995).

In Britain, animal rights activists helped pass the Martin Act in 1822 titled "Act to Prevent Cruel and Improper Treatment of Cattle." A few years later the Royal Society for the Prevention of Cruelty to Animals (RSPCA) was founded. In 1829, New York State passed a law to protect domestic animals such as horses, oxen, and other cattle. By the end of the nineteenth century, Britain passed stricter animal protections laws. Feminist and animal activist Frances Power Cobbe became more active in opposing vivisection. Late in the 1800s, all experiments with animals in Britain required a yearly license. During the same period, Darwin's theories on evolution convinced many people that human beings are descended from animals, which heightened sensitivity to subjecting animals to pain. The animal rights movement in England remained dormant until the 1960s. In 1972, Richard Ryder, Ronald Lee, and Clifford Goodman founded the Animal Liberation Front (ALF) in England.

In the United Sates, the National Academy of Sciences and the American Medical Association formed a coalition to push for legislation in the late 1890s to disarm anti-vivisectionists and promote medical progress. The animal rights movement was practically nonexistent in the United States until the 1960s and the advent of the civil rights movement. This was followed by Singer's book *Animal Liberation* (1975 [1990]). In the 1980s, two strident activists, Alex Pacheco and Ingrid Newkirk, formed PETA which has since engaged in numerous high-profile activities to highlight the plight of animals in scientific experiments. Although PETA has not condoned the use of violence, many people have alleged that the organization has used unethical and illegal means to reach its goals (Oliver 1999).

THE ARGUMENT FOR USING ANIMALS IN RESEARCH

Why do scientists use animals in research? The main argument can be understood in starkly utilitarian terms (Cohen 1986, Botting and Morrison 1997): Animal research produces important basic and applied knowledge that promotes human health and well-being. An added benefit of animal research is that it can yield knowledge that improves the health and welfare of animals. The argument is utilitarian because it holds that the ends (e.g., promoting human health and well-being) justify the means (animal research). Those who criticize this argument argue either (a) that the ends do not justify the means or (b) that the means are not effective at achieving the ends. The first type of critique raises moral objections to animal research; the second raises scientific or technical objections. Below we examine both of these critiques as well as replies to them.

MORAL CRITIQUES OF USING ANIMALS IN RESEARCH

To understand moral arguments against using animals in research, it is useful to draw an analogy with human research. Two of the guiding principles in human subjects' research are beneficence and respect for persons (National Commission 1979). We can use humans in our research provided that we take steps to promote and respect their inherent moral worth. The phrase "human guinea pig" literally means treating a research subject (or person) like an animal, that is, treating the research subject as something that can be sacrificed for a greater good (Jonas 1980). (Chapters 12–15 discuss human research in more depth.) Moral objections to animal research assert that research protections pertaining to human beings should be extended to animals: We should take steps to promote and respect the moral worth of animals (LaFollette and Shanks 1996). If we are not morally justified in performing an experiment on a human, then we should also not perform that experiment on an animal. Animals, like human beings, have inherent moral worth and should not be sacrificed for a greater good. There are two very different ways of supporting the idea that animals have moral worth: a utilitarian perspective defended by Singer (1975 [1990]), and a rights-based perspective defended by Tom Regan (1983).

According to Singer, the central question in our treatment of animals is whether they deserve moral consideration. Singer does not believe that all organisms deserve moral consideration; for example, he would not claim that bacteria deserve moral consideration. Merely being a living creature is not a sufficient reason for special moral status. Singer also thinks that the question of whether animals can reason or communicate is not the only relevant issue in deciding whether they deserve moral consideration—what matters is that animals have the ability to suffer. Singer cites Bentham on this point. In discussing the moral status of black slaves, Bentham argued that the color of their skin was irrelevant to determining their moral status. And Bentham extended this argument to animals: The most important question in determining the moral status of a being is not whether the being can think, talk, or reason, but whether it can suffer (Bentham 1789 [1988]). According to Singer, many animals can suffer and therefore pass the key test for determining whether they deserve moral consideration. This view implies that human beings have a moral obligation to refrain from causing animals to suffer.

Most people would likely accept most of this argument: It is certainly wrong to inflict *needless* suffering on animals, and one should take steps to minimize animal suffering. But most people would say that animal suffering could be justified to promote important causes, such as improving human health and well-being. Animals have some moral worth, but human beings have a higher moral status or worth (Cohen 1986, Frey 1994). Thus, we should consider animal suffering in deciding how to treat animals, but we should give far more weight to human suffering.

However, Singer does not believe that we should give more weight to human suffering. According to Singer, all beings that deserve moral consideration deserve equal moral consideration. Thus, we should give equal weight to human and animal suffering. Our refusal to give equal consideration to animals is a form of bias that Singer calls "speciesism." Singer equates speciesism with racism and sexism because these "isms" discriminate between different classes of beings based on what he believes to be morally irrelevant characteristics. Just as skin color and gender are not relevant to a person's moral standing, species membership is also not relevant. According to Singer: "Speciesism... is a prejudice or attitude of bias toward the interests of members of one's own species. It should be obvious that the fundamental objections to racism and sexism made by Thomas Jefferson and Sojourner Truth apply equally to speciesism" (Singer 1975 [1990], p. 7).

Singer's view does not prohibit animal experiments on species not considered to merit moral consideration, including plankton, worms, and many other lower species. Nor does it imply that experiments on animals with moral worth can *never* be morally justified. Because Singer is a utilitarian, he believes that animal experiments could be justified if the experiments promote the greatest good for the greatest number of beings that deserve moral consideration—humans and animals. However, his view implies that most of the animal experiments that are performed today should be stopped and that we should consider performing experiments on human beings that lack the ability to suffer, such as human beings with severe brain damage, instead of performing those experiments on animals.

Many writers have attempted to refute Singer's charge of speciesism by arguing that there are morally significant differences between human beings and other animal species (Caplan 1983, Cohen 1986, Carruthers 1992, Frey 1994). The charge of speciesism is unsound because, although there are no significant moral differences among races, there are significant moral differences between humans and other species. Typical, adult humans have the following qualities:

- *Rationality:* the ability to reason and solve problems, to engage in abstract thinking
- *Linguistic communication:* the ability to communicate information using a language
- *Emotion:* the ability to feel and respond to emotions such as empathy, compassion, mercy, guilt, anger, fear, and shame
- *Morality:* the ability to formulate and follow moral rules, to make moral choices
- *Creativity:* the ability to come up with new ideas, behaviors, solutions, and expressions
- *Spirituality:* the ability to form a concept of higher power or being, the need to find meaning in life and a connection to the whole universe
- *Self-consciousness:* the awareness of one's self in one's environment, of one's own beliefs, emotions, desires, and attitudes
- *Self-determination:* the ability to make deliberate choices in controlling one's own behavior
- *Consciousness:* the awareness of sensations, including the capacity to feel pain

Admittedly, not all human beings have all of these characteristics all of the time. Children develop these characteristics as they mature, people with severe brain injuries no longer have all of these characteristics, and people born with mental disabilities may never have some of them. Nevertheless, the typical adult member of the species *Homo sapiens* has all of these characteristics, while the typical, adult animal used in an experiment may have only some of these characteristics. A laboratory rat probably has consciousness, emotion, and perhaps even some degree of self-awareness, but it probably does not have reasoning, linguistic communication, morality, self-determination, spirituality, and other qualities. Some animals have more of these human characteristics than rats. For example, chimpanzees have consciousness, emotion, self-awareness, creativity and rudimentary reasoning, linguistic ability, and self-determination. Accordingly, chimpanzees should be treated differently than rats because they have more human qualities. (We'll return to this point later.)

An additional critique of Singer's view is that his use of the term "suffering" is somewhat naive and simplistic. It would appear that Singer uses the term "suffer" as a substitute for "feel pain," but suffering is not the same thing as feeling pain (Cassell 1991). There are many different types of suffering: unrelieved and uncontrollable pain; discomfort, as well as other unpleasant symptoms, such as nausea, dizziness, and shortness of breath; disability; and emotional distress. However, all of these types of suffering involve much more than the awareness of pain: They also involve self-consciousness, or the awareness that one is aware of something. For a creature to experience unrelieved pain, the creature must be aware that it is in pain and that the pain is not going away.

If we think of suffering in this fashion, then it may not be at all obvious that animals suffer, because we do not know the extent to which animals are self-conscious. Although one might argue that it is also difficult to prove that animals feel pain, most people find this idea easier to accept (based on behavioral and neurological similarities) than the claim that animals are self-conscious. However, once again, a great deal depends on what species of animal we have in mind. Monkeys, dogs, and cats probably have enough self-consciousness to experience suffering. But what about fish, frogs, and mice? Perhaps Singer has chosen the wrong word to describe animal experiences. If he had said that the key point is that animals can feel pain, rather than suffer, then perhaps his claims would be less contentious (Rollin 1989).

Regan, like Singer, also believes that animals have moral status but, unlike Singer, argues that animals have rights. This seems like an outrageous claim to those who believe that only moral agents—beings who can formulate and follow moral rules—have rights. Rights and responsibilities go hand in hand: One cannot have a moral right unless one can also accept moral responsibilities. Because animals are incapable of following moral rules and making moral choices, they are not moral agents. Thus, they have no moral rights (Fox and DeMarco 1990).

To make sense of the idea that animals have rights, Regan (1983) draws a distinction between moral *agents* and moral *patients*. Regan argues that moral communities include members who have moral rights but not moral responsibilities, such as young children, mentally retarded adults, and permanently comatose adults. We grant moral rights to these people because they still have interests even if they do not have responsibilities. Moral patients do not have all the rights accorded to moral agents. For instance, children do not have the right to vote, the right to enter a contract, or the right to marry, but they do have some basic rights, such as the right to life and the right to health.

Animals, according to Regan (1983), are moral patients because they have inherent value. By this he means that animals are capable of valuing their own experiences and their own lives. They can prefer pleasure to pain, freedom to captivity, and life to death. They have perception, memory, and a sense of their own past and future. Because animals have inherent value, one should treat them just like other beings that have inherent value, because inherent value does not come in degrees. Indeed, it would be a form of speciesism to insist that humans have more inherent value than do animals. Thus, animals should be accorded the same rights that we grant to other moral patients, such as children.

According to Regan, and many other writers, the purpose of rights is to serve as moral "trump cards" to protect and promote individual interests.

For example, when we say that a person has a right to vote, we imply that this right should not be taken away in order to promote a greater good. Animal rights also function as moral trump cards that forbid us from sacrificing animals for some "greater good." In particular, we should not use animals in experiments that are not designed to promote their interests. Because more than 99% of animal experiments yield no benefits for the experimental subject, almost all animal experiments are immoral (Regan 1983). The only kind of animal experiment that could be justified would be an experiment that is designed to benefit the animal or promote its interests. Here we can draw an analogy with the ethics of experimentation on children: According to most research ethics regulations and policies, we should conduct research on children that is more than minimal risk only if the benefits to the subjects outweigh the risks. Risky research on children should promote that subject's best interests (Glantz 1998).

Many different writers have criticized Regan's view. Some of the critiques address the theoretical underpinnings of his position and argue that he has not adequately explained how animals can have rights (Carruthers 1992). First, one might ask whether the claim that animals *really* value their own experiences and their own lives has any plausibility. The word "value" connotes more than having a preference, want, or desire: To value something, one must make a judgment about the worth of that thing. Wanting a drink of water is not the same thing as a valuing a drink of water: A person who values a drink of water also makes a judgment that the water is good in some respect, for example, good for quenching thirst. It is not at all clear that animals have the cognitive capacity to make judgments about value even if they have desires or preferences.

Second, according to most accounts of rights, rights promote or protect interests (Feinberg 1973). An interest, on this view, is something that one *needs* in order to promote one's overall well-being. For example, people need food, shelter, freedom, companionship, freedom from pain, and many other things that promote well-being. Regan's view implies that animals also have various interests, such as interests in living, in freedom of movement, in food and water, and in freedom from pain. But how can we make sense of the interests of animals? If we are not careful, we may find ourselves accepting the idea that plants have interests, if we maintain that animals have interests. Plants need water, sunlight, and soil to grow and flourish. But do plants have interests? This view seems patently absurd. So, there must be some difference between an interest and a *biological* need. One might argue that interests are different from biological needs in several ways. First, beings with interests are aware of those interests. Second, beings with interests can communicate those interests. Although a plant may need water, it is not aware of this need, nor can it communicate it (as far as we know). Although laboratory mice may be aware of their biological needs in some sense, they cannot communicate those needs.

Finally, other writers have objected to the practical problems with making sense of animal rights (LaFollette and Shanks 1996). One problem would be how we should resolve conflicts of rights among animals. For example, if a lion and a zebra both have a right to life, should we stop the lion from killing the zebra in order to protect the zebra's rights? How do we weigh the rights of a beaver to make a dam against the rights of those animals who will have their homes flooded if it does so? If we accept the idea that animals have rights, and that these rights are held equally, then there are no satisfactory solutions to these types of problems. To solve these issues, we would need to have some way of assigning value to various rights claims, for example, that the lion's right to food is more important than the zebra's right to life. But this opens the door to assigning greater value to the rights of human beings, which is a move that Regan wishes to avoid. If we say that animals and human beings both have rights but that human rights are more important, then we can also justify animal experimentation on the grounds that human rights to health and welfare outweigh animal rights to life, freedom from pain, and so on.

Although we think that Singer's and Regan's critiques of animal research have some serious flaws, they offer society and the research community some important lessons about our treatment of animals. In particular, these critiques clarify the importance of developing an account of the moral status of animals that is sensitive to both the similarities and differences between humans and animals (LaFollette and Shanks 1996). Singer and Regan, incidentally, do recognize that there are moral differences between humans and animals (Regan and Singer 1989). Singer (1975 [1990], 1985) holds that it is worse to kill a normal human adult than to kill a mouse, and Regan (1983) admits that if one must choose between saving the life of a human and saving the life of an animal, greater harm will occur if one does not save the human. But these concessions offer little comfort to those who place much greater value on human life than on animal life.

However, it is also important to remember that for many years many researchers held very little respect or consideration for animal pain or suffering. Some, such as the Cartesians and the behaviorists described above, adopted this stance based on their judgment that animals are unthinking and unfeeling beings, much like robots (Rollin 1992). We believe that the most sensible view lies somewhere between the extreme positions staked out by Singer and Regan on the one hand, and the Cartesians and behaviorists on the other. Most people, including most researchers, believe that we have moral duties toward animals and that we should respect and promote the welfare of animals (Bulger 1987). For example, most researchers favor extending the Animal Welfare Act (1966, 1996) to laboratory animals, including rats and mice (Plous and Herzog 2000). However, many people, including researchers, also recognize the importance of using animals in research (Botting and Morrision 1997). The only way to make sense of the competing claims is to adopt the view that animals have moral status (or moral value) but that human beings have a greater value (Frey 1980). There are degrees of moral value, and not all species have the same moral worth. We believe that the value of a species depends on its degree of similarity to the human species: Species that closely resemble humans, such as chimpanzees, have greater moral worth than species with little in common with humans, such as rats or cockroaches. (We do not explain why we consider human life to be valuable—we consider this to be basic assumption of ethics and morality.)

If animals have some moral value but less value than human life, how should we treat animals? To answer this question, it will be useful to distinguish between intrinsic and extrinsic value. Something has intrinsic value if people value it for its own sake; something has extrinsic value if it is valuable for the sake of something else. Some things can have both intrinsic and extrinsic value. For example, most people would agree that happiness has intrinsic value. The value of money, however, is extrinsic: Money is valued not for itself but for what you can buy with it. Education may be valuable for its own sake and also because it can help one to obtain employment.

Most people will agree that animals at least have extrinsic value because they are valuable as sources of food, clothing, labor, amusement, companionship, and so on. But saying that animals have extrinsic value provides very little in the way of any moral restriction of our conduct toward animals. One might appeal to the extrinsic value of animals in many different ways to justify restrictions on our conduct toward animals. For example, one might argue the following:

- It is wrong to torture a cat because people find this to be degrading or offensive.
- It is wrong to kill dolphins because people like dolphins.
- It is wrong to kill lions because people think lions are beautiful.
- It is wrong to exterminate prairie dogs because they play a vital role in the ecosystem of the Great Plains of North America.
- It is wrong to cause an animal species to become extinct because the species is an important source of biodiversity.
- It is wrong to harm an animal because a person who harms an animal is more likely to harm a human being.

These arguments, while important, are highly contingent, relativistic, and tentative because they depend on human wants, beliefs, and desires or features of the ecology. For instance, if no one liked dolphins, then it might be acceptable to kill dolphins; if prairie dogs did not play a vital role in the ecology, then it might be acceptable to exterminate them; if people were not offended by someone who tortures a cat, then it might be acceptable to torture a cat. For these reasons as well as others, it is important to show that animals have intrinsic value, not just extrinsic value (Taylor 1986). Thus, it is wrong to torture a cat because this act harms a valuable life, not just because people find it offensive.

So how might one *prove* that animals have intrinsic value? How does one prove that anything has intrinsic value, for that matter? We suggest that the process of assigning intrinsic value to something is not entirely rational, in that one does not arrive at judgments of intrinsic value based solely on empirical evidence or logical argument. In ethics and morality, judgments of intrinsic value are basic premises (or assumptions or axioms). More than 2,300 years ago, Aristotle argued that one cannot "prove" basic premises; one accepts basic premises and then makes arguments on the basis of those premises (McKeon 1947). Since that time, other philosophers have argued that it is not possible to conclusively "prove" that one should accept basic principles of deductive logic such as the rule of noncontradiction, basic principles of inductive logic such as Bayes's theorem, or basic axioms of Euclidean geometry such as the parallel postulate (Quine 1969, Putnam 1981). (As noted above, we hold consider the premise "human life has intrinsic value" to be a basic axiom of ethics and morality.)

So how does one come to accept basic premises, assumptions, or axioms? Arguments can be helpful, but experience is also important. For example, consider how one develops an appreciation for a piece of classical music, such as Mozart's *Eine Kleine Nachtmusik*. You might start to like this piece after someone convinces you by means of an argument that this is good music. Someone might argue that the music is well ordered, creative, intelligent, expressive, lively, and so on. But you will probably only come to appreciate this piece after listening to it. You experience the value or worth of the music after becoming familiar with the music. Some might even describe this psychological phenomenon as a "conversion" experience or a Gestalt shift: You simply come to "see" the value of the music much as you come to "see" that a picture of a witch is also a picture of a young woman (a classic gestalt figure). We suggest that the same psychological mechanisms apply to coming to appreciate the intrinsic value of animals (and humans): Arguments can play a role in helping us to accept the value of animals, but we also must have some experience with those animals in order to fully appreciate their value. For instance, if you have never had a pet pig, then an argument may convince you of the worth of the pig, but you will not truly appreciate the worth of the pig until you have come to know the pig, much in the same way that one would come to know the worth of Mozart's music.

Although we admit that it is not possible to *conclusively prove* that animals have intrinsic value, we do believe that *arguments by analogy* can play a role in helping people come to appreciate the value of animals. As noted above, we assume that people already accept the claim that human beings have intrinsic value. To show that animals also have intrinsic value, one may construct an argument by analogy with human beings: Animals have value insofar as they are like human beings. As the analogy increases in strength, the value of animals should also increase (LaFollette and Shanks 1996). An argument by analogy does not conclusively establish basic premises about moral worth, but it can help people become more willing to accept those premises. To develop the analogy between humans and animals, one must consider the above list of characteristics of human life and ask whether animals are like us in relevant ways.

Many of these questions are very difficult questions to answer and require a great deal of scientific investigation and research. After many years of treating questions about animal cognition and emotion as "unscientific," we are just now beginning to understand in greater detail important aspects of animal consciousness, experience, emotion, self-consciousness, and communication (Bonner 1980, Rollin 1989, Griffin 1992, De Waal 1996). Evidence that can help us answer these questions can come from the study of animal behavior, evolution, genetics, physiology, neurology, or endocrinology. Because different animal species have different behavioral, physiological, neurological, genetic, and biochemical traits as well as different evolutionary histories, it is likely that some species will be more like humans than others. As mentioned above, given what we know about chimpanzees and mice, we have reasons to believe that chimpanzees are more like humans than are mice. Thus, chimpanzees have greater moral worth than mice have. One can make similar comparisons for other species, such as dogs, cats, elephants, dolphins, monkeys, and birds. The more the species is humanlike, the greater its moral value.

ETHICAL ANIMAL EXPERIMENTATION AND SCIENTIFIC VALIDITY

We now consider arguments dealing with objections to the scientific validity of animal experiments. Before deciding whether to conduct an experiment on animals, we must decide whether it is morally acceptable to use the animal in the experiment, given the purpose of the study, the experimental design, the methods used, and the species of the animal. If we decide that the animal has some moral value and that the experiment would harm the animal in some way (e.g., by causing pain, suffering, disability, or death), then we must argue that the experiment can be ethically justified, given the expected benefits. Researchers can conduct experiments on animals, provided that they provide a sufficient moral justification for the experiments (LaFollette and Shanks 1996). Because many different animal species may have some degree of moral value, the moral burden of proof rests with researchers who plan to conduct experiments on animals; researchers do not simply have a moral "free ticket" or "blank check" regarding animal experimentation. Moreover, because animal species may differ with respect to their moral worth, an experiment can be morally acceptable in one species but not be morally acceptable in a different species. For example, it may be morally acceptable to create a transgenic mouse that is prone to various forms of cancer (i.e., an oncomouse), but it may not be morally acceptable to create an "oncochimp" or an "oncomonkey," because chimpanzees and monkeys have greater moral value than mice by virtue of their higher degree of similarity to human beings.

Because animal species have some degree of intrinsic value, our view implies a commonly accepted policy known as the "three Rs" of animal experimentation (Russell and Birch 1959):

- *Replacement:* When it is possible to answer a research question without using an animal, replace the animal with a methodology that does not use animals, such as cell studies or computer modeling. When it is possible to answer a scientific question using a morally "lower" species of animal, replace the "higher" species with a lower one.

- *Reduction:* When it is possible to answer a research question using a smaller number of animals, reduce the number of animals used.
- *Refinement:* Wherever possible, refine research methods, techniques, concepts, and tools to reduce the need for animals in research and to reduce harms to animals.

The three Rs can be justified on the grounds that they minimize harm to animals and promote animal welfare within the context of animal experimentation. Of course, these three Rs make sense only if one believes that the research protocols are likely to yield results with scientific, medical, or social value. Thus, a fourth R should also apply to animal research:

- *Relevance:* Research protocols that use animals should address questions that have some scientific, medical, or social relevance; all risks to animals need to be balanced against benefits to humans and animals.

Finally, a fifth R is also important (and plays a key role in U.S. animal research regulations):

- *Redundancy avoidance:* Avoid redundancy in animal research whenever possible—make sure to do a thorough literature search to ensure that the experiment has not already been done. If it has already been done, provide a good justification for repeating the work.

Avoiding redundancy is important to avoid using animals unnecessarily and wasting research resources.

This approach to the ethics of animal experimentation suggests that researchers also need to address the five Rs relating to the scientific validity of every research protocol they propose, because these issues can affect the ethical soundness of the research. This idea is conceptually similar to examining scientific validity of human experimentation: Good experimental design is a moral requirement for human subjects research because a poorly designed experiment may not yield generalizable knowledge and may therefore impose unnecessary risks on human subjects (Levine 1988). Some of the scientific issues that researchers need to address are as follows:

1. *The scientific necessity of the experiment:* If the experiment has already been done, it may or may not be worth repeating. Although it is often important to repeat new experimental findings, unnecessary repetition should be avoided.
2. *The appropriateness of the animal model:* Animal research protocols specify the species used to answer a research questions. These protocols should therefore provide a rationale for the particular species chosen; they should describe how the experiment would provide evidence that is relevant to the research question. For example, if one is interested in learning about the conduction of nerve signals along axons, then a species of squid may be a good animal model because it has long axons that are easy to study. Moreover, knowledge about

squid axons may be generalized to other species. If one wants to know whether a specific chemical is likely to be toxic or carcinogenic in humans, then it is important to use an animal species that is metabolically similar to the human species. For many years, researchers have more or less assumed that toxicity studies in mice can be generalized to humans, but this assumption is now being questioned (Anonymous 1993, LaFollette and Shanks 1996). For example, researchers at one time thought that saccharin can cause bladder cancer in humans based on studies in which mice were fed huge doses of the saccharin and then formed tumors in their bladders. However, we know that laboratory mice have a mechanism of waste elimination that is different from the human mechanism. When mice eliminate saccharin, they build up uric acid crystals in their bladders. Human beings, on the other hand, do not form these crystals. Thus, conclusions about the carcinogenic effects of saccharin in laboratory mice probably do not apply to human beings (Cohen 1995).

3. *The number of animals used:* The principle of reduction implies that researchers should reduce the number of animals used, whenever this does not affect their ability to obtain useful data, but researchers must also make sure that they use enough animals to obtain statistically significant data and results. Otherwise, the experiment causes unjustifiable harm to the animal subjects. Here, good statistical design and ethical practice go hand in hand (see the discussion of statistics in chapter 3).

4. *Efforts to promote animal welfare and reduce animal harm:* Because animal research should promote animal welfare and minimize animal harm wherever possible, researchers need to take measures to minimize pain, suffering, disability, and death. Researchers must consider the appropriate use of analgesia and anesthesia; humane forms of euthanasia, when this is required for pathological findings; appropriate and sterile surgical procedures; disease control; and living conditions, such as nutrition, living space, and exercise. The United States and many other countries have many regulations addressing these issues (discussed below). However, although regulations govern this important aspect of research, we want to stress that these regulations have a sound moral justification; they are not needless rules or "red tape" (LaFollette and Shanks 1996).

5. *Alternatives to the animal model:* As mentioned above, the principle of replacement implies that researchers should find alternatives to animal models, wherever possible. In the last few decades, scientists have made greater strides in developing alternatives to animal models, such as cell and tissue cultures and computer simulations (Office of Technology Assessment 1986, LaFollette and Shanks 1996, Stokstad 1999). We encourage further developments in this direction. However, it is unlikely that researchers will be able to

completely eliminate the need for animal subjects, because many complex physiological, behavioral, and developmental phenomena can be understood only within the context of a whole organism. For example, to understand how a vaccine protects an organism against infection, one must eventually use the vaccine in whole organisms. Inferences about the vaccine from tissue cultures or computer models simply will not provide relevant knowledge about whole organisms. Most research questions related to animal behavior, such as research on Parkinson's disease, obesity, aggression, and addiction, will probably require whole organisms.

CHIMERAS

One of the important techniques researchers have developed in the last decade is using human–animal chimeras to study human diseases. The word "chimera" comes from Greek mythology. A chimera was a monster composed of parts from a lion, human, snake, and goat. In modern biology, a chimera is simply an organism composed of parts from different species or different parts from the same species. Biomedical science and technology have helped to create many different types of chimeras, such as mice with human genes, cells, or tissues containing cells with different genomes. Some humans have pig heart valves to replace their defective valves. Chimeras are playing an increasingly important role in biomedical research, since they can help scientists to study basic biological processes, such as embryonic development and cell differentiation, and to model human diseases, such as cancer, obesity, heart disease, and Parkinson's disease (Robert 2006).

Chimeras raise some interesting and important ethical issues. Animals with human parts could be like human beings in many important ways. It is possible, in theory at least, that an animal–human chimera might be created that is "subhuman." For example, suppose that researchers transfer human genes or stem cells into chimpanzees, which cause the chimpanzee to develop sophisticated linguistic and reasoning skills. How should we treat this chimpanzee? As a chimpanzee or as a human being? Would it have moral or legal rights? What kind of impact would the creation of subhuman have on the idea that human beings have inherent worth and value? Would the creation of subhumans degrade the value we place on human life (Robert and Baylis 2003)? The possibilities are so novel and fantastic that most people have not had time to think clearly about the ethics of creating human–animal chimeras. For this reason, research proposals that aim to create human–animal chimeras need careful review and oversight (Moreno and Hynes 2005).

REGULATIONS

Here we briefly review some of the regulations that govern research on animals in the United States. Many of these rules embody the ethical and scientific considerations mentioned earlier in this chapter. The two main federal laws that govern animal research in the United States are the 1966 Animal Welfare Act (AWA), since revised several times (1970, 1976, 1985, 1996), and the U.S. Public Health Service (PHS) *Policy on the Humane Care and Use of Laboratory Animals* (Rollin 1992, Public Health Service 2000b). The PHS policy applies only to animal research conducted using PHS funds, but the AWA applies to all animal research. Many research institutions and professional societies have their own guidelines for the humane care and use of animals in research (Bennett 1994). There are also private organizations, such as the Association for the Assessment and Accreditation of Laboratory Animal Care (AAALAC), that accredit research institutions. Institutions can legally conduct animal research without AAALAC accreditation, but accreditation stills plays an important role in shaping research practices, because many research sponsors require AAALAC accreditation (Association for the Assessment and Accreditation of Laboratory Animal Care 2007). States also have their own laws pertaining to animal welfare, but we do not review them here because these laws all allow animal research that conforms to the federal laws.

The PHS policy applies to all vertebrate animals used in research supported by PHS funds. The NIH is one of the units of the PHS and the largest one supporting research with animals. The PHS policy requires compliance with AWA and the National Research Council (NRC) *Guide for the Care and Use of Laboratory Animals* (1996). The NIH Office for Laboratory Animal Welfare (OLAW) is responsible for monitoring recipients of PHS funds for compliance with the regulations, which takes the form of a written assurance of compliance by the research institution. OLAW can inspect animal care facilities "for cause."

The AWA, originally titled the "Laboratory Animal Welfare Act," promulgates rules that apply to the commercialization and use of animals, including the use animals in research. Animals originally covered by the AWA included (nonhuman) primates, dogs, cats, guinea pigs, hamsters, and rabbits (Bennett 1994). In 1970 the AWA was expanded to cover all warm-blooded animals, but in 1976 the Secretary of Agriculture excluded rats, mice, birds, farm animals, and horses from being covered under the AWA. The U.S. Department of Agriculture (USDA) oversees compliance with AWA. In 2000 the USDA announced a plan, later suspended, to once again include rats, mice, and birds under the AWA (Malakoff 2000b). Although some researchers have objected to including rats, mice, and birds under the AWA, polls suggest that many researchers would welcome this change (Plous and Herzog 2000).

Both the AWA and the PHS policies stipulate rules and guidelines for the humane care and use of animals in research. These regulations address a variety of issues, including living conditions, such as cage sizes, temperature, environmental stimulation, food, and exercise; efforts to reduce pain and discomfort, including analgesia and anesthesia; surgical procedures, including antisepsis and euthanasia; veterinary care and disease prevention and control; procurement and transportation; and the qualifications, health, and safety of personnel. The NRC's *Guide for the Care and Use of Laboratory*

Animals (1996) spells out many of the standards for animal experimentation contained in AWA and PHS regulations. The NRC established the Institute for Laboratory Animal Research to study, prepare, and distribute documents on the care and use of animals in the research community. The AAALAC also considers the NRC guide when deciding whether to accredit an institution. All researchers and students are advised to consult the NRC guide when preparing animal experimentation protocols and proposals (Bennett 1994).

The AWA and PHS regulations also address institutional responsibilities for ensuring and promoting the humane care and use of animals in research. These rules require institutions that conduct animal research to establish an institutional animal care and use committee (IACUC) to review and approve animal research protocols and to monitor animal research. IACUCs are also charged with educating researchers about scientific and ethical aspects of animal research and relevant regulations. IACUC members should include a veterinarian, a scientist, and a community representative. Many IACUCs also include an ethicist or a lawyer. IACUCs have the authority to stop any animal research protocol that does not meet the standards set forth in relevant policies. IACUCs also conduct inspections of laboratories on a biannual basis. IACUCs function much like institutional review boards (IRBs) in that both of these research committees are charged with ensuring institutional compliance with relevant regulations and with protecting research subjects.

QUESTIONS FOR DISCUSSION

1. How can we know whether an animal can feel pain or suffer?
2. What is the difference between the ability to feel pain and the ability to suffer? Do you agree with this distinction? Why or why not?
3. Do you agree with Singer that speciesism is like racism? Do you think speciesism can be justified? Why or why not?
4. What is the moral status of nonhuman animals? Do they have rights? Should we promote their welfare?
5. What criteria should we use to decide whether a being has moral status, that is, whether it is a member of the moral community?
6. Are there some types of animal experiments that you regard as unethical? Why?
7. Do you think that animal research always has important benefits for human beings? Are there some types of research that are not very beneficial?
8. Do you think animal research is overregulated, underregulated, or is regulated just about the right amount? Why?
9. Are there some ways in which we take better care of animals in research than we do humans?
10. Do you think it would be ethical to create human–primate chimera for research purposes?

CASES FOR DISCUSSION

Case 1: Brain Injury Experiments on Baboons

In the spring of 1984, head injury experiments on baboons conducted by investigators at the University of Pennsylvania drew protests from animal rights groups, intense media coverage, and a review by Congress and the NIH, the federal agency sponsoring the research. The primary aim of the experiments was to develop and test a model for brain injuries that occur in automobile accidents and some sports, such as football or boxing. Secondary aims were to measure the effects of the experimentally induced trauma on brain tissue. In the experiments, baboons were outfitted with helmets, which were fastened to their heads with cement. The baboons were sedated with a tranquilizer. A hydraulic piston would hit the helmet with a force two thousand times the force of gravity. The piston was not designed to cause a penetrating wound to the animal's head, but to produce trauma to the soft tissue of the brain. The experiments resulted in brain injuries, paralysis, and coma. The baboons were kept alive for two months following the experiments and then euthanized for pathological analysis. Although the animals were sedated prior to the experiments, some animals were awake when the piston hit their helmet. Some animals received more than one blow to the head (Orlans et al. 1998).

On May 18, 1984, five members of Animal Liberation Front (ALF), an animal rights organization, entered the Head Trauma Research Center at the University of Pennsylvania and stole 30 videotapes the researchers had made of the experiments. They also vandalized equipment in the laboratory and wrote "ALF" on the walls. The ALF gave the videos to PETA, which released excerpts to the media. The video excerpts showed the piston hitting the heads of the baboons, with some of them apparently writhing in great pain and distress. These grotesque images shocked researchers, convinced many people to support the animal rights movement, and instigated public debate about animal research. Researchers and leaders at the University of Pennsylvania defended the experiments as important studies with potential benefits for human health. They also maintained that the experiments were in full compliance with federal regulations. A congressional committee held hearings on the experiments, which included testimony from officials from the NIH, the FDA, and PETA. The NIH, the FDA and the University of Pennsylvania continued to maintain that the experiments were ethical and in compliance with the law. The NIH even renewed the $300,000 grant it had awarded to Thomas Gennarelli, a professor of neurosurgery at the University of Pennsylvania who was the principal investigator for the baboon experiments (Orlans et al. 1998).

PETA and other animal rights groups continued to protest against the experiments. They held peaceful demonstrations at the University of Pennsylvania and the NIH. PETA put together a 20-minute video from 60 hours of stolen videotape titled *Unnecessary Fuss*, after a remark Gennarelli had made about the experiments. In the summer of 1985, animal activists petitioned members of Congress to stop the experiments. Sixty members of Congress signed a letter to the director of the Department of Health and Human Services, Margaret Heckler, asking her to stop funding the experiments. When Heckler refused to stop the experiments, animal activists staged a sit-in at the NIH and occupied several offices. Four days later, Heckler suspended the funding for the study. The NIH then conducted a full investigation of the experiments and found that the University of Pennsylvania had failed to comply with the NIH's animal welfare policies. The investigation determined that the researchers had failed to provide adequate anesthesia, analgesia, nursing care, training, and supervision. The University of Pennsylvania reprimanded Gennarelli and Thomas Langfitt, chair of the Department of Neurosurgery, who was associated with the study. The USDA fined the university $4,000 for violating the Animal Welfare Act. The NIH eventually removed its funding restrictions, but by that time the university had abandoned its primate head trauma program (Orlans et al. 1998).

- Can we justify this type of experiment?
- Did the NIH and the FDA act ethically the first time? Second time?
- How would you have designed the experiment?

Case 2

You are a graduate student conducting an experiment for Dr. Beeson on repairing spinal cord injuries in rats. During the experiments, you sever the spinal cord, allow the rat to heal, determine the degree of paralysis, and then inject neural stem cells into the site. You will use 25 rats in your protocol. In one of the experiments, one of the rats died after surgery, and Dr. Beeson replaced the rat. You asked him what he planned to do in response to this death, and he said nothing. You prod him a bit further, and he says that he knows why the rat died—he gave the rat too much anesthetic—and he doesn't think the death will have any effect on the experiment's results. He also does not want the IACUC to know about the death.

- How should you respond?
- What should the IACUC do if it finds out about this incident?

Case 3

A researcher is planning to test a new analgesic medication in dogs. The medication may prove useful in relieving pain associated with severe burns. To conduct the experiment, she will use two groups of 12 dogs each, an experimental group and a control group. Both groups will receive a burn on the back. Both will receive treatment for the burn, but only one group will receive the medication under study. She will also attempt to measure their degree of pain and discomfort by touching the burn to evoke a response, such as behavioral cues, heart rate, and blood pressure. Would you approve of this experiment? Why or why not?

Case 4

Surgeons are developing a robotic device for performing heart surgery. The device, if successful, would be a safer and more reliable way to perform heart surgery than current human techniques. One of the main benefits of the robotic system is that it minimizes unstable movements. It is also expected to reduce the risk of infection. In developing the device, the surgeons are planning to test it on dogs, cats, and pigs. They also plan to use animals to teach surgeons how to use the device. For each animal, the protocol will include a surgical procedure followed by a postoperative recovery period, followed by euthanasia and pathological analysis. One of the research protocols also involves using the device for emergency surgery to respond to heart trauma, such as gunshot or knife wounds. For this protocol, animals will be shot or stabbed and prepared for surgery. All animals will be given appropriate analgesia and anesthesia. Would you approve of this protocol? Why or why not?

Case 5

Researchers are developing an animal model for a rare genetic disorder known as osteogenesis imperfecta. Children born with this disorder have a genetic mutation that prevents them from forming strong bones. Neonates usually have many broken bones at birth and usually die within 6 months, or within a year at most. They usually also have severe brain damage, organ damage, and a variety of infections. Researchers plan to develop a transgenic mouse with a genetic defect that results in a condition similar to osteogenesis imperfecta. They will study the etiology of the disease in mice as well as a variety of treatment modalities. Do you have an ethical problem with this experiment? Why or why not?

Case 6

A researcher is conducting experiments on tumor suppression with transgenic oncomice. In his experiments, he allows the mice to develop cancerous tumors, and he then treats them with different gene therapy vectors designed to inhibit cell cancer growth. Two graduate students and two lab technicians work with him on his project. Last week, one of his graduate students informed the IACUC chair that several mice had developed tumors that impeded their free movement in the cages. The tumors were growing around their front legs and on their chests. This was a violation of the researcher's animal protocol. The graduate student also reported that the researcher has been out of town consulting with biotech companies and has not been paying close attention to work being conducted in his lab. How should the IACUC handle this problem?

Case 7

Dr. Murphy is studying aggression in rats. She is experimenting with a variety of environmental and hormonal influences on aggression in order to study the relationship between aggression and overcrowding, stress, food deprivation, and levels of sex hormones. In some of her experiments, rats will be placed in crowded cages; in others they will be deprived of food; in others they will receive high doses of testosterone. She will attempt to minimize pain and discomfort to the rats, but they are likely to fight and injure each other. Would you allow Dr. Murphy to do this experiment?

12

The Protection of Human Subjects
in Research

> This chapter discusses the history of human experimentation, giving special attention to cases that have helped to shape ethical guidelines and policies. It discusses important ethics codes and provides an overview of U.S. federal regulations. The chapter also addresses some key concepts and principles in human research, such as informed consent, risk/benefit ratios, minimal risk, and research versus therapy.

The use of human subjects in research came into sharp focus during the Nuremberg war crimes trials, when the world discovered the atrocities committed by Nazi doctors and scientists on tens of thousands of prisoners held in concentration camps. While these tribunals were unfolding, the American Medical Association (AMA) was developing a set of principles to be followed in experiments using human subjects (Advisory Committee on Human Radiation Experiments 1995). After the tribunals concluded in 1947, the research community adopted the world's first international code for research on human subjects, the Nuremberg Code (Advisory Committee on Human Radiation Experiments 1995, p. 103). The code emphasized the importance of fully informed and voluntary consent of research subjects, minimization of harms and risks to subjects, scientific validity of research design, and the social value of the research. Since Nuremberg, many documented cases of unethical or questionable research have also been conducted in the United States and other countries (Capron 1989, Advisory Committee on Human Radiation Experiments 1995, Beauchamp and Childress 2001, Washington 2006). There have also been many ethical controversies in human subjects research, such as the use of placebos, the nature and practice of informed consent, and research on vulnerable populations, such as children, prisoners, economically disadvantaged people, and the mentally ill (Shamoo and Irving 1993, Egilman et al. 1998a,b, Washington 2006). As a result, various federal agencies and scientific and professional associations have developed regulations and codes governing human subjects research, and there has been a great deal of discussion and debate about ethical standards that should govern the use of humans in research (Levine 1988, Pence 1996). This chapter reviews these regulations after providing a historical and philosophical perspective on these issues.

HUMAN EXPERIMENTATION BEFORE WORLD WAR II

Alexander Morgan Capron (1989, p. 127) has observed that "the darkest moments in medical annals have involved abuses of human research subjects."

A brief survey of the history of human subjects research supports this view. Before the Scientific Revolution (ca. 1500–1700 A.D.), medicine was an observational rather than experimental science. Medical research was based on the teaching of Hippocrates (460–377 B.C.), the father of scientific medicine. Hippocrates developed theories and principles that explained diseases in terms of natural rather than supernatural causes. According to his teachings, health was a state of balance among the four humors of the body, blood, phlegm, yellow bile, and black bile. Disease occurs when the body becomes out of balance as the result of too much or too little of one or more humors. The goal of medicine is to use various treatments and therapies to restore the body's proper balance. For example, Hippocratic physicians believed that bloodletting could restore health by eliminating excess blood.

Hippocrates' method was observational rather experimental because he did not use controlled interventions (or experiments) to obtain medical knowledge. Instead, Hippocrates gathered knowledge through careful observation of disease conditions, signs, symptoms, and cures. He also developed detailed case histories. Hippocratic physicians believed in the body's ability to heal itself, and they tended to prescribe nonaggressive and noninterventional therapies, such as special diets, herbal medications, exercise, massage, baths, and prayer. The Hippocratic School developed a code of medical ethics that emphasized the importance of promoting the welfare of the individual patient. Two of the Hippocratic Oath's key tenets, which evolved hundreds of years after Hippocrates' death, are to keep patients from harm and injustice ("do no harm") and to benefit the sick. Although Hippocratic physicians sought to improve medical knowledge, their code of ethics and their philosophy of medicine implied that medical advances would occur slowly and would not sacrifice the welfare of the individual patient for scientific progress (Porter 1997).

This conservative approach to medical research began to change during the Scientific Revolution, as physicians such as Paracelsus (1493–1542), Andreas Vesalius (1514–1614), and William Harvey (1578–1657) (see Harvey 1628 [1993]) challenged medical dogmas and sought to apply the new experimental method to medicine. However, these physicians still did not conduct many controlled experiments on human beings. Although Paracelsus, Vesalius, and Harvey dissected human bodies, they did not gain their knowledge of anatomy from experiments on living human beings. While Harvey conducted some experiments on human beings, his experiments were relatively benign and noninvasive. For example, he used a tourniquet to demonstrate the direction of the flow of blood in human veins, and he measured pulse and blood pressure. He conducted his more invasive procedures, such as vivisections, on animals. As physicians began to apply the experimental method to medicine, experiments on human beings became more common and more risky. One famous eighteenth-century experiment conducted by the English physician Edward Jenner (1749–1823) illustrates some recurring ethical concerns. Jenner observed that dairymaids who developed cowpox did not develop smallpox. He hypothesized that cowpox provided an inoculation against smallpox. To test his hypothesis, he inoculated James Phipps, an eight-year-old boy, with some material from a cowpox pustule. The boy developed a slight fever but suffered no other ill effects. Six weeks after this inoculation, Jenner exposed Phipps to the smallpox virus and he did not develop the disease (Porter 1997).

During the nineteenth century, experiments on human beings became even more common. For example, William Beaumont (1785–1853) treated Alexis St. Martin for a bullet wound in the stomach. The wound healed but left a hole in the stomach. Beaumont hired Martin as a servant and used him as an experimental subject, because he could observe the process of digestion through the hole in Martin's stomach (Pence 1995). During the twentieth century, physicians began to accept the germ theory of disease developed by Louis Pasteur (1822–1896) and Robert Koch (1843–1910). Despite Pasteur's unquestioned place in science, there are now historical studies indicating that his behavior was not above ethical reproach. For example, Pasteur treated a patient for rabies without first ensuring the safety of the treatment in animal experiments (Geison 1978). The surgeon Joseph Lister (1827–1912) performed a variety of experiments to develop and test antiseptic methods in medicine. For instance, Lister observed that carbolic acid was effective at reducing infections among cattle, and he hypothesized that this compound has antiseptic properties. To test his idea, he applied lint soaked in carbolic acid and linseed oil to a boy's wound. He also took measures to prevent germs from entering the wound. The boy, James Greenlees, did not develop an infection. Lister applied his method to dozens of other cases of compound fractures and amputations and published his results in *Lancet* in 1867 (Porter 1997).

One of the most disturbing experiments in the United States before World War II took place in 1874 in Cincinnati, when Robert Bartholomew inserted electrodes into the brain of Mary Rafferty, a 30-year-old "feeble-minded" patient who was dying of terminal cancer, which had spread to her scalp. Bartholomew saw a research opportunity and for several hours electrically stimulated Rafferty's brain and recorded her responses, which were often cries of pain (Lederer 1995).

Many of the human experiments were inspired by the work of Pasteur and Koch, who developed vaccines for bacterial infections. To implement this methodology, researchers needed to establish a link between a pathogen and a disease, isolate a disease pathogen, develop a vaccine, and then test the vaccine. In 1895, Henry Heiman, a New York pediatrician, infected two mentally retarded boys, 4 and 16 years old, with gonorrhea. In 1897, the Italian researcher Giuseppe Sanerilli injected yellow fever bacteria into five subjects without their consent in order to test its virulence. All five subjects became severely ill, although none died (Lederer 1995). Many physicians, including William Osler (1849–1919), condemned this experiment. In his textbook *The Principles and Practice of Medicine* (1898), Osler discussed Sanerilli's experiments as well as some other studies of yellow fever.

U.S. Army physician Walter Reed and his colleagues in Cuba conducted his famous yellow fever experiments around 1900. Yellow fever had become a major health problem for military operations in Cuba, the Caribbean, and Central America. At the time, researchers hypothesized that yellow fever was transmitted to humans by the *Aedes aegypti* mosquito. Because there were no animal models for the disease, human subjects were required to study its transmission. The risks to human subjects were great, because medicine had no cure for the disease, which often resulted in death. Two investigators working with Walter Reed, James Carroll and Jesse Lazear, allowed themselves to be bitten by mosquitoes in order to test the hypothesis. Reed had also agreed to participate in these experiments, but he was in Washington when his colleagues exposed themselves to the disease. Both colleagues contracted yellow fever, and Lazear died from the disease. After Lazear died, Reed decided not to use himself as an experimental subject, but he continued experimenting on human beings to develop a vaccine. Because the risks of participating in these experiments were so great, Reed and his colleagues had volunteers sign written contracts stating that they understood the risks of the experiment and that they agreed to participate. Volunteers were also given $100 in gold and free medical care for their participation. Although other researchers obtained undocumented informed consent from subjects, this is believed to be the first case of the documentation of informed consent in research. Research subjects who participated in subsequent yellow fever experiments came to be regarded as heroes and martyrs. Surviving volunteers (all soldiers) received gold medals and government pensions (Lederer 1995).

There have also been unethical experiments on vulnerable African Americans. Harriet Washington's 2006 book *Medical Apartheid* cites many examples of such experiments. For instance, in the early 1800s, 250 out 251 subjects in an experiment testing inoculations of smallpox vaccine were African Americans. In 1846, Dr. Walter F. Jones of Virginia poured boiling water on patients with typhoid pneumonia. Washington describes many dangerous and humiliating experiments on African-American slaves. The author recognizes that progress has been made in research with black populations. She ends her book with grave concerns regarding high-risk experiments that have moved recently to Africa to exploit vulnerable blacks.

Before World War II, physicians and surgeons had ambivalent attitudes toward human experimentation. On the one hand, most physicians accepted the Hippocratic idea that they should not harm their patients. Claude Bernard (1813–1878) restated the principle in his *Introduction to the Study of Experimental Medicine* (1865 [1957]). According to Bernard, physicians should never perform on humans an "experiment, which might be harmful to him to any extent, even though the result might be wholly advantageous to science" (p. 101). On the other hand, physicians regarded many risky and untested interventions as therapeutic and believed that it was sometimes necessary to try these treatments in order to benefit the patient. While physicians condemned many of the unethical experiments that were brought to their attention, they also had a strong commitment to medical research and experimentation and did not want to place any burdensome restrictions on research. Most physicians thought that self-experimentation was noble and virtuous, but they did not think that informed consent was always necessary. Indeed, most physicians at the time thought that it was more important to avoid harming the research subject than to obtain the subject's consent. For several decades, the AMA considered adopting a code of ethics for research on human subjects, but it did not do so until 1946 (Lederer 1995).

In 1900, Prussia was the first nation in the world to formalize the prohibition of medical interventions other than for therapeutic purposes (Capron 1989). The Prussian directive required that consent be given and that prospective subjects be informed of adverse consequences. It also excluded minors from research. These directives were not given in a vacuum or without a cause: They came as a reaction to numerous and repeated abuses of patients in medical research. For example, the discoverer of the bacillus strain that causes leprosy, Amauer Hansen (1841–1912), carried out an appalling experiment on an unwitting 33-year-old woman when he twice pricked her eye with a needle contaminated by nodules of a leprous patient (Bean 1977). Hansen was later merely reprimanded.

The conduct of Walter Reed and his colleagues stands in sharp contrast to some egregious examples discussed above. Reed was one of the medical officers during the construction of the Panama Canal. The project required a great deal of manpower, but numerous workers were dying from yellow fever. Reed developed a vaccine for yellow fever and tested it on the canal workers. None of Reed's 22 subjects who contracted yellow fever died from it, and thousands of workers were able to avoid the infection. Reed demonstrated a genuine concern for the welfare of his fellow human beings and for informed consent (Bean 1977). Reed himself prepared English and Spanish versions of a contract signed by each individual volunteer. His informed consent document and the process he followed were exemplary and are considered by some to be as good as today's methods and documents. Although some scholars claimed that the ethical/legal doctrine of informed consent evolved in the 1950s and 1960s (Advisory Committee on Human Radiation Experiments 1995), Reed's work shows that he followed this paradigm before it became more broadly accepted.

In the early 1900s, the eugenics movement flourished in Europe and in the United States (for further discussion, see chapter 14). In the 1930s, one Canadian province and 28 U.S. states passed laws requiring the sterilization of the criminally insane, presumed "feeble-minded," psychopathic personalities, and the mentally ill (Proctor 1988, Ollove 2001). By the late 1930s, California alone had sterilized 13,000 persons, and the U.S. total is estimated at between 30,000 and 100,000 persons (Proctor 1988, Ollove 2001). The State of Virginia in the early twentieth century was a leader in sterilization efforts. *Baltimore Sun* reporter Michael Ollove chronicled the ordeal of a Virginian who was sterilized for being "feeble-minded." Later, this Virginian became a soldier, winning the Purple Heart, the Bronze Star, and Prisoner of War honors during World War II (Ollove 2001). The eugenics movement helped provide impetus for the Nazi atrocities committed in World War II. Hitler was a strong advocate of eugenics, and he believed it was necessary to control human breeding in order to prevent the Aryan race from being corrupted by inferior races, such as the Jews and Gypsies (Proctor 1999).

HUMAN EXPERIMENTATION AFTER WORLD WAR II

The Nazi experiments conducted on human beings have been regarded by many as the worst experiments ever performed on human subjects. None of the subjects gave informed consent, and thousands of subjects were maimed or killed. Many of the experiments were not scientifically well designed or conducted by personnel with appropriate scientific or medical qualifications. Moreover, these experiments were planned, organized, and conducted by government officials. Subjects included Jews, homosexuals, convicted criminals, Russian officers, and Polish dissidents. Some of the experiments included the following (Proctor 1988, Caplan 1992, Müller-Hill 1992, Pence 1995):

- Hypothermia studies where naked subjects were placed in freezing cold water
- Decompression studies where subjects were exposed to air pressures equivalent to the pressures found at an altitude of 70,000 feet
- Wound healing studies, where subjects were shot, stabbed, injected with glass or shrapnel, or otherwise harmed to study how their wounds healed
- Vaccination and infection studies, where subjects were intentionally infected with diseases, such as typhus, staphylococcus, malaria, and tetanus, in order to test the effectiveness of vaccines and treatments
- Josef Mengele's (1911–1979) experiments designed to change eye color, which resulted in blindness
- Mengele's human endurance experiments, where subjects were exposed to high levels of electricity and radiation
- Mengele's twin studies: exchanging blood between identical twins, forcing fraternal twins to have sex to produce children, creating conjoined twins by sewing twins together at the back, placing children in virtual isolation from birth to test the role of nature and nurture in human development

Although historians and ethicists have focused on Germany's horrific experiments with human subjects during World War II, less attention has been given to Japan's atrocities during this era. From 1932–1945, Japanese medical researchers killed thousands of human subjects in medical experiments. Most of the experiments took place in China while this country was under Japanese occupation. The experiments included intentionally wounding and operating on human beings for surgical training, vivisection of live humans, infecting humans with pathogens, exposing subjects to extremes of temperature, and biological and chemical warfare research. Most of the human subjects were people of Chinese ancestry, but victims also included Allied prisoners of war. At the end of the war, the U.S. government made a deal with Japan to gain access to the data from chemical and biological warfare experiments. In exchange for this data, the U.S. government agreed to not prosecute Japanese physicians and scientists for war crimes. As a result of this cover-up, the Japanese atrocities were not widely known until the 1990s, and Japanese political leaders have been reluctant to acknowledge that these crimes against humanity occurred (Tsuchiya 2008).

By the mid-twentieth century, human experiments, ethical and otherwise, were becoming more common, but the research community had not put a great deal of thought into the ethics of research on human subjects. Although some physicians, most notably Bernard and Osler, had written about the ethics of human experimentation, and the AMA had drafted some documents on human experimentation, there were no well-established ethical codes for experimentation on human subjects before 1947. This is one reason that the Nuremberg Code has such an important place in history: It was the first internationally recognized code of ethics for human research.

Although the Nuremberg Code did help to define and clarify some standards for the ethical conduct of human experiments, many abuses took place after the code was adopted. Some of these ethical problems in research were discussed by Henry Beecher (1904–1976) in an exposé he published in the *New England Journal of Medicine* in 1966. Beecher described 22 studies with ethical violations, including the now well-known Tuskegee syphilis study, the Willowbrook hepatitis experiments on mentally disabled children (see chapter 13), and the Jewish chronic disease case study.

The Tuskegee study took place from 1932 to 1972 in a public health clinic in Tuskegee, Alabama. The purpose of the study was to follow the natural progression of later-stage syphilis in African-American men. Six hundred subjects were enrolled in the study, which was funded by the U.S. Department of Health, Education, and Welfare (DHEW), the precursor to the Department of Health and Human Services (DHHS). The subjects were divided between an "experimental" group of 399 subjects with untreated syphilis and a "control" group of subjects without syphilis. The initial plan was to conduct the study for one year, but it lasted nearly 40 years. The subjects who participated in the study were not told that they had syphilis, that they were not receiving a medically proven treatment for syphilis, or that they were participating in an experiment. Subjects with syphilis only knew that they had "bad blood" and could receive medical treatment for their condition, which consisted of nothing more than medical examinations. Subjects also received free hot lunches and free burials. An effective treatment for syphilis, penicillin, became available in the 1940s, but the subjects were not given this medication. In fact, there is evidence that the study investigators took steps to prevent subjects from receiving treatment for syphilis outside of study. The study not only had ethical problems, but also had scientific flaws: Key personnel changed from year to year, there were no written protocols, and records were poorly kept. Even though Beecher brought the study to the attention of the public, it was not stopped until Peter Buxton, who worked for the U.S. Public Health Service (PHS), reported the story to the Associated Press. The story soon became front-page news, and a congressional investigation followed. In 1973, the U.S. government agreed to an out-of-court settlement with families of the research subjects, who had filed a class-action lawsuit (Jones 1981, Pence 1995). In 1997, the Clinton administration issued an official apology on behalf of the U.S. government.

The Jewish chronic disease case study took place in Brooklyn, New York, in 1964. In this case, researchers introduced live cancer cells into 22 unsuspecting patients (Faden and Beauchamp 1986). The purpose of the study was to learn more about the transplant rejection process. Previous studies had indicated that healthy subjects and subjects with cancer have different immune responses to cancer cells: Healthy subjects reject those cells immediately, whereas cancer patients have a delayed rejection response. Researchers claimed that they obtained informed consent, but they did not document the consent. They claimed that there was no need for documentation because the procedures they were performing were no more dangerous than other procedures performed in treating cancer patients. Investigators also did not tell the subjects that they would receive cancer cells, in order to avoid frightening them unnecessarily (Levine 1988).

Human radiation experiments took place in the United States from 1944 to 1974, during the cold-war era (Advisory Committee on Human Radiation Experiments 1995). These experiments were funded and conducted by U.S. government officials or people associated with government institutions on more than 4,000 unsuspecting citizens and military personnel. Many of these experiments violated standards of informed consent and imposed or risks on the subjects. Most of these experiments were conducted in order to aid U.S. cold-war efforts by providing information about how radiation affects human health. Most of these studies used radioactive tracers and did not result in significant harm to the subjects. However, several of the studies that involved children exposed them to an increased lifetime cancer risk, and several studies caused death shortly after the administration of radiation.

In 1994, the Clinton administration began declassifying documents related to these experiments and appointed a commission to develop a report on this research, which it issued in 1995. Although the commission openly discussed some ethical problems with the research, it also found that most studies contributed to advances in medicine and public health (Advisory Committee on Human Radiation Experiments 1995, Moreno 2000, Beauchamp 1996, Guttman 1998). It also judged the experiments by the standards that existed at the time that they were conducted: According to the commission, most of these experiments did not violate existing ethical or scientific standards. Nevertheless, as Welsome (1999) observed: "Almost without exception, the subjects were the poor, the powerless, and the sick—the very people who count most on the government to protect them" (p. 7). Some of the more noteworthy studies that came to light that may have violated the existing ethical standards included the following:

- Researchers at Vanderbilt University in the late 1940s gave pregnant women radioactive iron to study the effects of radiation on fetal development; a follow-up study found that children from these women had a higher than normal cancer rate.

- In Oregon State Prison from 1963 to 1971, researchers X-rayed the testicles of 67 male prisoners, who were mostly African Americans, to study the effects of radiation on sperm function.
- During the late 1950s, researchers at Columbia University gave 12 terminally ill cancer patients radioactive calcium and strontium to study how human tissues absorb radioactive material.
- Researchers released a cloud of radioactive iodine over eastern Washington State to observe the effects of radioactive fallout.
- From the 1940s to the 1960s, researchers injected encapsulated radium into the nostrils of more than 1,500 military personnel; many developed nosebleeds and severe headaches after exposure.

Perhaps the most troubling aspect of these studies is that most of them took place after the international community had approved the Nuremberg Code. It is ironic that the U.S. government, which had been so outspoken in its criticism of Nazi research, would also sponsor human experiments that many would consider unethical (Egilman et al. 1998a,b).

The Jesse Gelsinger case, mentioned several times in this book, has had a significant impact on the ethics and regulation of research with human subjects. Gelsinger was an 18-year-old youth with a genetic liver disease, ornithine transcarbamase deficiency (OTD). People with OTD lack a functional copy of gene that codes for the liver enzyme ornithine transcarbamase. Although most people with OTD die in infancy, Gelsinger's disease was well managed through strict dietary control and pharmacotherapy. James Wilson, a gene therapy researcher at the University of Pennsylvania, asked Gelsinger to participate in an experiment to transfer a functional copy of the gene that codes for ornithine transcarbamase into patients with OTD. Gelsinger agreed to participate in the study. The study was a phase I trial, so there was little chance that Gelsinger would receive a medical benefit from participation. In the experiment, the investigators infused a maximum tolerable dose of the andenovirus vector into Gelsinger's liver. Unfortunately, Gelsinger died from a massive immune reaction shortly after receiving this infusion (Marshall 2000a, Greenberg 2001, Gelsinger and Shamoo 2008).

Gelsinger's death made national headlines and resulted in a lawsuit against Wilson and the University of Pennsylvania and investigations from the Office for Human Research Protection (OHRP), the National Institutes of Health (NIH), and the U.S. Food and Drug Administration (FDA). These investigations found several ethical and legal problems with the experiment. First, the investigators had not adequately informed Gelsinger about the risks of the study, because they had not told him that several monkeys had died after receiving an adenovirus infusion. The investigators had also failed to adequately inform Gelsinger about their financial interests in the research. Wilson had more than 20 gene therapy patents and owned 30% of the stock in Genovo, a small startup company that sponsored the research.

The University of Pennsylvania also had stock in the company. Second, Gelsinger's death, which was an adverse event, was not reported to the proper authorities in a timely fashion. Investigators and sponsors in clinical trials are legally and ethically obligated to report adverse events (Marshall 2000a, Greenberg 2001). Moreover, under pressure from industry, the FDA did not create a system to track and share adverse events of gene therapy subjects with other industry stakeholders (Gelsinger and Shamoo 2008). It was found, after Gelsinger's death, that there have been 700 adverse events among gene therapy subjects. The FDA actually received about 95% of the reported adverse events without sharing them with others. In response to Gelsinger's death, the NIH Recombinant DNA Advisory Committee revaluated phase I gene therapy studies, and the FDA, OHRP, and NIH tightened rules for reporting adverse events. The FDA, OHRP, PHS, and other organizations issued new guidelines concerning conflicts of interest (COIs) in research (see chapter 8).

Besides these important cases from the history of biomedical research, there have also been some noteworthy cases in social science research. One of the methodological problems with social science experiments, known as the Hawthorne effect, is that research subjects may change their behavior as a result of knowing that they are participating in an experiment. As a result, the experiment may be biased. To get around this bias, many social science researchers believe that it is sometimes necessary to deceive human subjects about the experiments in which they are participating. This is precisely what Stanley Milgram did in his 1960s experiments relating to obedience of authority. In one of these experiments, Milgram used two subjects, a "teacher" and a "learner." The teacher was led to believe that the purpose of the experiment was to test the effects of punishment on learning. The teacher provided the learner with information that the learner was supposed to recall. If the learner failed to learn the information, the teacher was instructed to give the learner an electric shock. The severity of the shock could be increased to "dangerous" levels. In some designs, teachers and learners were in different rooms; in other designs, teachers and learners could see each other. In reality, the learners never received an electric shock, and they faked agony and discomfort. Milgram was attempting to learn about whether the teachers would obey the authorities, that is, the researchers who were telling them to shock the learners (Milgram 1974). At the end of each session, Milgram debriefed his subjects and told them the real purpose of the experiment. Many of the teachers said that they suffered psychological harm as a result of these experiments because they realized that they were willing to do something that they considered immoral (Sobel 1978).

Another noteworthy case of deception in social science research took place in Wichita, Kansas, in 1954. During these experiments, investigators secretly recorded the deliberations of six different juries in order to gain a better understanding of how juries make their decisions. The judges of the Tenth Judicial Circuit and the attorneys in the cases approved of the study, although the litigants were not told about the study. When this study came to light, the integrity of the jury system was cast in doubt. In 1955, a subcommittee of the Senate Judiciary Committee held hearings to assess the impact of this research on the jury system. As a result of these hearings, Congress adopted a law forbidding the recording of jury deliberations (Katz 1972).

ETHICAL DILEMMAS IN RESEARCH WITH HUMAN SUBJECTS

Research on human subjects creates many different ethical dilemmas. We discuss a number of these below.

The Good of the Individual versus the Good of Society

Human experimentation raises an ethical dilemma addressed by moral philosophers since antiquity—the good of the individual versus the good of society. According to all of the moral theories discussed in chapter 2, human beings have moral worth and moral rights. On the other hand, most theories also stress the importance of promoting social welfare. Scientific and medical research can promote many important goals that enhance social welfare, such as human health, education, control of the environment, and agriculture. It is important to use human subjects in research in order to gain scientific and medical knowledge, but this also places people at risk and may violate their dignity or rights. Thus, a central ethical question in all research with human subjects is how to protect the rights and welfare of individuals without compromising the scientific validity or social value of the research. Different authors have responded to this dilemma in different ways. Those who take a very conservative approach argue that we should never compromise human dignity, rights, or welfare for the sake of research (Jonas 1992). Others take a more progressive approach and argue that although it is important to protect human subjects, to advance research we must sometimes give the interests of society precedence over the interests of individuals (Lasagna 1992).

This tension between the individual and society occurs in many different aspects of research ethics. For instance, most people would hold that it is acceptable to use an individual in research if he or she gives informed consent. Informed consent respects the rights and welfare of the individual. But it is often difficult for patients to understand the benefits and risks of participating in a study, or the nature of the study and its scientific importance. A variety of influences may also affect the subject's ability to make a perfectly free choice, such as the hope of a cure, money, and family expectations. Those who place a high priority on protecting individuals will insist on a high standard of informed consent in research, while those who do not want to undermine research may be willing to accept standards of informed consent that are less demanding, claiming that they are more realistic. We may have no research subjects at all if we insist that subjects be *completely* informed and uninfluenced by any coercive factors (Inglefinger 1972).

In some types of research, such as social science experiments, it may not be possible to achieve unbiased results if subjects are informed of the nature of the research, because the subjects may change their behavior if they know the hypothesis that the experiment is trying to test. In his study on obedience to authority, Milgram maintained that he needed to deceive his subjects in order to obtain unbiased results. Although his experiments may have obtained socially important results, they caused some psychological harm to the individual subjects. One might also argue that these experiments violated the dignity of the subjects because they were manipulated and deceived (Sobel 1978).

The good of the individual versus the good of society is also an issue in the use of experimental and control groups in research. Suppose a clinical trial is testing a new type of HIV treatment regimen. Subjects are randomized into two groups: one that receives the standard HIV regimen, which includes a combination of several antiretroviral drugs, and another that receives a different combination of drugs, which includes a new drug. If the study starts to yield positive results, at what point should researchers stop the study and make the new treatment regimen available to all patients? Stopping the trial sooner than planned may save the lives of some of the patients in the study. However, stopping the trial too soon may keep the study from yielding statistically significant data, which would prevent researchers from saving the lives of many other patients.

The good of the individual versus the good of society is also an issue in the clinical trials that use placebo control groups: Using placebo control groups may enhance the study's scientific validity at the expense of the welfare of those individuals who receive placebos, because the individuals will not be receiving an effective therapy. (Chapter 15 discusses placebos in research in more depth.) Randomization, double-blinding, strict adherence to the protocol, and many other issues related to the methodology of clinical trials involve conflicts between the good of the individual and the good of society (Kopelman 1986, Schaffner 1986, Friedman 1987, Menikoff 2006).

Weighing Risks versus Benefits

Conflicts between the good of the individual and the good of society frequently arise in weighing the benefits and risks of a research protocol. In assessing a research protocol involving human subjects, one must ask whether the benefits of the protocol to the individual and society outweigh the risks to the individual (Levine 1988). Risk is a function of both the *probability* and the *magnitude* of the harm: A study that has only 2% probability of causing death may be deemed very risky due to the severity of this harm, whereas a study that has a 60% probability of causing a side effect, such as dizziness or headache, may be deemed not very risky due to the relatively benign nature of the harm. In contrast, most benefits are not usually analyzed in probabilistic terms. We believe the concept of "opportunity" could fill this conceptual role: An opportunity is a function of the probability and magnitude of a benefit. A study that has only a small probability of a great benefit, such as a cure from a fatal illness, may be judged as presenting a great opportunity, whereas a study that has a high probability of a minimal benefit may be judged as presenting only a moderate opportunity.

Although assessments of statistical probabilities are empirical or scientific judgments, evaluations of harms or benefits are ethical or value judgments (Kopelman 2000a,b). Therefore, ethical decision making is essential to all calculations of risk versus benefit. For example, phase I clinical trials often raise important risk/benefit questions (Shamoo and Resnik 2006b). In phase I drug studies, investigators are measuring the pharmacological and toxicological effects of a drug to estimate a maximum tolerable dose. Although some phase I studies are conducted on very sick patients, such as cancer patients (see the Jesse Gelsinger discussion), the overwhelming majority of the subjects in phase I trials are healthy subjects who receive no medical benefits from participating. Usually, healthy research subjects are paid for their participation, so they derive some financial benefit. Subjects also provide their informed consent, so this is not usually the main issue. The issue often boils down to this: What level of harm should a subject withstand in the name of research? Most would agree that phase I studies on healthy subjects should not be conducted if there is even a small probability of death, but what about other harmful effects, such as cancer, kidney failure, or allergic shock?

A recent lapse in ethical conduct of a phase I trial occurred in the TeGenero trial in England. Six healthy volunteers received the active drug, and two received a placebo. The active drug was a monoclonal antibody known as TGN1412, a presumed stimulator of the production of the T-lymphocytes. The six volunteers who received TGN1412 developed a severe immune reaction and became very ill, with multiple organ failures. Three of these subjects nearly died. There were serious lapses in the design and safety considerations in this trial. For example, the investigators did not wait long enough before administering doses to new volunteers. When testing a drug on a human being for the first time, the usual procedure is to wait and see how the first volunteer reacts to the drug before administering it to other volunteers. The researchers in this study did not observe this safety rule (Shamoo and Woeckner 2007).

Pesticide experiments on human subjects have generated considerable controversy. In these experiments, private companies have exposed healthy adult human subjects to pesticides to generate safety data to submit to the U.S. Environmental Protection Agency (EPA). The EPA has had to decide whether to accept this type of data for regulatory purposes. The agency has formulated new regulations pertaining to "third-party" research, or research sponsored by private companies to submit to the EPA. One of the key issues in pesticide experiments is whether the benefits to society of the research, such as better knowledge of pesticides for regulatory and public health purposes, outweigh the risks to human subjects. Environmental groups have opposed all pesticide research involving human subjects, but a report by the Institute of Medicine (IOM) determined that some types of low-risk pesticide experiments are acceptable, if they meet stringent ethical and scientific standards (Robertson and Gorovitz 2000, Resnik and Portier 2005, Krimsky and Simocelli 2007).

In medicine, there is also a long and honorable history of self-experimentation. One could view self-experimentation as a highly altruistic act and morally praiseworthy. However, are there limits on the risks that a researcher may take in the name of science? A few years ago, a group of researchers said they would test an HIV/AIDS vaccine on themselves (Associated Press 1997). Although this is certainly a worthy cause, one might argue that these researchers should not be allowed to take the risk of contracting HIV/AIDS. Even when risks are not likely to be great, controversies can still arise if the subjects are not likely to derive any benefits from research participation and cannot provide adequate informed consent. Also, there can be design problems with self-experiments, such as a small sample size and bias (Davis 2003).

Just Distribution of Benefits and Harms

Human experimentation also raises important dilemmas in the distribution of benefits and harms. Although these issues have existed since human experiments first began, they have not received a great deal of attention in the literature so far (Kahn et al. 1998). The benefits of research participation may include access to new medications or treatments, knowledge about one's physical or mental condition, money, and even self-esteem. The harms include many different physical effects, ranging from relatively benign symptoms, such as dizziness or headache, to more serious effects, such as renal failure or even death. Research also may involve legal, financial, or psychosocial risks. For many years, researchers and policy makers have focused on the need to protect human subjects from harm. This attitude was an appropriate response to the harmful research studies discussed above, such as the Tuskegee study, and many research policies and practices that did not protect some vulnerable subjects from harm.

The policy of protecting vulnerable subjects from harm has also led to the exclusion of children, women, and other groups from research, often to their detriment. In the 1980s, women's interest groups objected to policies that excluded women from research on the grounds that these policies were harmful to women's health. They lobbied Congress and the NIH, which changed its policies in response to this political pressure (Dresser 1992, 2001). Groups representing the interests of children have also demanded that drug companies sponsor more pediatric clinical trials—90% of children's prescriptions are prescribed on an off-label basis because there have not been enough studies on how drugs affect children. The idea that one can extrapolate pediatric drug dosing from adult dosing is a faulty assumption, because children often react differently to drugs than adults (Friedman Ross 2006).

Research versus Practice

An important conceptual issue that has a direct bearing on many of the difficult dilemmas in human experimentation is the distinction between research and practice. Health care professionals often perform interventions on patients that are innovative or unproven. For example, a surgeon may try a new surgical technique in performing an emergency tracheotomy, or a general internist may tailor her medication orders to meet the needs of her particular HIV patients and may use nonstandard doses or combinations of drugs. If these interventions were conducted in order to benefit the patient, they have a reasonable chance of benefiting the patient, and they are not part of a research study, then many would hold that these interventions are not research—they are simply innovative practices. As such, they do not need to conform to standards of research ethics, but they should be based on standards of acceptable medical practice. If, on the other hand, interventions are conducted to develop generalizable knowledge, then they may be regarded as research (National Commission 1979, President's Commission 1983a,b).

Although distinctions between research and practice make sense in the abstract, they become blurry in concrete cases. For example, consider the case of Baby Fae, an infant born with a defective heart who received a baboon heart when no human hearts were available (Pence 1995). The cross-species transplant (or xenograft) was conducted despite very low odds of success. She lived with a baboon heart from October 26 to November 15, 1984. Clearly, this was a highly innovative procedure that probably benefited transplant science much more than it benefited the patient. Or consider the birth of Louise Brown, the world's first test tube baby, in 1978. No one had performed in vitro fertilization in human beings before she was conceived. An obstetrician, Patrick Steptoe, performed this procedure in order to benefit his patient, Lesley Brown, who was suffering from fertility problems (Pence 1995). Health care quality improvement studies also straddle the border between research and practice. For example, suppose several hospitals plan to study procedures for reducing medication errors. Their study will compare different procedures for reducing errors. The hospitals plan to collect and analyze the data and publish it in a journal. Although the goal of this study is to determine the best way to reduce medication errors, one might consider it to be a form of research (MacQueen and Buehler 2004).

One might argue that innovations in practice, such as the cases of Baby Fae and Louise Brown, should be viewed as research and should be regulated as such. To protect patients/subjects from harm or exploitation, these innovations in practice should be developed into formal research protocols. Indeed, some research ethics codes, such as the Helsinki Declaration (World Medical Association 2004), encourage clinicians to perform their innovations under research protocols. However, because innovations in medical practice are fairly commonplace and often benefit patients, and clinicians would rather avoid the administrative burden of developing a protocol and submitting it to an institutional review board (IRB) for review, it will continue to be difficult to compel clinicians to meet this ethical standard (King 1995).

BALANCING ETHICAL PRINCIPLES

The issues we have just discussed are but a sampling of the many different ethical dilemmas and questions that arise in research with human subjects.

Below we discuss some ethics codes and regulations that deal with some of these issues. But before we examine these, let us first examine the moral or ethical principles these codes and regulations are founded on. The Belmont Report, an influential document authored by the National Commission for the Protection of Human Subjects of Biomedical and Behavioral Research, provides an answer to these questions (National Commission 1979). According to the Belmont Report, there are three basic ethical principles for research with human subjects: respect for persons, beneficence, and justice. Respect for persons requires researchers to protect the autonomy and privacy of competent research subjects and to provide protections for subjects who cannot make their own decisions. Beneficence requires researchers to minimize risks to subjects, maximize benefits to subjects (and society), and ensure that the balance of benefits to risks is reasonable. Justice requires researchers to ensure that the benefits and burdens of researcher are distributed fairly and to ensure that vulnerable subjects are not taken advantage of in research. According to the Belmont Report, one should carefully weigh and balance these different principles when making an ethical decision. This is similar to the approach to ethical decision making we defended in chapter 2.

In an influential review article published in the *Journal of the American Medical Association*, Emanuel et al. (2000) articulate and defend seven ethical principles for research with human subjects. Their principles are derived from the Belmont Report as well as various ethics codes, such as the Nuremberg Code. Other writers, such as Levine (1988) and Veatch (1987), have also defended principles of research with human subjects. Below is our list of eight principles, which we have derived from Emanuel et al. (2000), Levine (1988), and Veatch (1987):

1. *Scientific validity:* Research protocols should be scientifically well designed; experiments should yield results that are replicable and statistically significant. Researchers should strive to eliminate biases and should disclose and/or avoid COIs. Research personnel should have the appropriate scientific or medical qualifications. Research institutions should have the appropriate resources, procedures, and safeguards.

2. *Social value:* Researchers should conduct experiments that have scientific, medical, or social worth; they should not use human beings in frivolous or wasteful research.

3. *Informed consent:* Subjects should make an informed choice to participate in research. When subjects lack the capacity to make such decisions, researchers should obtain informed consent from a legally authorized representative. Subjects should be able to withdraw from a study at any time for any reason.

4. *Respect for persons:* Researchers should respect the privacy, dignity, and rights of research subjects.

5. *Beneficence:* Researchers should use various strategies and techniques to minimize harm to research subjects and maximize benefits to

subjects and society. The risks of research should be justified in terms of the benefits. Researchers should not conduct experiments that have a high probability of causing death, nor should they harm subjects in order to benefit society. Researchers should be prepared to end experiments in order to protect subjects from harm. Researchers should monitor data to protect subjects from harm.

6. *Equitable subject selection:* Researchers should promote a fair and equitable distribution of benefits and harms in research. Subjects should not be excluded from participation in research without a sound moral, legal, or scientific justification.

7. *Protection for vulnerable subjects:* Researchers should use appropriate safeguards to protect vulnerable subjects from harm or exploitation.

8. *Independent review:* An independent committee should review the scientific and ethical aspects of the research.

We believe that these eight principles can be applied to particular dilemmas that arise in research with human subjects. They also can provide an ethical foundation for different rules, regulations, and guidelines. As mentioned above, the Nuremberg Code was the first international ethics code. Since then, several other international codes have been developed, including the Helsinki Declaration (World Medical Association 2004), and the guidelines of the Council for International Organizations of Medical Sciences (2002). We discuss ethics codes and guidelines in more depth in chapter 15.

HUMAN RESEARCH REGULATIONS

Although ethical codes and principles are important in their own right and have a great deal of moral authority, they do not have the persuasive force of the regulations and laws that have been adopted by various countries and some states. We recognize that many different countries have laws and regulations dealing with research on human subjects, but we focus on the U.S. laws and regulations in this text. The laws and regulations adopted by other countries are similar to the U.S. laws and regulations. We also address only the U.S. federal laws, although we recognize that some states, such as California, have their own research ethics laws.

The first steps toward developing human research regulations in the United States took place in 1953, when the NIH opened an intramural research operation called the Clinical Center. This center oversaw human experiments conducted at the NIH's intramural campus in Bethesda, Maryland, and reviewed protocols in order to avoid "unusual hazards" to subjects before proceeding with experiments (Capron 1989, Advisory Committee on Human Radiation Experiments 1995, Hoppe 1996). Late in 1965, the National Advisory Health Council, at the prodding of then NIH director James Shannon, issued the first prior review requirement for the use of human subjects in proposed research (Capron 1989). In 1966, this action prompted the U.S. Surgeon General to generalize the prior peer-review requirement to all NIH-funded research on human subjects. The FDA in 1971 issued its own similar regulations for testing new drugs and medical devices.

In response to research scandals, most notably the Tuskegee Syphilis Study, the United States enacted the National Research Act in 1974, which required that the DHEW, a precursor to the current DHHS, to codify all of its policies into a single regulation, now known as the Common Rule, which is codified in the Code of Federal Regulations at Title 45, Part 46, abbreviated as 45 CFR 46. These regulations required each research institution that conducts intramural or extramural research funded by the DHEW to establish or use an IRB to review and pass judgment on the acceptability of the proposed research according to the detailed requirements listed in the regulations. The regulations set forth rules for IRB composition, decision making, oversight, and documentation. IRBs should be comprised of people from different backgrounds, including scientific and nonscientific members, male and females members, as well as members from within the institution and members from the local community. In 1976, the NIH also developed the Office for Protection from Research Risks to provide oversight for research with human subjects. This office was later renamed the Office for Human Research Protection (OHRP) and relocated to report directly to the DHHS, in order to provide it with stronger, more independent, and broader governing authority.

In 1979, the first presidentially appointed commission on human experimentation, the National Commission for the Protection of Human Subjects of Biomedical and Behavioral Research, known simply as the National Commission, issued the Belmont Report, as mentioned above. The 1974 National Research Act mandated the formation of the National Commission. The Belmont Report provided a conceptual foundation for major revisions of the federal research regulations. In 1978, the DHEW revised its regulations to add additional protection for pregnant women, fetuses, in vitro fertilization, and prisoners. From 1981 to 1986, changes in U.S. regulations included revisions to DHEW's regulations for IRB responsibilities and procedures, changes in the FDA regulations to bring them in line with DHEW regulations, further protections for children, and a proposed federal common policy for the protection of human research subjects (Advisory Committee on Human Radiation Experiments 1995, p. 676).

In 1991, the U.S. government issued its final federal policy—the Common Rule, 45 CFR 46—which was adopted by 16 agencies and departments (Office of Science and Technology Policy 1991). It applies to research that is funded by or conducted by federal agencies. However, three federal departments, including the EPA and FDA, never adopted the Common Rule. The EPA adopted the Common Rule for EPA-sponsored research and has developed a different set of rules for privately funded research submitted to the EPA (Resnik 2007b). The FDA adopted rules similar to the Common Rule that apply to privately funded research conducted to support applications for new drugs or medical devices submitted to FDA. However, these rules have some potential regulatory loopholes or gaps pertaining to research with human subjects. For example, privately funded medical research that is not conducted in support of an application for a new drug or medical device is not covered by any existing federal regulations. This is in contrast to the Animal Welfare Act (1966, 1996), which covers use of all animals in research. In order to close this regulatory gap and provide uniform protection for human subjects, Jay Katz was the first to suggest in 1973 that the United States adopt a law to govern the use of *all* human subjects in research (U.S. Department of Health, Education, and Welfare 1973, Katz 1993, 1996, Shamoo and O'Sullivan 1998, Shamoo 2000).

In the fall of 1995, President Clinton appointed the National Bioethics Advisory Commission (NBAC) to examine bioethical issues in human research and medical practice. Several federal agencies, members of Congress, and various patient and consumer advocacy groups had called for the formation of such a commission to address ongoing issues and concerns. Since its formation, the NBAC has issued reports on human cloning, embryonic stem cell research, gene therapy research, and research on people with mental disorders. In 2001 the NBAC issued its most comprehensive report addressing the most pressing problems in research with human subjects (National Bioethics Advisory Commission 2001). The primary recommendation was that the entire oversight system needs an overhaul. The NBAC recommended, as advocacy groups such as CIRCARE (Citizens for Responsible Care and Research 2007) had also urged, that all research with human subjects be regulated. Shamoo and Schwartz (2007) argued strongly for a universal and uniform regulation for human subjects protections. The NBAC recommendations emphasized greater participation of noninstitutional members in IRB membership, better informed consent processes, better management of COIs, better adverse events reporting, mandatory education and training for researchers and IRB members, and compensation for research-related injuries. Most of these recommendations were also recommended by many groups and scholars. Although the NBAC's recommendations have had an influence on the overall discussion and some of the implementations of protections for human subjects, most of its recommendations were not enacted into laws or regulations. NBAC's charter expired in 2001. Moreover, a report in 2002 from the IOM (Institute of Medicine 2002b) also recommended federal oversight for all research with human subjects. Various bills have come before Congress to close the loopholes in the federal research regulations, but none have passed so far.

In late 2000, the secretary of the DHHS created the National Human Research Protections Advisory Committee (NHRPAC). One of us (Shamoo) was a member. The NHRPAC attempted to deal with the applications of NBAC's recommendations and any new incoming ethical concerns (see the OHRP Web site, http://www.hhs.gov/ohrp/, for reports and recommendations). NHRPAC issued reports and documents addressing COIs, the status of third parties in research, public data use, genetics research, people with impaired decision-making capacity in research, and confidentiality of data. The George W. Bush administration allowed the NHRPAC's charter to expire in August 2002. The NHRPAC's reports and recommendations have had some influence in the application of the federal regulations, but like the NBAC's recommendations, none of them have been enacted into laws or regulations.

In 2003, the new secretary of the DHHS created the Secretary's Advisory Committee on Human Research Protections (SACHRP). The advisory committee clearly had greater representation from performers of research with human subjects than did the NHRPAC, such as members from industry, academic medical centers, and private for-profit IRBs. Moreover, the committee had a better representation of the social and behavioral field. SACHRP tackled important questions such as whether to modify the Common Rule and its various subparts. It has also handled issues involving children, especially those who are wards of the state. Finally, SACHRP dealt with issues of involvement of prisoners in research and the requirement for accreditation (Association for the Accreditation of Human Research Protection Programs 2008) for IRBs. In all of these cases, SACHRP recommended some improvements but left most of the application of its suggestions to institutions and their IRBs. SACHRP has avoided the serious reforms that NBAC recommended and that advocates have been calling for, such as universal oversight of all research with human subjects in research, mandatory training and education of researchers and IRB members, and better uniform adverse events reporting.

In recent years, researchers from the social and behavioral sciences, journalism, and oral history have argued that the U.S. research regulations are excessively burdensome and are better suited to biomedical research (Hambruger 2005, American Association of University Professors 2006). One of us (Shamoo), while a member of NHRPAC (U.S. Department of Health and Human Services 2002), has spoken and written (Shamoo 2007) on reducing regulatory requirements for low-risk research, such as social and behavioral research.

Despite the many national commissions and committees addressing research with human subjects over the years, there are serious gaps and flaws in the overall system of the protections of human subjects in research that can be summarized into five categories: (a) gaps in regulation where a segment of human subjects are not covered, (b) lack of mandatory training and education for investigators, (c) no universal and uniform system of adverse event reporting, (d) inadequate regulations concerning children, and (e) lack of adequate oversight and monitoring of research in foreign countries (Shamoo and Strause 2004, Shamoo and Katzel 2007).

INSTITUTIONAL REVIEW BOARDS

Although the federal human research regulations have been revised many times since 1978, their basic structure has not changed much. These regulations all require IRB oversight and review of research protocols. Although research institutions are charged with assuring to the federal government that human research meets specific requirements, IRBs are charged with fulfilling these institutional responsibilities. IRBs have the full authority to review and approve or disapprove research protocols, to require modifications of protocols, to monitor informed consent, to gather information on adverse events, to stop experiments, to examine COIs, and to require adherence to federal and local, institutional requirements (45 CFR 46.109).

Most IRBs consist of 15–25 members who are employees of the research institution and one or more nonscientist members from the community. Critics have claimed that these situations can raise COI issues for IRB members (Cho and Billings 1997, Shamoo 1999), as discussed above and in chapter 10. The IRB or its members may face some pressure to accept a proposal so that the institution can obtain a grant or other source of funding. IRBs often face enormous workloads and unrealistic deadlines. Demands to accelerate or streamline the review process can affect the quality of IRB review. To do its job well, an IRB often must take some time to carefully review protocols and discuss key issues, but this can slow down a process that researchers and institutions would like to accelerate. IRB members could face liability for negligence or other legal wrongs if the IRB does not do its job well (Resnik 2004d). IRBs also often have trouble recruiting qualified and committed members, because membership on an IRB can place heavy demands on a person's time. In response to these and other problems, many commentators, as well as the NBAC, have recommended some structural changes in IRBs, but no significant changes have occurred to date (Moreno et al. 1998, U.S. Department of Health and Human Services 1998, Mann and Shamoo 2006).

To ensure that the rights and welfare of human subjects are adequately protected and to ensure that institutions comply with regulations, IRBs usually consider the following questions and issues when reviewing research proposals.

1. *Is it research?* Research is defined by federal regulations as a systematic investigation that contributes to generalizable knowledge (45 CFR 46.102(d)). To determine whether a proposed study qualifies as research, IRBs must consider the primary intent or purpose of the study (or procedure). If the primary intent of the study is to advance human knowledge, then the study probably qualifies as research. If the primary goal of the study is something other than the production of systematic knowledge, then it may not be research. Medicine includes many innovative or experimental therapies that are not treated as research, because their primary aim is to benefit the patient. It can be difficult to determine whether an activity should be classified as research when the activity has research and nonresearch goals, such as student projects, which aim to teach students about research but also involve research, and health care quality improvement and public health interventions, which aim to promote human health but also involve the systematic collection of data.

2. *Is it human research with human subjects?* The regulations define a human subject as a "living individual about whom an investigator…obtains data through interaction or intervention with the individual or identifiable private information" (45 CFR 46.102(f)). Research on stored human tissue is not explicitly covered by current federal regulations and presents IRBs with some important ethical and policy dilemmas. Several organizations, including the NBAC, have made some recommendations concerning research on stored human tissue. The IRB's main concern should be not with the tissue itself, but with the person whom the tissue came from. In evaluating research with stored human tissue, IRBs should pay careful attention to how the tissue was obtained, how it will be stored and protected, and how it will be used (Clayton et al. 1995, Weir et al. 2004). We explore these issues in more depth in chapter 14.

3. *Does the research require IRB approval?* The federal regulations treat some research involving human subjects as "exempt," meaning that the regulations do not apply to it. Some categories of exempt research include some types of survey or educational research; research relating to existing, publicly available data if subjects cannot be identified directly or through links to the data; research that evaluates public benefit programs; and food quality research (45 CFR 46.101(b)). If research is exempt, then it does not require IRB review. However, the IRB, not the investigator, should make the determination of whether research qualifies as exempt.

4. *Does the research require full IRB review?* If the research is classified as minimal risk, then it can be reviewed on an expedited basis by the IRB chair or a designee (45 CFR 46.110). Minor changes to IRB-approved protocols can also be reviewed on an expedited basis. "Minimal risk" is defined as follows: "The probability and magnitude of the harm or discomfort anticipated in the research are no greater in and of themselves than those ordinarily encountered in daily life or in routine physical or psychological examinations or tests" (45 CFR 46.102(i)).

5. *Are risks minimized?* The IRB must determine that investigators have taken appropriate measures to minimize risks (45 CFR 46.111(a)). There are many different strategies for minimizing risk, including rules for excluding people from a study who are likely to be harmed, clinical monitoring, data monitoring, extra protections for confidentiality, follow-up with subjects, referrals for health problems, and availability of counseling.

6. *Are the risks reasonable in relation to the benefits?* Risks are reasonable if they can be justified in terms of the benefits of the research for the subjects or society. Benefits to subjects include medical as well as psychosocial benefits (participation itself may be a benefit). Benefits to society may include advancement of knowledge. Federal agencies do not treat money as a benefit to the individual, because this could lead to the involvement of research subjects in highly risky research with money as the only benefit.

7. *Is selection of subjects equitable?* To satisfy the demands of justice as well as the federal regulations, the IRB needs to ask whether there are sound ethical and scientific reasons for including or excluding some types of subjects in

research. A good scientific reason for excluding a type of subject would be that the subject cannot provide data relevant to the question being studied. For example, there is no good reason for including men in a study of ovarian cancer. A valid moral rationale for excluding subjects would be to protect those subjects or a third party from harm. For example, pregnant women may be excluded from a clinical trial in which the drug is likely to harm the fetus and the medical benefit of the drug to the mother is not likely to be significant.

8. *How will subjects be recruited?* IRBs address this question to promote equitable subject selection and to avoid potential coercion, undue influence, deception, or exploitation. IRBs usually review advertisements to ensure that are not deceptive or coercive. Informed consent should take place under circumstances that minimize the chances for coercion or undue influence (45 CFR 46.116). IRBs need to pay careful attention to the payments offered to subjects as recruitment incentives to ensure that these payments are appropriate (Grady 2005).

9. *Is confidentiality protected?* Federal regulations require researchers to take adequate steps to protect privacy and confidentiality (45 CFR 46.111(a)). There are different strategies for protecting confidentiality, including removing personal identifiers from the data, restricting access to the data, keeping the data in a secure place, and obtaining a Certificate of Confidentiality from the DHHS, which allows researchers to refuse a request to share data as part of a legal proceeding. Although privacy and confidentiality are important, some types of situations that occur in research may require researchers to break confidentiality or privacy. For example, if a pediatric researcher discovers that one of her subjects is the victim of child abuse, she should report this to the proper authorities.

10. *Will informed consent be obtained?* The federal regulations require that consent be obtained from the research subject or the subject's legally authorized representative (45 CFR 46.116). Researchers who are obtaining consent must determine whether the subject is capable of making an informed choice. Moreover, the consent process should avoid therapeutic misconception that subjects can easily perceive (Appelbaum et al. 1987). If he or she cannot, then the researchers must find an appropriate representative, such as a parent or guardian, spouse, or other family member. In cases where consent is given by a representative, the IRB may also require that the subject be allowed to assent to the research, if it is in the subject's interests to provide assent. There are a couple of exceptions to the consent requirement. Investigators do not have to obtain informed consent for research on public benefit programs, for example, research evaluating the effectiveness of a new Medicare benefit. IRBs may also waive some of the informed consent requirements if obtaining informed consent would invalidate or bias the research and waiving the consent requirement will not adversely affect the subject's rights or welfare (45 CFR 46.116(d)). Some types of social science experiments may qualify for this exception. Additionally, the FDA has adopted its own special regulations waiving informed consent requirements in emergency medical research (21 CFR 50.24).

11. *Will informed consent be documented?* Federal regulations require that informed consent be documented except in cases where the research is judged to be minimal risk, the principal harm arising from the research would be potential breach of confidentiality, and the only record linking the subject to the research would be the informed consent document itself, or where the research is minimal risk and documentation of the research procedure is not usually required outside of a research context (45 CFR 46.117). In cases where the research subject is not able to sign an informed consent form, the legally appointed representative must sign the form. The IRB may also decide to require that the signature be witnessed.

12. *Is informed consent process adequate?* IRBs take a careful look at the *process* of obtaining informed consent. IRBs are concerned about the quality and integrity of the process as well as fairness and potential exploitation or coercion. Although IRBs often spend a great deal of time examining the informed consent form to make sure that it covers everything that needs to be discussed, and that it is understandable and well written, informed consent should be much more than just signing a form. It should be an ongoing conversation between researchers and subjects about the benefits and risks of research participation (Veatch 1987). Federal regulations provide details about information that must be included in informed consent (45 CFR 46.116), including information about the purpose of the research, its duration, and relevant procedures; the number of subjects enrolled; foreseeable risks, costs, or discomforts; possible benefits; alternative treatments; confidentiality protections; any compensation for injuries; payment for participation (if any); ability of the subjects to refuse or withdraw at any time without any penalty; ability of the researchers to withdraw subjects to protect them from harm; potential disclosure of significant findings to subjects or their representatives during the study; and whom to contact regarding the research. Also, IRBs must be satisfied that enrolled subjects can comprehend the information given to them. If the subjects have impaired decision-making capacity, then arrangements should be made for determining their legally appointed representatives.

13. *Are any safeguards required to protect vulnerable subjects?* Safeguards that may help protect vulnerable subjects could include provisions to promote adequate informed consent, such as using legally appointed representatives if the subjects cannot give consent. IRBs may also need to include members with some knowledge of the needs and interests of these vulnerable populations. For example, to review research on prisoners, an IRB must have a member who is qualified to represent the interests of prisoners. The federal government has other protections for vulnerable subjects, which are discussed in chapter 13.

14. *Is the protocol scientifically valid?* The IRB should also ensure that research proposals meet appropriate standards of scientific validity, because poorly designed studies expose subjects to research risks without the promise of obtaining useful results (Levine 1988). An IRB may consider statistical issues, such as sample size, sampling biases, and surrogate end points. The IRB may consider

conceptual issues related to research design, such as the merits of quantitative versus qualitative research and using the "intent to treat" model in clinical research. Although it is very important for research protocols to meet the highest standards of scientific rigor, this requirement may sometimes conflict with other ethical principles, as discussed above. Randomized, controlled clinical trials have become the "gold standard" for clinical research, but they have generated a great deal of controversy relating to the use of placebo controls as well as rules for starting or stopping clinical trials. The IRB may also need to consider whether there is adequate empirical data for proceeding with a study, such as data from previous animal or human studies. If the research is part of a clinical trial to generate data to support an application to the FDA, then the IRB will want to know whether the study is a phase I, phase II, or phase III study, whether it is an expanded access or human use trial, and so on.

15. *Will the data be monitored to protect subjects from risks and ensure the validity of the results?* It is very important for researchers to monitor clinical trials to learn about potential benefits, problems with the protocol, and adverse events, such as unanticipated harms, including death. Adverse events need to be reported to the sponsor and the IRB. Most clinical trials use data and safety monitoring boards to handle some of these tasks. Although the IRB does not monitor research, it is responsible for ensuring that adequate processes and mechanisms are in place for monitoring (DeMets et al. 2005).

16. *Are there any COIs?* Chapter 10 discusses COIs in more detail. COIs can have an adverse impact on the integrity and trustworthiness of clinical research, as demonstrated by the Gelsinger case, described above. Psaty et al. (2004) have shown that COIs may have contributed to the suppression of adverse events reporting for research on the cholesterol drug cervastatin. Beside COIs, many other factors may contribute to the underreporting of adverse events, such as the lack of knowledge by both the subjects and investigator as to what to report (and to whom) and lack of uniformity of reporting requirements (Shamoo 2005, Shamoo and Katzel 2008). Moreover, in randomized clinical trials, there is evidence of bias in publishing more favorable results than unfavorable results (Ridker and Torres 2006, Chou and Helfand 2005). Although a variety of rules require disclosure of COIs (see chapter 10), IRBs often do not learn about COIs. Clinical researchers may have a variety of financial interests in research, including private funding, stock, salary, and economic incentives to recruit subjects into clinical trials—to encourage investigators to recruit research subjects, companies may pay them a fee for each subject recruited or patient care costs over and above what it costs to provide care for the patient (Spece et al. 1996). When IRBs obtain information about COIs, they must decide what to do with it. Should they require investigators to disclose COI information to potential research subjects? Should they require investigators to take steps to minimize the impact of COIs on their research? The research community is just beginning to explore these sorts of questions (Resnik 2004a).

17. *Are there any collaborating institutions?* A great deal of research today involves collaboration among many different researchers at different institutions. The IRB needs to ensure that this collaboration does not harm subjects and that other institutions abide by the relevant regulations and guidelines. International research collaboration has become especially controversial, because researchers in different countries may abide by different ethical standards (Macklin 1999a,b).

MISCONDUCT IN CLINICAL RESEARCH AND CLINICAL TRIALS

As discussed in chapter 8 on scientific misconduct, all research supported by the federal government must follow the federal policy on scientific misconduct. Therefore, the policy applies to all federal agencies. Each agency provides an oversight for compliance with the scientific misconduct policy, including when human subjects are involved. The PHS provides much support for clinical research. Scientific misconduct within the context of clinical research provides a unique challenge to agencies because it involves human subjects. The Office of Research Integrity (ORI) has drafted Guidelines for Assessing Possible Research Misconduct in Clinical Research and Clinical Trials, which attempt to adapt the scientific misconduct policy to clinical research (Office of Research Integrity 2001). The ORI emphasizes that the guidelines are to be used as a supplement to the ORI's model for responding to allegations of research misconduct.

In brief, the guidelines require that any investigation into misconduct in clinical research and trials should follow these elements:

1. Coordination of human subject protections with the IRB
2. Informing physicians and patients when significant public health issues are involved
3. Informing the Data and Safety Monitoring Board
4. Informing the affected journals
5. Coordination of the investigation with OHRP and FDA, if appropriate

The investigation should include the assessment of discrepancy of the data submitted to federal agency, publications, or other forms, from the original clinical data in the medical records.

Under the Freedom of Information Act, Shamoo (2001) obtained the institutional incident reports for NIH-supported research that contain adverse events, including deaths, for the years 1990–2000. Only eight deaths and 386 adverse events were reported throughout the 10 years, even though tens of millions of human subjects took part in these experiments. This likely represents an underreporting of adverse events in this large population. This high level of underreporting casts doubts as to the integrity of some clinical data.

QUESTIONS FOR DISCUSSION

1. Is the media paying too much, just enough, or too little attention to questionable research with human subjects? Why do you think that is so?
2. Do you think freedom of inquiry should surpass other values? Why?
3. What risks would you and members of your family take for public good? Why?
4. In your opinion, should there be a difference between informed consent for volunteering as a human subject and informed consent for medical treatment? What would those differences be, if any? Why?
5. Do you think IRBs are doing a good job of protecting human subjects? Can they? How would you improve the system, if you believe improvement is needed?

CASES FOR DISCUSSION

Case 1: Migrants and Phase I Trials

Oscar Cabanerio is a 41-year-old immigrant living in the United States without any legal documentation. He is poor and needs cash to send to his family in Venezuela. SFBC International Inc. is a large contract research organization (CRO) testing drugs in human subjects for drug companies in Miami, Florida. Cabanerio agreed to be in a trial to test Oros Hydromorphone, made by Alza Corporation. The study paid each research subject $1,800. The subjects are instructed to swallow the tablets and not chew them, since chewing can cause overdosing. Adverse reactions include heart attack, allergic reaction, and even death. Informed consent for this study is usually very quick. The subjects are eager to earn money, so they just look over the document and sign it. Many of them have limited English-speaking abilities.

- What are some ethical problems with this study?
- Are there any problems with the selection of subjects, the consent process, or safety?
- Is there fair subject selection in these studies and why?
- How would you run such a facility ethically?

Case 2

An announcement in the newspaper and radio encourages people to enroll in research protocols to test a new anti-flu medication. The announcement emphasizes that subjects will receive a free physical exam, free health care for 60 days, and $400 compensation. The new drug is very promising in either stopping the full-blown symptoms of the flu or preventing it altogether. The protocol has already been approved by an IRB.

- What questions would you ask if you were a potential subject?
- Should the IRB have approved the protocol? Why?

Case 3

A hospital associated with a research university has a policy that every new employee should give a blood sample. The employee is told that the blood sample will be frozen for a long time. The hospital's purpose in collecting the blood sample is to reduce their liability in case anyone contracts HIV. From the frozen sample they can determine whether the patient had the HIV virus prior to employment. This will reduce the hospital's liability. A few years later, a researcher at the hospital is developing an HIV diagnostic instrument directly from the blood. The instrument, if it works, would advance HIV screening. The researcher wants to use the samples without any names attached to them. He just wants to test different samples from different people. The researcher designed the protocol such that once the samples are obtained, no one would know which sample belonged to which person.

- What concerns would you have if your blood sample was included in this research?
- What should the IRB do?
- What should the informed consent form contain?

Case 4

Many clinical faculty members at research universities receive 20–90% of their salary from research grants and contracts, a large percentage of which consist of clinical trials sponsored by pharmaceutical companies.

- Do you see potential problems with this arrangement?
- Can you suggest some solutions?
- Should a university engage in routine clinical trials?

Case 5

Subjects for a research study will be recruited from private pain treatment clinics and the medical school's pain service. Preliminary studies have shown that the drug thalidomide may provide some relief for migraine headaches, arthritis, and neuropathy conditions. Because thalidomide's harmful effects on fetuses are well known, women of childbearing age will be excluded from this study.

- Are there social/scientific benefits from the study?
- If you are a member of the IRB, what questions would you ask?
- What should the risk/benefit analysis include?

Case 6

A company is developing a safer pesticide for use on a variety of crops, including tomatoes, corn, apples, green beans, and grapes. The company plans to use healthy subjects (its employees) to test the pesticide for toxic effects. Subjects will be paid $500 and will be monitored carefully for three days. They will report toxic effects, such as dizziness, nausea, headache, fatigue, shortness of breath, and anxiety. If the pesticide proves to be safe, it may replace many existing pesticides commonly used on crops.

- What are the risks to subjects?
- What are the benefits to subjects?
- Would you put any conditions on the protocol before going forward? What would they be?
- Do you see any COIs? Can they influence the outcome of the study?

Case 7

A nontenured assistant professor at a medium-sized university is a member of her university's IRB. One of the human subject protocols the IRB is reviewing is from a world-renowned professor at another department at her university. This world-renowned professor is a member of the promotion and tenure committee. The assistant professor's package for promotion to tenured associate professor will go to the committee in six months. The assistant professor has a great deal of concern about the proposed protocol. She feels that the risks are watered down and the benefits or potential benefits are exaggerated.

- What should the assistant professor do? What you would do?
- How should the university handle the problem?

Case 8

A research proposal and its informed consent forms were submitted to an IRB of an independent nonprofit research facility in San Francisco. The protocol deals with 300 drug addicts, of whom 20% are also suspected of having HIV. The protocol is a survey of social habits of these addicts. The surveyor will follow the addicts around in their daily routine for one week to register their food intake, drugs used, sexual habits, and so forth. The researcher considered the study to be a minimal risk study and said so on the proposal submitted to the IRB.

- What should the IRB do?
- What should the informed consent form contain?
- Should confidentiality of information be dealt with in the informed consent form?
- Is this research of minimal risk? Why?

13

Protecting Vulnerable Human Subjects in Research

This chapter discusses ethical issues and policies relating to the protection of vulnerable subjects in research. It reviews the history of human experimentation with such vulnerable human subjects as people with impaired decision-making capacities, children, fetuses, and subordinate populations (i.e., prisoners, students, employees, and soldiers). The chapter discusses the definition of vulnerability as well as the criteria for decision-making capacity, examines the federal regulations and the special safeguards, or the lack thereof, currently available to protect vulnerable populations, discusses different types of research with children, and examines the concept of minimal risk.

HISTORICAL BACKGROUND

In chapter 12, we recounted some of the historical abuses of human subjects in research in the past century, such as the Tuskegee Syphilis Study and the Nazi experiments. An important theme of human experimentation before World War II is that many of the subjects were from vulnerable populations: Children, mentally ill people, poor people, prisoners, minorities, and desperately ill people were often harmed or exploited in research. As noted in chapter 12, Mary Rafferty, Robert Bartholomew's subject for electrical stimulation of the brain, was "feeble-minded." In 1911, Hideyo Noguchi of the Rockefeller Institute used orphans and hospital patients as experimental subjects for developing a diagnostic test for syphilis. Noguchi injected his subjects with an inactive solution of the causative agent of syphilis, which produced a skin reaction in subjects with syphilis but no reaction in neutral controls. Many leaders of medical research were outraged by Noguchi's use of healthy subjects in his experiment. In 1915, Joseph Goldberger, an investigator for the U.S. Public Health Service, attempted to induce the disease pellagra in male prisoners in Mississippi by placing them on a diet of meat, cornmeal, and molasses. He was able to prove pellagra is a nutritional deficiency resulting from this diet. The governor of Mississippi granted the prisoners pardons for their participation (Lederer 1995).

After World War II, vulnerable subjects continued to suffer harm or exploitation in research. The Willowbrook hepatitis experiments were conducted at Willowbrook State School in Willowbrook, New York, from 1956 to 1980. A team of researchers, led by Saul Krugman and Joan Giles, began a long-range study of viral hepatitis in this institution for mentally retarded children. Viral hepatitis was endemic at Willowbrook: Most children who entered Willowbrook became infected within 6–12 months of admission. Although the disease is usually not life threatening, it can cause permanent liver damage. Victims of the disease usually have flulike symptoms, such as fever, fatigue, and nausea. The disease is transmitted orally through contact with feces or body secretions. In their research, Krugman and Giles infected healthy subjects with viral hepatitis. This allowed them to study the natural progression of the disease, including its incubation period, and to test the effectiveness of gamma globulin in preventing or treating the disease. They collected more than 25,000 serum samples from more than 700 subjects. The two researchers justified their study on the grounds that it offered therapeutic benefits to the subjects: The children in the study would receive excellent medical care, would avoid exposure to other diseases, and would acquire immunity against more potent forms of hepatitis. Krugman and Giles obtained written informed consent from parents, although some critics have charged that the parents did not understand the nature of the study. Krugman and Giles also obtained appropriate approvals for their study from the New York State Department of Mental Hygiene, the New York State Department of Mental Health, and the human experimentation committees at the New York University School of Medicine and the Willowbrook School (Munson 1992).

As discussed in chapter 12, human radiation experiments took place in the United States from 1944 to 1974, during the cold-war era (Advisory Committee 1995). These experiments, which were government funded, involved more than 4,000 unsuspecting citizens and military personnel as human subjects. Several of these radiation studies involved children. Because children are in the developing years of their lives, exposure to radiation could greatly increase their lifetime risk of cancer. Several noteworthy studies involving children and other vulnerable populations may have violated existing ethical standards, included the following:

- Researchers at Vanderbilt University in the late 1940s gave pregnant women radioactive iron to study the effects of radiation on fetal development; a follow-up study found that children from these women had a higher than normal cancer rate.
- From the 1940s to the 1960s, researchers injected encapsulated radium into the nostrils of more than 1,500 military personnel; many developed nosebleeds and severe headaches after exposure.
- Prior to 1973, pharmaceutical companies conducted about 70% of their phase I clinical trials on prisoners (Moreno 1998).
- Thousands of military personnel were used in LSD experiments in the 1950s (Moreno 1998).

During the 1990s, the research community learned about a variety of ethically questionable studies on mentally ill patients. The national media also covered many of these stories. As a result, the National Bioethics Advisory Commission (NBAC) issued a report recommending changes in federal regulations on research on people with mental disorders (Shamoo 1997a, National Bioethics Advisory Commission 1998). Many of these problems originally came to light through a series of papers delivered at a conference held in 1995 (Shamoo 1997c) and a series of articles published in journals (Shamoo and Irving 1993, Shamoo and Keay 1996, Shamoo et al. 1997). This was followed by a major series of articles in the *Boston Globe* (see Kong and Whitaker 1998). Many of these research projects involved "washout" periods as part of their protocol for testing a new drug. It requires a subject to stop taking a medication until it no longer has significant pharmacological effects, to reduce biases and complications due to drug interactions in the clinical trial. After the washout period, the protocol randomly assigns the subject to receive either an existing treatment or a new drug. The protocols may also include a placebo control group. In some washout studies, the harms to subjects are fairly minimal, especially if the washout period is short and subjects are carefully monitored under inpatient settings, but in others the harms may be substantial due to the absence of necessary treatment during the washout period.

In the studies that many people regarded as unethical, the subjects were taking medications for depression, schizophrenia, and other serious mental disorders. One study of washout research with schizophrenia patients found that many subjects suffered the effects of withdrawal from medications and experienced relapses, which could include increased psychosis or rehospitalization (Crow et al. 1986, Wyatt 1986, Baldessarini and Viguera 1995, Gilbert et al. 1995, Wyatt et al. 1999). As a result, more than 10% of subjects dropped out of these studies (Shamoo and Keay 1996, Shamoo et al. 1997) for a variety of reasons. Because 10% of schizophrenics commit suicide, a relapse of this disease can be very dangerous. In 1991, Craig Aller, a patient with schizophrenia at the University of California, Los Angeles, and his family argued that he suffered permanent brain damage due to a relapse caused by a medication washout as part of his participation in a research protocol (Aller and Aller 1997). Another patient in this study allegedly committed suicide (Aller and Aller 1997). In some of these studies, researchers asked the subjects to consent, but critics questioned whether the patients were capable of giving informed consent (Shamoo and Keay 1996). Many of these experiments did not even give the subjects the opportunity to consent. Other experiments that were criticized included studies in which mentally ill subjects were given ketamine to induce psychosis and delusions, to study the mechanism of the disease, and healthy children 6–12 years old were given fenfluramine (an obesity drug) to test whether they were prone to violence (Sharav and Shamoo 2000)—children were selected for these studies because their siblings were incarcerated. We describe the fenfluramine case in more details as case 1 below.

During the 1990s, the research community also became more aware of ethical issues related to another type of vulnerable group: people living in developing nations. Chapter 15 discusses this research in more detail.

THE CONCEPT OF VULNERABILITY

As we have seen, historically, vulnerable subjects have been harmed and exploited in research. Many of the rules and policies dealing with vulnerable subjects were developed to protect vulnerable subjects from harm and exploitation (Mastroianni and Kahn 2001). The net effect of these policies is that some vulnerable groups, such as children and pregnant women, were frequently excluded from research, which adversely affected the welfare of these groups. For example, as result of policies designed to exclude children from research, very few drugs that are prescribed to children have actually been tested on pediatric populations. Physicians prescribe these drugs to pediatric populations on an "off-label" basis by extrapolating from their effects on adult populations. This practice assumes that children are physiologically similar to adults, which is often a faulty assumption. Since the 1990s, researchers and advocacy groups have urged researchers to include more children in research (Tauer 1999, Friedman Ross 2006). Women were also routinely excluded from clinical trials as a result from exclusionary policies and methodological biases. In the mid-1980s, feminist activists and politicians pressured the National Institutes of Health (NIH) to include more women in research studies. As a result, the NIH now has policies for the inclusion of women and minorities in research (Dresser 2001). The exclusion of mentally ill people from research has also had an adverse impact on our understanding of mental illness (DuBois 2008).

There is an ethical tension, then, between including and excluding vulnerable subjects in research. It is important to include vulnerable subjects in research to improve their medical and psychiatric treatment and learn more about how to enhance their welfare (National Bioethics Advisory Commission 1999). But it is also sometimes necessary to exclude vulnerable subjects to protect them from harm and exploitation. The key is to strike a reasonable balance between these two ethical imperatives. Research regulations and policies that we discuss in this chapter attempt to achieve such a balance. Do the current regulations and policies provide not enough protection, too much protection, or just the right amount of protection? This is an important question for the reader to consider as we discuss protecting vulnerable subjects in research.

One of the most fundamental questions relating to the protection of vulnerable subjects in research is a conceptual one: What is a vulnerable research subject? Most commentators have focused on two criteria for defining vulnerable subjects: (a) compromised decision-making capacity (DMC): Vulnerable subjects have difficulty making decisions; and (b) lack of power: Vulnerable subjects lack the ability to promote their own interests (Macklin 2003). As result of their compromised DMC, lack of power, or both, vulnerable subjects may have a diminished ability to provide consent. They may have difficulty understanding information that is presented to them, or they may be susceptible to coercion or undue influence. Some vulnerable subjects, such as children or mentally disabled adults, may not be able provide consent at all. Since vulnerable subjects are more easily harmed or exploited than others, there is an ethical imperative, based on the principles of justice and respect for persons (discussed in chapter 12) to protect vulnerable subjects from exploitation or harm (Macklin 2003).

The U.S. federal regulations (45 CFR 46.111(b)) require that research protocols include special protections for vulnerable subjects, where appropriate: "When some or all of the subjects are likely to be vulnerable to coercion or undue influence, such as children, prisoners, pregnant women, mentally disabled persons, or economically or educationally disadvantaged persons, additional safeguards have been included in the study to protect the rights and welfare of these subjects." International research guidelines, such as the Helsinki Declaration (World Medical Association 2004) and the Council for International Organizations of Medical Sciences (CIOMS) guidelines (Council for International Organizations of Medical Sciences 2002) also state that research protocols should include special protections for vulnerable subjects. In addition, subparts of the federal regulations include specific protections for children, prisoners, and fetuses (via rules pertaining to pregnant women). In what many commentators consider to be a significant omission, the federal regulations (as of the writing of this book) do not include any specific protections for mentally disabled research subjects. International ethics codes, such as the CIOMS guidelines, also include specific protections for children, fetuses, mentally disabled people, and prisoners (Council for International Organizations of Medical Sciences 2002).

In addition to the categories of vulnerable subjects just discussed, other types of vulnerable subjects may include desperately ill people (e.g., people with cancer or AIDS), people in extreme poverty (e.g., people in developing nations), people who have limited language abilities (e.g., non-English-speaking people in the United States), and subordinates (e.g., students, employees, and soldiers). Some writers have argued that the definition of vulnerability is so broad that nearly everyone might be classified as a vulnerable subject (Levine et al. 2004). While we think this is a potential problem with the definition of vulnerability, we think that researchers and institutional review board (IRB) members can deal with it by carefully considering the capabilities of research subjects that will be recruited for a particular protocol and determining whether they need extra protections and what those protections might be.

A variety of safeguards can be used to protect vulnerable subjects. Some of these include a clear and convincing rationale for using the vulnerable population (rather than a nonvulnerable population) in research, rationale for the risk limitations for the subjects, modification and assessment of informed consent, provisions for assent when consent is not appropriate, use of legally authorized representatives in the consent process, translation of documents, use of interpreters, independent monitors of consent, independent assessment of subjects' DMC, witnesses to the consent process, and including members on the IRB who are familiar with the needs and interests of the vulnerable group. To protect vulnerable subjects, it is especially important to make sure that the population is being used because there is likely to be some definite benefit to the population (or its members) and that the population is not being used for reasons of convenience, availability, or cost savings (Moreno 1998). For example, studies involving prisoners should provide direct benefits to the research subjects, such as research on methods of rehabilitation, or should provide benefits to prisoners as a group, such as research on prison safety or the causes of incarceration and criminal behavior. As noted above, in the 1960s and 1970s, prisoners were often used in phase I drug trials in which neither the subjects nor the prison population would receive a direct benefit.

Having discussed some general concerns with research involving vulnerable subjects, we now consider some specific issues relating to particular vulnerable groups.

RESEARCH SUBJECTS WITH DECISION-MAKING IMPAIRMENTS

Throughout history, mentally ill or mentally disabled people have received unfair and even inhumane treatment, including being subjected to forced hospitalization and sterilization. Nazi eugenics and racial purification policies began with programs to prevent mentally ill or disabled people from reproducing. In the past, mentally ill research subjects have not been accorded the same protection as other research subjects (Shamoo and O'Sullivan 1998, Shamoo 2000). Both the Belmont Report (National Commission 1979) and the President's Commission for the Study of Ethical Problems in Medicine and Biomedical and Behavioral Research (1983a,b) recognized that mentally disabled people should be treated as vulnerable or potentially vulnerable research subjects. (We distinguish here between mental illness and mental disability to emphasize that mental illnesses, such as depression, are different from mental disabilities, e.g., dyslexia or mental retardation. A mental illness may or may not cause a patient to be mentally disabled.)

There are three basic types of research subjects with decision-making impairments: (a) subjects who are born with or acquire permanent cognitive or emotional disabilities, such as people with Down's syndrome, autism, or Parkinson's dementia; (b) people who have a mental illness, such as schizophrenia, depression, or substance abuse; and (c) people who are temporarily mentally disabled, such as someone recovering from head trauma due to an automobile accident. It is important to distinguish among these types of subjects in research because a person with a mental illness, such as depression, may be perfectly capable of making decisions when the illness is well managed, whereas a person with a severe form of autism may never be capable of making some types of decisions. Someone with Alzheimer's dementia may become progressively disabled during a research study, whereas someone with head trauma may gain decision-making abilities during a research study as his or her condition improves.

In thinking about mentally disabled research subjects, it is important to distinguish between two different but related concepts: competency and capacity for decision making (sometimes called "decisional" capacity). Competency is a legal notion: Competent people have the full range of rights available to citizens. In the United States, children (or minors) are deemed to be incompetent, and parents or guardians make decisions for them. The age at majority varies from jurisdiction to jurisdiction, but most set it at age 18.

A court may grant a minor the right to make decisions, as an "emancipated" minor, if the court decides that it is in the minor's best interests to make decisions. A court may also grant a mature minor a limited right to make decisions, such as the right to refuse medical treatment. Adults are deemed to be legally competent unless they are adjudicated incompetent by a court. If an adult is adjudicated incompetent, the court will appoint a guardian to make decisions for that adult. A court may also stop short of adjudicating an adult as completely incompetent and allow the adult to make some decisions but not others (Buchanan and Brock 1990, Berg et al. 2001).

Decision-making capacity (again, DMC) is a clinical or ethical concept. Persons have DMC if they have the ability to make sound decisions for themselves. A person could be legally competent yet lack DMC, such as when a patient with advanced Alzheimer's dementia has not been adjudicated incompetent. Conversely, a person could have DMC but be legally competent, such as a mature minor who has not been emancipated by a court. In clinical medicine, physicians assess patients' DMC to determine whether they are capable of making medical decisions. If a person is not capable of making a medical decision, then the physician will use a surrogate decision maker. Although there is some variation in different legal jurisdictions, most states recognize the following order of preference for surrogate decision making for people lacking DMC: guardian (including a parent if the person is a child), health care power of attorney, spouse, sibling or adult child, other family members. It also worth noting that DMC is not an all-or-nothing characteristic: People may have more or less DMC, depending on their circumstances. In general, the greater the risk and complexity associated with the decision, the greater the amount of DMC that is required. For example, a person with dementia would probably have sufficient DMC to make a simple decision, like deciding which shirt to wear, but probably could not make a more complex and risky decision, such as deciding whether to participate in a phase II clinical trial. More DMC is required for complex and risky decisions to protect people with diminished DMC from making harmful choices.

The ability to make a decision depends on a number of different cognitive and emotive functions. To determine whether a person has adequate DMC to make a particular choice, one should assess the following (Grisso and Appelbaum 1998, Berg et al. 2001, Appelbaum 2001, Buchanan et al. 2001, DuBois 2008):

- *Consciousness:* Is the person conscious, awake, alert, or coherent?
- *Emotional stability:* Is the person emotionally stable?
- *Memory:* Can the person remember information needed to make a decision?
- *Expressing choice:* Can the person express a choice?
- *Understanding:* Does the person understand the information being presented?
- *Appreciation:* Does the person appreciate the significance of the information presented, that is, what it means for him or her?

Mentally disabled patients fail to meet some or all of these conditions. The more severe and chronic the mental disorder, the more likely the impairment of DMC. Patients with schizophrenia, manic depression, major depression, and anxiety disorders may represent the most likely group with problems related to DMC. When mentally ill or disabled subjects are confined in an institution, DMC can be compromised even further due to potential coercive factors in that environment. Assessing a person's DMC can be a difficult task, especially when their cognitive and emotive abilities wax and wane over time and often depend on circumstances. In difficult cases, it may be necessary for different people to observe a person closely, over time, in different circumstances.

If we apply this framework to the research context, it is important to distinguish between competence and DMC among research subjects (or potential subjects). Children do not have the legal right to consent to research, unless a court grants them this right via emancipation or under the mature minor doctrine. However, this does not mean that children lack the ethical right to consent to research. Many commentators have argued that children should be allowed to consent to research when they have sufficient DMC to make a decision, usually when they are age 14 or older. Difficult ethical questions can arise when parents and teenage children disagree about research participation. The parents may have the legal right to enroll a child with cancer in a clinical trial, but what should be done if the child refuses (Friedman Ross 2006)?

Adults have the legal right to consent to research participation unless a court declares them to be legally incompetent for that task. Adults who have been adjudicated incompetent will have a court-appointed guardian who can make decisions concerning research participation. For adults who have not been declared legally incompetent but have diminished DMC, it may be advisable for a surrogate decision maker, such as a spouse or close family member, to help the adult with the consent process and to receive the information that the adult receives. Having a surrogate decision maker assist in this way can help protect the adult from harm or exploitation. Also, if the adult's DMC declines, the surrogate decision maker will be available to make decisions (National Bioethics Advisory Commission 1998). In some cases, an adult may use a document, such as living will or health care power of attorney form, to express a desire to participate in research if he or she loses the ability to make decisions. Researchers should honor the wishes expressed in these documents (Berg et al. 2001).

One of the more controversial issues relating to the assessment of DMC is whether there should be an independent assessment of DMC for research participation (Shamoo 1994b, National Bioethics Advisory Commission 1998). The argument in favor of an independent assessment is that researchers have a potential conflict of interest when it comes to assessing DMC, because it would be in the researcher's interests to find that a person has sufficient DMC to participate in research to meet enrollment goals. The argument against independent assessment is that this can be very burdensome, expensive, and time-consuming and is not necessary in low-risk research. A compromise position, which we recommend, is that the need for an independent assessment of prospective subjects' DMC should vary with the benefit/ risk ratio of the research. When the risks of research are more than minimal and the subjects will receive no direct benefits, an independent assessment of prospective subjects' DMC should be required. For example, a phase I clinical trial of a new drug on healthy subjects should have independent assessment of prospective subjects' DMC. When the risks of research are more than minimal but the subjects are likely to benefit from participation, an independent assessment of DMC is advisable but not required. For example, an independent assessment of prospective subjects' DMC would be advisable in a phase II clinical trial for a new chemotherapy agent. When the risks of research are minimal, an independent assessment of DMC is not required. For example, there would be no need for an independent assessment of prospective subjects' DMC for a study that only requires subjects to fill out a health information questionnaire and have 5 ml of blood drawn every two years for ten years.

Researchers must take great care in obtaining informed consent from mentally disabled research subjects to avoid mistakes and manipulations. Informed consent rests on four pillars: (a) adequate DMC, (b) sufficient information, (c) understanding of information, and (d) freedom from coercion or undue influence (Buchanan and Brock 1990, Berg et al. 2001). As we just described, assessing a person's DMC can be difficult because the subject is susceptible to manipulation by the researcher. Additionally, mentally disabled people, such as schizophrenics or people with dementia, may have difficulty understanding the information that is presented to them. Some mentally disabled people may be institutionalized or living under other conditions where they do not feel empowered to make their own choices. For many, saying no to their physician or psychiatrist may be difficult. These individuals can easily fall prey to real or perceived duress or coercion. Some mentally disabled people, such as people with bipolar disorder, may be poor at assessing risks and may therefore take risks not normally taken by the average person.

Although both the National Commission for the Protection of Human Subjects of Biomedical and Behavioral Research and the President's Commission recommended extra protections for the mentally disabled, the federal government did not issue special regulations on the mentally disabled as a separate subpart of 45 CFR 46 when it revised this document in 1991, due to insufficient advocacy for the mentally ill at that time (Shamoo and Irving 1993). The Common Rule discusses some added protections for the mentally disabled as part of the IRB composition: "If an IRB regularly reviews research that involves a vulnerable category of subjects, such as children, prisoners, pregnant women, or handicapped or mentally disabled persons, consideration shall be given to the inclusion of one or more individuals who are knowledgeable about and experienced in working with these subjects" (45 CFR 46.107(a)).

At present, what constitutes an acceptable degree of risk when mentally disabled individuals are enrolled in research as subjects can be interpreted differently by various IRBs. The continuing controversies regarding this vulnerable group are due in part to this lack of clarity in the guidelines. For example, some researchers argue that the washout studies described above pose an acceptable degree of risk, while others argue that these studies expose subjects to unacceptable research risks.

The NBAC spent two years studying ethical issues in research on the mentally ill or disabled and filed a report in 1998. In its executive summary (National Bioethics Advisory Commission 1998), the report recommends the following:

1. IRBs that review protocols for research on mentally ill or disabled subjects should include at least two members familiar with these disorders and with the needs and concerns of the population being studied.
2. For research protocols with greater than minimal risk, IRBs should require independent, qualified professionals to assess the potential subject's decision-making ability.
3. For research that involves greater than minimal risk and offers no prospect of direct medical benefits to subjects, the protocol should be referred to a national Special Standing Panel for a decision, unless the subject has a valid legally appointed representative to act on his/her behalf.

The NBAC concept of minimal risk is borrowed from the regulation on children (discussed below). As noted above, the federal government has not issued any new regulations based on the NBAC's recommendations.

CHILDREN

The use of children in research poses unique and challenging ethical dilemmas. As far back as 1931, German law prohibited research with children if it endangers the child (Wendler 2006). Most people would agree that it is ethically acceptable to expose children to risks in research if they are likely to receive direct benefits, such as medical treatment. For example, a phase III clinical trial of a new treatment for asthma that enrolls asthmatic children would offer the children the benefit of treatment. Enrolling children in this study would be acceptable as long as the risks to the children can be justified in terms of the expected benefits. More troubling is the use of children in research that offers them no direct benefit (Kodish 2005). For example, consider a proposed study of diabetes in children:

> The long-term aim of the proposed study is to increase understanding about the metabolic changes that precede the development of Type 2 diabetes in children, and the influence of Asian ethnicity on the diabetes risk. In this longitudinal study, it is proposed that a cohort of 450 healthy children be enrolled who are 8 to 10 years of age. Three hundred children will be of Japanese ancestry, while the other 150 children will be their Caucasian cousins. Each subject will have

two evaluations (two years apart, one at baseline and the other at the two-year follow-up). At each evaluation, the subjects will undergo the following procedures: (i) medical and family history; (ii) physical activity and dietary assessment; (iii) physical examination including pubertal staging; (iv) blood drawn by venipunture to measure the level of multiple hormones, lipids, and other serum factors and proteins; (v) intravenous glucose tolerance test which involves placement of an intravenous catheter, infusion of glucose, and serial measurements of serum levels of glucose, and insulin, and other factors; and (vi) measurement of body composition by means of DEXA and intra-abdominal fat determination by MRI. (U.S. Department of Health and Human Services 2001)

This study has an important goal: to advance our understanding of metabolic changes that lead to diabetes in children and the role of Asian ancestry in diabetes risk. Knowledge gained from this study could definitely help to improve children's health. However, the study would administer an intravenous glucose tolerance test to healthy children, which would expose them to a significant amount of risk and discomfort without the prospect of any direct benefit, such as medical treatment. Is it acceptable to expose these children to significant risks for the sake of other children? We allow adults to choose to risk their own health and well-being for the common good, because adults have a right to autonomy. We usually protect children from risks, however, because they lack autonomy (i.e., DMC). The best interest standard, a principle of ethics and jurisprudence for making decisions for children and incompetents, holds that we should make decisions that are in the child's best interests (Kopelman 2000a). If it is not in a child's interests to participate in a research project, because the project risks the child's health, then the child should not be in that project.

Strict adherence to the best interest standard has an unfortunate implication: It would seem to imply that all research on children that poses risks without offering benefits in return should be halted. Many would find this situation to be ethically unacceptable because it would stop research on vaccines and many new treatments for children. Even low-risk research, such as research that involves only surveys or questionnaires or research that involves only observations of children in their natural environment, would be unethical under the best interest standard. If we agree that pediatric research is important and should go forward, we need to ask what is the level of risk that a child may be exposed to in research when the child is not expected to benefit from the research (Kopelman 2000a). Ethicists have been debating this issue since the 1970s, when philosopher Richard McCormick (pro-children's research) and theologian Paul Ramsey (anti-children's research) wrote on opposite sides of the debate. Since then, three basic approaches have crystallized (Friedman Ross 2006):

Position 1: Children cannot participate in risky research unless they are likely to receive direct benefits that justify the risks.
Position 2: Children, as members of society, should make some sacrifices for the common good, such as participating in research that entails risks but no benefits.

Position 3: Children should be able to participate in research that imposes risks but offers no direct benefits (e.g., treatment) but can offer indirect benefits, such as learning the value of contributing to society or enhanced self-worth.

We think that position 3 is the most reasonable one. Parents frequently enlist their own children in risky activities that benefit society, such as working at a soup kitchen or building a home under the Habitat for Humanity program, to teach them moral lessons. Research is no different. Children can participate in research when these indirect benefits offset the risks. The problem with position 1, which we have already alluded to, is that it is too restrictive and would stop important research involving children. The problem with position 2 is that it is not restrictive enough and would seem to place no limits on the risks that children may be exposed to for the common good. Position 3 is the most reasonable one because it allows children to participate in risky research but places limits on those risks. The risks must be offset by the indirect benefits that the children may receive, such as learning the importance of making a contribution to society. Under position 3, participation by children in risky research would not be morally obligatory, but it would be morally acceptable and even commendable.

As noted above, one of the consequences of the restrictive children's research policies is that there has not been enough drug testing on children. Before 1998, all drugs were tested on children on a voluntary basis. The pharmaceutical industry avoided extensive research with children because of federal policies (discussed earlier) that limited the risks that children could be exposed to in research and because of fear of liability. As a result, 70–80% of drugs approved by the Food and Drug Administration (FDA) were not labeled for pediatric use (Food and Drug Administration 2003), and most of the drugs prescribed for children were used on an "off-label" basis. "Off-label" prescribing of drugs to children can be risky, because it requires physicians to make guesses about appropriate doses and schedules.

The U.S. government took some steps to encourage more drug testing on children. In 1998, the FDA mandated that the pharmaceutical industry test drugs and biological products on children if they are to be used on children. The logic behind this regulation is the same as that requiring the inclusion of women and minorities in research trials (Tauer 1999). In 2002, Congress passed the Best Pharmaceuticals for Children Act (BPCA), which gives pharmaceutical companies an additional six months of market exclusivity for new drugs if they test those drugs on children (Food and Drug Administration 2002). The BPCA was renewed in 2007.

Federal Pediatric Research Regulations

Earlier in this chapter we noted that children are not legally competent to consent to research, unless a court grants them this right. The federal research regulations do not define the age of majority, but leave this to each state. The Common Rule (45 CFR 46) includes a special set of regulations (subpart D) for the protection of children in research. Subpart D applies to research funded by the federal agencies that have adopted the Common Rule. It does not apply to U.S. Environmental Protection Agency (EPA)–sponsored research, since the EPA has adopted its own regulations pertaining to EPA-funded research involving children, fetuses, or pregnant women (Resnik 2007a). The main difference between the EPA regulations and the Common Rule is that the EPA regulations do not allow any intentional exposure studies involving children. Although the FDA has not officially adopted subpart D, it endorses the extra protections contained in this regulation. So, privately funded pediatric research submitted to the FDA should conform to the requirements of subpart D. Subpart D also does not apply to some types of research that have been exempted from the federal regulations, such as research on educational practices or instructional strategies in normal educational settings (45 CFR 46.101(b)).

Subpart D distinguishes between four types of research that are approvable under this regulation:

- Category I: Research not involving greater than minimal risk (45 CFR 46.404)
- Category II: Research involving greater than minimal risk but presenting the prospect of direct benefit to the individual subjects (45 CFR 46.405)
- Category III: Research involving greater than minimal risk and no prospect of direct benefit to individual subjects, but likely to yield generalizable knowledge about the subject's disorder or condition (45 CFR 46.406)
- Category IV: Research not otherwise approvable that presents an opportunity to understand, prevent, or alleviate a serious problem affecting the health or welfare of children (45 CFR 46.407)

The first two categories are the least controversial of the four. Category I is not very controversial because it involves only minimal risk. For this category to be approvable, the IRB must find that the research has no more than minimal risks for the subjects and that the investigators have made adequate provisions for soliciting informed consent of the parents/guardians and the assent of the research subjects (if appropriate) (45 CFR 46.404). Category II is not controversial because it encompasses research that is likely to be in the child's best interests. For this category to be approvable, the IRB must find that the risks to the subjects are justified by the anticipated benefits to the subjects, the relationship between benefits and risks is at least as favorable to the subjects as alternative available approaches, and investigators have made adequate provisions to obtain consent from parents/guardians and the subject's assent, where appropriate (45 CFR 46.405).

Categories III and IV are more controversial because they involve research that is more than minimal risk and does not offer the subjects any prospect of direct benefit. Category III research is approvable if the IRB finds that the risks are only a minor increase over minimal risk; the research interventions or procedures are similar to what the subjects experience or are likely to experience in their medical, dental, psychological, social or educational situations; the research is likely to yield important knowledge about the subjects' disorder or condition that is important for understanding the disorder or condition; and the investigators have made adequate provisions to obtain consent from parents/guardians and assent from the subject, where appropriate (45 CFR 46.406). Category IV research is approvable only if the IRB finds that the research represents an important opportunity to understand or alleviate a serious problem affecting the health or welfare of children, and the Secretary of the U.S. Department of Health and Human Services (DHHS) convenes a panel of experts from science, medicine, education, ethics, and law, who determine, following a public comment period, that the research satisfies the conditions that it represents an important opportunity to understand or alleviate a serious problem affecting the health or welfare of children and it will be conducted in accordance with sound ethical principles (45 CFR 46.407). Because category IV involves considerable administrative burdens, such as referring the research to the secretary of the DHHS and convening an expert panel, it is seldom used. Most investigators and IRBs try to make pediatric research protocols fit into one of the other three categories (Friedman Ross 2006).

The concept of minimal risk plays an important role in categories I and III in the pediatric research regulations, because category I refers to minimal risk and category III refers to a minor increase over minimal risk. There has been considerable debate and controversy surrounding minimal risk. Minimal risk is defined in the federal regulations as follows: "The probability and magnitude of harm or discomfort anticipated in the research are not greater in and of themselves than those ordinarily encountered in daily life or during the performance of routine physical or psychological examinations or tests" (45 CFR 46.102(i)). The definition consists of two ways of defining minimal risk. Minimal risk is either a risk not greater than risks ordinarily encountered in daily life or a risk not greater than routine physical or psychological tests. Since the federal regulations do not define risks "ordinarily encountered in daily life," there has been considerable disagreement about what this phrase might mean. There is also evidence that different investigators and IRBs interpret this phrase differently. In one study, 23% of IRB chairpersons classified allergy skin testing as minimal risk, 43% classified it as a minor increase over minimal risk, 27% classified it as a more than a minor increase over minimal risk, and 7% answered "don't know" (Shah et al. 2004).

There are two ways of interpreting risks "ordinarily encountered in daily life": a relativistic interpretation and an absolute one. According to the relativistic interpretation, daily life risks can vary according to the population and circumstances. For example, a child living in a ghetto probably encounters more risks than a child living in the suburbs. A child with a serious, chronic disease encounters more risk than a healthy child. A child who rides horses encounters more risks than one who does not. Kopelman

(2000b) argues against the relativistic interpretation on the grounds that it would lead to inconsistencies and injustices. The relativistic interpretation would lead to inconsistencies because different IRBs could classify different studies as minimal risk, depending on the population and circumstances. The relativistic interpretation would lead to injustices because some populations might be required to bear a greater burden of research risks than other populations, because they already encounter higher risks in their daily lives. To avoid these ethical problems, an absolute interpretation should be used (National Bioethics Advisory Commission 2001, Institute of Medicine 2004). Minimal risk should be the risk that a typical, healthy child ordinarily encounters or the level of risk encountered in routine medical examinations (Wendler et al. 2005).

In addition to distinguishing among different categories of approvable pediatric research, subpart D also states requirements for informed consent by the parents/guardians and the child's assent. The regulations require the IRB to determine whether the subjects in a research study will be capable of providing assent, taking into account the age, maturity, and psychology of the children (45 CFR 46.408). If the IRB determines that the subjects will be capable of assent, then the investigators must make adequate provisions for obtaining assent. The IRB may determine that assent is not necessary if the subjects are not capable of providing assent or the prospect of direct benefit to the subjects is very great and can only be obtained by participating in research (45 CFR 46.408). Note that assent is not the same as consent. A subject's assent merely indicates that they have been informed about a study and they approve it. It does not indicate that they fully understand the study or have freely chosen to participate in it. Despite the good intent of the regulations requiring assent, assent can be manipulated. For example, in some cases playing high school sports has been conditional upon assenting to urine drug tests for research (Shamoo and Moreno 2004).

The IRB shall also determine whether the investigator has made adequate provisions for obtaining consent from the parents/guardians. If the research is category I or II, the consent of one parent is needed. If the research is category III or IV, the consent of both parents is required, unless one parent is deceased, unavailable, or incompetent. The IRB may also waive parental consent requirements if parental consent would not be reasonable and waiving parental consent is consistent with state law (45 CFR 46.408). Parental consent might not be reasonable if the child is abused or neglected, for example.

PREGNANT WOMEN AND FETUSES

Involving pregnant women in research presents investigators and institutions with some difficult ethical and legal questions. On the one hand, pregnant women have a right to decide whether to participate in research, and they can also benefit from research that provides them with medical or psychological therapy. Additionally, it is important to learn about prescribing drugs during pregnancy and how to treat medical problems during pregnancy. On the other hand, including pregnant women in research may expose the fetus to risks. Even if one does not consider the fetus to be a human being with full moral or legal rights, one must still be concerned about the harms that may occur to the child in utero. The thalidomide tragedy of the 1950s and 1960s provides a stark reminder of the dangers of exposing the fetus to drugs. Thousands of children (mostly in Europe) were born with severe birth defects (e.g., missing or deformed limbs) as a direct result of in utero exposure to thalidomide, prescribed as a treatment for morning sickness (Stephens and Brynner 2001). Investigators and research sponsors have been wary of including pregnant women (or even women who could become pregnant) in research out of fear of the legal liability resulting from birth defects related to research. Concerns about how research procedures and interventions might affect the fetus are one reason that women were routinely excluded from research for many years (Dresser 2001).

Although pregnant women are not vulnerable subjects per se, fetuses are. To protect the fetus from harm, it is necessary to craft rules that apply to pregnant women. The federal research regulations and the CIOMS guidelines (Council for International Organizations of Medical Sciences 2002) both include extra protections for pregnant women and fetuses, which strike a balance between the rights and welfare of pregnant women and the welfare of the fetus (and rights of the future child). Subpart B of the Common Rule includes additional protections for pregnant women, human fetuses, and neonates in research. (We do not discuss research on neonates in this text.) The regulations allow pregnant women or fetuses to be involved in research if (a) the risk to the fetus is minimal and the purpose of the research is to develop important biomedical knowledge, or (b) if the risk to the fetus is more than minimal, and the research has the prospect of direct benefit to the woman, her fetus, or both (45 CFR 46.204). The regulations stipulate that only the woman's consent is required, unless the research is likely to benefit only the fetus, in which case the father's consent must also be obtained. The father's consent need not be obtained if he is unavailable, incompetent, or incapacitated or the pregnancy resulted from rape or incest (45 CFR 46.204). The regulations also require that the investigators not offer an inducement to terminate a pregnancy and that they are not involved in any decisions relating to the termination of the pregnancy. Investigators also must play no role in determining the viability of the fetus (45 CFR 46.204).

SUBORDINATE POPULATIONS

A subordinate population is one that is under the control of some institution of authority. Subordinates may lack power, have compromised DMC, or both, due to their position. They may face coercion, manipulation, intimidation, duress, or other circumstances that can interfere with their ability to make free choices. Prisoners are the extreme example of a subordinate population, because they are institutionalized and their lives are under tight control of authority figures. Prisoners may feel pressured to participate in research to reduce their sentence or gain prison privileges. They may face harassment or intimidation from guards or other inmates (Pasquerella 2002). Military personnel have more freedom and power than do prisoners, but their lives are still controlled by rules, protocols, and superior officers. Soldiers who disobey an order can face severe punishments, ranging from a reprimand to imprisonment. Soldiers have been ordered to participate in experiments on the grounds that their participation was necessary to promote military goals (Moskop 1998). Employees and students are also in positions where they are under the power of an authority and may face coercion or duress (Levine 1988). An employer may threaten to fire an employee who does not participate in research, and a teacher may threaten to give a student a bad grade for refusing to participate in an experiment. Because they are susceptible to harm and exploitation, subordinate populations require extra protections in research.

Prior to the National Commission's report in 1979, the use of prisoners in research was common. Current federal regulations reflect the National Commission's recommendation for special restrictions on the recruitment and use of this population as human subjects in research. Prisoners are compensated only for discomfort and time spent in research. The normal compensation package for adults outside the prison could be regarded as coercive and exploitative in the prison environment because most prisoners would prefer research participation to the daily boredom of prison life. One might argue that most prisoners would not participate in research if they were not in prison. There are also problems with maintaining confidentiality in the prison environment. In his book *Acres of Skin*, Hornblum (1998) chronicles how in the 1960s and 1970s researchers used the skin on the backs of prisoners to test numerous drugs and perfumes for toxicity and carcinogenicity. The experiments were conducted at Holmesburg Prison in Philadelphia, by University of Pennsylvania researchers. The records of these experiments are "lost." Several ethical issues came to light: Subjects received payment, housing for the experiment was better than that provided for other prisoners, the human contacts during the experiments may have been unduly coercive, and informed consent was barely informative.

Subpart C of the Common Rule provides additional safeguards for prisoners when they are enrolled as research subjects. Research protocols may involve prisoners only if (a) the research is of minimal risk related to the causes, effects, and process of incarceration, or (b) the study is of the prison and inmates as an institution, or (c) the research is on conditions of prisoners as a class, or (d) the research has the potential to improve health of wellbeing of the research subject (45 CFR 46.306). When prisoners are used in a control group in which other prisoners in the study are likely to benefit, the Secretary of the DHHS can approve this research after consultation with experts and receiving public comments.

The definition of "minimal risk" is different for prisoners than for other populations addressed in the Common Rule. "Minimal risk" is defined in terms of daily life risks encountered by healthy people or the risks of routine physical or psychological examinations (45 CFR 46.303). The intent of this definition is to make the standard of minimal risk for prisoners no different from the standard of minimal risk for nonprisoners. By making the reference to the risks of "healthy people," it avoids the relativistic interpretation of minimal risk mentioned above. A prisoner is defined as any person involuntarily confined to a penal institution. This includes people awaiting trial, arraignment, or sentencing and people who are punished under alternatives to incarceration, such as home confinement (45 CFR 46.303).

In addition to these regulations, the Common Rule also requires that an IRB that reviews research on prisoners must include at least one member who is a prisoner, a prisoner representative, or someone with appropriate background to speak for prisoners. The majority of the IRB members must have no formal association (e.g., an employment relationship) with the prison involved in research (45 CFR 46.304). The regulations also include provisions to deter undue inducement, coercion, or duress in the consent process.

Note that an advisory group convened by the Secretary of the DHHS has been contemplating possible revisions to the Common Rule, including revisions to the regulations pertaining to prisoners. There is a general consensus among researchers and ethicists that the rules in subpart C are overly restrictive and outdated. It is possible that the subpart C will have been revised before our book appears in print.

As we have mentioned in several chapters, military personnel have been used in ethically questionable experiments, such as human radiation experiments. Because most military research is classified, there is probably a great deal more ethically questionable research that the public does not know about (Moreno 2000). Even so, the military's research record is not entirely bleak. Walter Reed's yellow fever experiments, described in chapter 12, were an example of responsible conduct of research by the military. Also, the Department of Defense was first among all federal agencies to adopt the Nuremberg Code in 1953. Despite the Department of Defense's (DOD) adoption of the Nuremberg Code, aberrations occurred (Moreno 1998, 2000). Both the Common Rule and FDA regulations apply to military research. In addition, military branches have their own rules and regulations.

One of the most controversial episodes of military research with human subjects occurred during the first Gulf War (1990–1991), when the DOD obtained an informed consent waiver from the FDA to administer anthrax vaccines to thousands of soldiers in the war without their consent. The military wanted to vaccinate soldiers against anthrax because it was thought that Iraq had developed and stockpiled biological and chemical weapons, including weapons-grade anthrax dust. The vaccine was an investigational new drug. There were no published studies of the safety or efficacy of the vaccine in humans prior to the war. The military also deviated from recommending dosing schedules. The military's rationale for giving the vaccine without informed consent is that it would not be feasible to obtain informed consent in a battlefield setting. Also, if soldiers refused, they could endanger other soldiers and military operations if they contracted anthrax. Some soldiers did refuse the vaccine, and they were court-martialed and punished. Many soldiers suffering from Gulf War illness claim that their mysterious disease was caused by exposure to the anthrax vaccine. An FDA review of the military's procedures found that they deviated from the FDA's approved plan for testing the vaccine. For example, the military convened a second IRB to approve the experiment after the first one determined that it was unethical (Moreno 2000, Cummings 2002).

Employees are sometimes asked to participate in research studies conducted by their employers. Although employees have much more freedom and power than do prisoners and soldiers, they still may face coercion or intimidation during the consent process. Although employees are always free to quit their jobs if they do not want to do what their employers ask them to do, employees may need to keep their jobs out of economic necessity. In chapter 1, we mentioned that Woo Suk Hwang used his employees (lab technicians) as egg donors for his stem cell experiments. Another case of an unethical experiment involving employees occurred when Novartis asked company managers to ingest the pesticide diazinon as part of an experiment (Resnik and Portier 2005). For employer-sponsored experiments involving employees to be ethical, great care must be taken to safeguard the employees' ability to consent (or refuse to consent) to research as well as the employees' privacy and confidentiality. One way to do this is for the employer to hire an independent contractor to conduct the study. The contractor, not the employer, would have access to the names of people who volunteer for the study. Since the employer would not know who participates (or does not participate) in the research, the employer will not be able to reward employees for participating or penalize employees for not participating. Employees who volunteer for this type of research should be assured that their participation will in no way affect their employment status, salary, and so on.

Finally, students often participate in research conducted by their professors. These studies range from filling out self-administered surveys of behaviors, attitudes, or opinions distributed in psychology or sociology classes, to providing biological samples (e.g., blood, tissue, urine) for chemical or genetic analysis, to participating in controlled behavioral experiments, such as Stanley Milgram's obedience of authority experiments in the 1960s (see chapter 12) (Moreno 1998). Students, like employees, may face coercion, undue inducement, or intimidation during the consent process. To protect the rights and welfare of student research participants, the student's privacy must be protected, and the student must be free to consent (or refuse to consent) to participate in research. Students should not be punished for refusing to participate in research. Participation in research should not be a part of the course grade, unless the professor gives the student an alternative to research participation that takes the same amount of time and effort, such as writing a short essay.

QUESTIONS FOR DISCUSSION

1. How would you design a system to ensure enrollment of mentally ill individuals into research studies while protecting them from harm? Why?
2. Should there be decisional capacity assessment for the mentally ill? All of them? Why? When and how?
3. Should we ever do research on children? Infants? Newborn? Fetuses? How?
4. Should we conduct surgical research on infants? How?
5. What safeguard would you design for a professor who wants to conduct research on his students?
6. Should the military conduct research on blood substitutes with soldiers who are wounded on the battlefield and potentially could die with no access to fresh blood? If not, should you do such research on civilian trauma patients?

CASES FOR DISCUSSION

Case 1: Fenfluramine Challenge Studies on Children in New York City

In four studies funded by the NIH and conducted at Columbia University in New York City during the early 1990s, 68 healthy children and 66 children diagnosed with attention-deficit/hyperactivity disorder (ADHD) were enrolled in a nontherapeutic experiment (with no potential direct medical benefit). The purpose of the study was to investigate possible correlations between serotonergic activity and aggression. There are some data to indicate that an increased secretion of serotonin may be associated with aggression. Some of these children were identified as having aggressive behavior. The rest of the children were identified as having potential aggressive behavior because they had siblings who were incarcerated or had family, social, genetic, or environmental factors associated with aggressive behavior. The rationale for this experiment is that administering a dose of fenfluramine could greatly increase the rate of secretion of serotonin in the experimental group compared with the control group, which could reveal an association between serotonergic activity and aggressive behavior.

Fenfluramine is one of the active components of the "Fen-Phen" anti-obesity drug combination, which has significant known risks, such as cardiac damage. ADHD itself is recognized as a risk factor for aggression. All children were African Americans or Latinos from very poor environments, and all were receiving drug treatment. The children were given a gift certificate of $25, and the parents were given $100 to $150 each. Incarcerated siblings were identified from sealed court records without permission from the courts or the city administration. Afterward, the researchers contacted the subjects and their siblings at their homes and identified themselves as persons conducting research funded by the federal government. The study was approved by several IRBs. NIH study sections also approved the original research proposal.

The procedure required subjects to undergo a battery of tests, including genetic tests. The subjects then underwent an overnight fast and were administered a challenge dose of fenfluramine, with blood samples collected for several hours. Samples of spinal fluid (by spinal taps) were collected every 3 hours for 12 hours. Blood levels of serotonin were monitored continuously through a vein catheter.

Because the study offered no potential direct medical benefits, it could not be approved under the federal designation 45 CFR 46.404 or 46.405 (category I or II). It could only be approved under designations 46.406 or 46.407 (category III or IV; see definitions above).

In the fall of 1997, the human rights organization, Citizens for Responsible Care and Research (CIRCARE), cofounded by one of the authors (Shamoo), called for an investigation by the federal oversight agency, the Office of Protection from Research Risks (OPRR), the predecessor of the Office for Human Research Protections. CIRCARE also alerted the media and other advocacy groups. The determination letter from OPRR claimed that using healthy subjects was appropriate, whereas using ADHD subjects was not appropriate. Also, OPRR said that the informed consent document did not adequately inform the parents. Unfortunately, the OPRR determination was based on technical matters such as whether there was a properly convened meeting of the IRB, not on substantive matters concerning the ethics of the study (Shamoo and Tauer 2002, Koocher 2005).

- Were the risks to subjects minimized?
- Were the benefits sufficient to justify the study?
- What kind of informed consent process could be adequate?
- Was the selection of subjects equitable?
- Was confidentiality of subjects, family, or their siblings protected?
- What can we learn from this episode?

Case 2: The Kennedy Krieger Lead Abatement Study

Lead poison is a serious problem in houses built prior to 1980. There are data to indicate that lead poisoning in children causes serious deficits in the neurodevelopment. Between 1995 and 1998, investigators from the Kennedy Krieger Institute (KKI) at Johns Hopkins University enrolled 50 poor families with very young children living in Baltimore for a lead abatement experiment. The families were randomly assigned to receive one of three different abatement programs to reduce lead, or to be in one of two control groups. The three different abatement programs, which were less expensive than complete lead abatement, ranged in cost from $1,650 to $7,000. The two control groups included one group of families with full lead abatement and one living in housing without lead paint. The recruited families had children ranging in age from 6 months to 7 years. The investigators collaborated with the landlords on this experiment. Landlords were encouraged to recruit families with young children and in some cases were told that they must have young children. The investigators helped the landlords apply for grants for lead abatement. The lead level in each house was measured by wipe tests and air samples. Each house was tested before and after abatement and while the families were living in these houses. The blood level of lead in each child was also measured at different time periods. The informed consent document stated that once the investigators know the child's blood lead level, "we would contact you to discuss a summary of house test results and steps that you could take to reduce any risks of exposure." The study was sponsored by the EPA and the Maryland Department of Housing and Community Development.

Families of two children sued the KKI for negligence. The suit alleged that the investigators failed to inform the parents in a timely fashion about dangerous lead levels and failed to provide adequate informed consent (*Grimes v. Kennedy Krieger Institute, Inc.* 2001). One of the children, Erika Grimes, was ten months old when she first lived in the house, and the other child, Myron Higgins, was three and a half years old. The acceptable lead level for a child is 10 µg/dl. KKI found that Erika Grimes's blood level was higher than normal. The allegations are that KKI waited nine months to inform the Grimes family. Her original blood level was 9 µg/dl and subsequently her blood level was 32 µg/dl and 22 µg/dl. Medical treatment or intervention is indicated at 25 µg/dl. Erika then was moved out of the house. The defendants moved to dismiss the case, on the grounds that the researchers had no legal duties to the subjects that could justify a negligence claim. The circuit court agreed. The plaintiffs appealed the case to the Maryland Court of Special Appeals. The appellate court overruled the lower court's decision on the grounds that the investigators had legal duties to the research subjects because they have a "special relationship" with the subjects. The appellate court remanded the case to the circuit court for finding consistent with its opinion, and opined that children can only participate in research that poses no risk or research in which they are expected to receive a direct medical benefit. This opinion sent shock waves through the research community, as it would invalidate (in the State of Maryland) research conducted in accordance with 45 CFR 46.405–407. In response to a swell of protest from research institutions, the appellate court issued a clarification that their order does not bar children's participation in nontherapeutic research with minimal risk. In discussing the case, the appellate court also stated: "There was no complete and clear explanation in the consent agreements signed by the parents of the children that the research to be conducted was designed, at least in significant part, to measure the success of the abatement procedure." The court went further and showed its dismay at the design of the study and stated that "researchers intended that the children be the canaries in the mines, but never clearly told the parents" (Shamoo 2002; see also Kopelman 2002).

- What regulatory rule would govern this study, and which section?
- Were the risks minimized?
- Were the risks justified in relation to benefits?
- Was the selection of subjects equitable?
- Was informed consent adequate?

Case 3

The aim of a study is to better understand condom use among adolescents and the psychosocial factors that increase or decrease condom use. The study is a survey of adolescent attitudes and beliefs about sexuality. It will include many different questions about sexuality as well as questions about alcohol and drug use, violence, musical tastes, and religion. The subjects will not be told the exact purpose of the study when they take the survey, but they will be told the results of the study. Subjects will be recruited from three local high schools. Personal identifiers will be removed for data analysis and publication. High school health education teachers will help administer the survey. Taking the survey will suffice as proof of consent. Subjects may refuse to take the survey without penalty. Parents will be notified about the study and may refuse to allow their children to participate.

- In what specific category of risk (discussed above) would you place this protocol? Why?
- How would you protect the privacy of subjects?
- Why should the parents be involved?
- Should the community be concerned?

Case 4

Subjects with Alzheimer's disease will be recruited from 10 nursing homes in the area. Subjects or their legally appointed representatives will give consent. Subjects will provide a blood sample for genetic testing. Personal identifiers will be removed from the samples, although researchers will retain the ability to link samples to subjects via a code. Researchers will develop a DNA database and attempt to find common genes associated with the disease, including variants of the *APOE* gene. Researchers will also compare the database with a database drawn from a matched cohort of patients without the disease.

- Do you have any concerns? If so, what are they?
- If your father or mother were one of the patients recruited, would you encourage your parent to enroll?

Case 5

A full professor/researcher at a major research university is a member of the National Academy of Sciences. In her research protocol, she describes briefly how she proposes to take 100 schizophrenia patients off their medications for four weeks (a washout period) while they are outpatients. The protocol calls for randomly selecting 50 as controls, so after washout, these 50 will receive placebo for the duration of the experiments (60 days), and the other 50 will receive a promising new drug that presumably will have fewer side effects. A colleague in a different department at the university raises some concerns during IRB deliberations.

- What you think the IRB member's concerns were?
- What are the risks/benefits of this protocol? Do the benefits outweigh the risks? Why?

- How would you deal with these concerns?
- Should the protocol proceed?

Case 6

An informed consent form describes a research protocol briefly. The protocol involves the washout of 60 patients from their current antidepressant drug. The research protocol dwells on potential worsening of the patients' condition if they are taken off their current medication but does not cite literature discussing that the washout may worsen prognosis in the future. The informed consent form makes the passing reference, "You may experience some symptoms of depression." The forms do mention alternative medications to take instead of their current medication or the experimental one.

- What are the risks and benefits? Do the benefits outweigh the risks? Why?
- How much should the informed consent form reflect the protocol? How much should it reflect the literature?
- What do you think of potentially scaring patients away from participation in the study?

Case 7

A research protocol calls for using 500 healthy young children to test a new chicken pox vaccine. The new vaccine's safety has been tested in adults, and all indications are that it is fairly safe, and probably safer than the current vaccine for the same disease. Half of the children will receive a placebo, and half will receive the new vaccine. All 80 children were screened to make sure that they have not yet received any vaccine for this disease and have not been exposed to chicken pox. Informed consent forms are very detailed on risks and benefits. However, the forms do not mention that those receiving the new vaccine will not be able to receive the old vaccine.

- Should the informed consent forms have mentioned the fact that the children can no longer obtain the old vaccine?
- Was the description of risks and benefits appropriate?
- Would you allow your children to enroll?

14

Genetics, Cloning, and Stem Cell Research

This chapter provides an overview of ethical, social, and policy issues related to research on human genetics and embryonic stem cells. It covers such topics as genetic engineering of animals and plants; genetic testing, privacy, and discrimination; storage of genetic samples; returning research results to subjects; family genetic testing; somatic gene therapy; germline manipulation; intellectual property concerns; embryonic stem cell research; and cloning.

EUGENICS AND HUMAN ENHANCEMENT

Since the topics covered in this chapter are so vast and varied, it is difficult to decide the best way to organize them. We have decided to approach these topics in (more or less) chronological order. Charles Darwin's (1809–1882) theory of evolution by natural selection is our starting place. In 1859 in *The Origin of Species*, Darwin produced evidence that species can change over time in response to the environment, and that all life forms, including humans, are descended from common ancestry. The initial reaction from many religious groups was to brand Darwin's theory as sacrilege because it contradicted the creation story in the Book of Genesis, removed man from the center of God's creation, and made him just another animal. In 1927, the famous John Scopes "Monkey Trial" pitted creationists against evolutionists over teaching Darwin's ideas in the public schools. This battle continues today, as evidenced by the debate in Kansas about teaching evolution in public schools (Dennett 1995, Dalton 1999).

Another idea that had a profound effect on science, social policy, and politics was the notion of improving our species through selective breeding. People began to realize that the same principles of selective breeding that apply to plants and animals might also apply to the human population. This led to the idea of eugenics, or controlled breeding of the human population to promote "desirable" or eliminate "undesirable" characteristics. Founders of the eugenics movement include Francis Galton, who argued for selective breeding in human beings, and Herbert Spencer, who defended the "struggle for existence" as a principle of social policy known as social Darwinism. According to this idea, society should allow its weakest members to perish

Note: This chapter includes material that presumes a basic undergraduate level of biological knowledge. We assume that readers are familiar with basic concepts in biology, such as cells, DNA, RNA, genes, genomes, proteins, chromosomes, embryonic development, evolution, mutation, genetic disease, organs, and tissues. We explain some technical terms as they occur. For readers who would like consult a biology textbook to have a better understanding of this chapter, we recommend Campbell and Reece (2004).

and should encourage the survival of its strongest members. This idea was used to justify laissez faire capitalism as an approach to social and economic policy, that is, capitalism with little government regulation and no social safety net. In the 1890s, the German philosopher Friedrich Nietzsche incorporated some ideas of Darwinism into his writings. Nietzsche described existence as "will to power" and envisioned the emergence of an *ubermensch* (superman) who would be superior to normal human beings. Nietzsche's writings influenced many writers, philosophers, and politicians, among them Adolf Hitler (1889–1945).

During the early twentieth century, many U.S. states and countries in Europe adopted eugenics laws and policies, including immigration quotas based on race and ethnicity and mandatory sterilization laws for the "feeble-minded." In the United States, the Supreme Court upheld mandatory sterilization laws in *Buck v. Bell* (1927). Nazi Germany extended eugenics ideas even further, with mandatory sterilization for people with schizophrenia, blindness, alcoholism, feeble-mindedness, and physical deformities. The Nazis also sterilized people of Jewish ancestry and enacted laws forbidding Aryans to breed with Jews, to promote "racial purity." They eventually exterminated millions of Jews and other "impure" or "inferior" people in the name of cleansing the Fatherland (Kevles 1995). The horrors of Nazi eugenics led U.S. states to repeal their eugenics laws. Today, no U.S. states have mandatory sterilization laws.

The specter of eugenics continues to haunt research on human genetics and reproduction. Many scholars, politicians, and organizations are concerned that research in these fields will be used to try to eliminate undesirable traits or promote desirable ones in the human population or to create "perfect" children (Kass 1985, 2004, Annas and Elias 1992, McGee 1997, Andrews 2000, Fukuyama 2003, President's Council on Bioethics 2003, 2004, Sandel 2007). Many of the new reproductive technologies give parents the ability to influence the traits of their children. Some of these include:

- *Sperm sorting:* This technique involves using a method to sort sperm by their Y or X chromosome. Since an egg contains a single X chromosome, if an X sperm fertilizes the egg, this will yield the combination XX, which will produce a female child. If a Y sperm fertilizes the egg, this will produce an XY combination, producing a male child. Sperm sorting has been used by couples to avoid giving birth to a child with a sex-linked genetic disease, such as hemophilia (President's Council on Bioethics 2003).
- *Gamete donation:* These techniques are useful for parents who have fertility problems because the man does not produce enough sperm or produces damaged sperm, or the woman does not produce eggs. In sperm donation, a male sperm donor provides a sperm sample that is used to fertilize a woman who is ovulating. Fertilization could also take place in vitro, using an egg harvested from a woman. The embryo can then be implanted in the womb. In egg donation, eggs

are harvested from egg donors, are fertilized in vitro, and then are implanted in a womb. If both the man and women have fertility problems, sperm and egg donation can be employed. Infertile couples can select the gamete donor based on his or her characteristics, such as height, eye color, hair color, cancer history, educational level, and IQ (President's Council on Bioethics 2003).

- *Prenatal genetic testing (PGT):* PGT is useful for parents who are concerned about giving birth to a child with a genetic disease that can be diagnosed in utero, such as Down's syndrome or Tay Sachs disease. Some PGT techniques include amniocentesis, which tests the amniotic fluid, and chorionic villi sampling, which tests a sample of the placenta. Both of these tests involve some risk to fetus. In the future, it will be possible to perform a genetic test on the fetus by taking a sample of blood from the mother. If parents conduct PGT and discover that the fetus has a genetic disease, they may decide to abort the fetus, provided that the law allows this option (President's Council on Bioethics 2003).

- *Preimplantation genetic diagnosis (PGD):* PGD is useful for parents who are concerned about giving birth to a child with a genetic disease that cannot currently be diagnosed in utero, such as Huntington's disease or cystic fibrosis. In PGD, parents create a number of embryos in vitro, which are tested for genetic abnormalities. Parents can choose to implant embryos that do not have the abnormalities being tested. Embryos with the defective gene(s) can be discarded (President's Council on Bioethics 2003).

In addition to these technologies, there are some that have been used successfully in animals but not in human beings (yet), such as the following (both are discussed in depth further below):

- *Cloning:* An infertile couple could use cloning techniques to procreate if neither parent has viable gametes (Pence 1998).

- *Germline manipulation (also known as germline gene therapy):* A couple could use recombinant DNA technologies to avoid giving birth to a child with a genetic disease influenced by multiple genes, such as cancer, alcoholism, or diabetes (Walters and Palmer 1997, Resnik et al. 1999).

While all of these different reproductive technologies have helped, or could help, infertile couples to procreate or to avoid giving birth to children with genetic diseases, they can also be used for nontherapeutic purposes. For example, a couple could use sperm sorting to increase their chances of having a female child because they have a male child and want one of each sex. A couple could use gamete donation to try to increase the chances that their child will be at least six feet tall or have good athletic ability. A couple could use PGT to abort a child that carries a gene for homosexuality, and a couple could use PGD to select an embryo of a particular sex or an embryo that is likely to have blue eyes. A couple with a child with leukemia could use PGD to create a child to be a bone marrow donor for the diseased sibling. A deaf couple could use PGD to select for a child that carries a gene for deafness. And the list goes on.

Many people are uncomfortable with nontherapeutic uses of reproductive technology because they think that medicine should be used for therapeutic purposes, that is, to diagnose, treat, or prevent disease. With the possible exception of creating a child to be a bone marrow donor, these other uses of reproductive technology would be for nontherapeutic purposes, or enhancement (Annas and Elias 1992, Andrews 2000, President's Council on Bioethics 2003, 2004, Kass 2004, Sandel 2007). Human enhancement is a broad topic that we do not cover in depth here. Instead, we briefly list some arguments against genetic enhancement and some in favor of it (for further discussion, see Parens 1999, Resnik et al. 1999):

- *Enhancement is unsafe:* It can cause physical or psychological harm to the child who is enhanced. Physical harm could occur as result of genetic abnormalities caused by reproductive technologies, such as cloning or manipulation. Psychological damage could occur if the child is not accepted by other members of society or feels constant pressure to live up to parental expectations and demands (Kass 1985, 2004). Even worse, genetic mistakes could be passed on to future generations.

- *Enhancement contributes to discrimination:* Parents could use enhancement to select against traits associated with specific racial or ethnic groups or to select against particular sexes. People with genetically based disabilities, such as blindness, deafness, or Down's syndrome, would face increased discrimination as they would be regarded as inferior beings who slipped through the "gene screen" (Parens and Asch 2000). Even "normal," nonenhanced people may face discrimination if they are regarded as imperfect.

- *Enhancement leads to social injustice:* Since most reproductive technologies are likely to be expensive for quite some time, only the rich will be able to afford them. They will be able to use reproductive technologies to produce children with "superior" traits, who will give birth to children, who will in turn also use these technologies. Over time, the rich will get richer, and the genetically "superior" people will increase their superiority. The socioeconomic and cultural divide will continue to grow (Silver 1998, Buchanan et al. 2001).

- *Enhancement is cheating:* Life is a series of competitive games such as competition for grades, jobs, athletic achievement, wealth, and so on. People who have artificially enhanced abilities and talents through drugs or genetic modifications gain an unfair advantage over others (Parens and Asch 2000).

- *Enhancement threatens human dignity:* As enhancement becomes widely used, parents will view their children as products or commodities, subject to manipulation or control, not as people to be loved and valued for their own sake (Fukuyama 2003, Kass 2004, Sandel 2007). Respect for the value, uniqueness, and importance of human life will also decline as it becomes easier to design human beings for specific purposes. Enhancement could also be misused to produce supersoldiers, slaves, monsters, and other horrors (Rifkin 1983).

- *Enhancement could threaten human genetic diversity:* If people use reproductive and genetic technologies to promote "desirable" trait, the genetic structure of the human population will change over time. For example, if everyone decides that blue eyes are better than other eye colors, the genes that code for other types of eye pigments will disappear from the human population. Over time, the human population could lose valuable genetic diversity as a result of selective breeding in the same way that maize (known as corn) has lost genetic diversity (Rifkin 1983, 1998).
- *Enhancement is "playing God":* We may acquire the technical skill to genetically engineer human beings, but we lack the wisdom. To assume that we could know how human beings should be designed is arrogant.

Those who defend enhancement have offered critiques of all of these arguments. In addition, they have proposed some arguments in favor of enhancement:

- *Enhancement honors human reproductive freedom:* Parents should be free to make reproductive choices as long as those choices do not harm other people. If a reproductive or genetic technology has been proven to be safe and effective, parents should be able to use it without interference from the state (Robertson 1994, Harris 2007).
- *Enhancement is an expression of human freedom and potentiality:* Far from "playing God," enhancement is a way of fulfilling our natural desire to strive for improvement.
- *Enhancement can improve human welfare:* Genetic and reproductive technologies could be used to create people with advanced intelligence, physical strength, agility, musical ability, disease resistance, and other talents and traits that could benefit humankind (Stock 2003, Silver 1998, Harris 2007). Enhanced humans could make significant contributions to science, medicine, technology, industry, and the arts.
- *Enhancement cannot be stopped:* It is very difficult to prevent human beings from obtaining things that they strongly desire. Attempts to criminalize marijuana and other drugs have created an enormous black market. Human organs are also bought and sold on the black market. In the United States, immigrants pour cross the border every day looking for work, and employers are eager to hire them. It is likely that technologies that enhance human traits will be highly desired and that people will find a way to obtain these technologies on the black market, if they are made illegal. Since a black market

pertaining to reproductive and genetic technologies would be very dangerous to children created by these technologies and to society, enhancement should be regulated, not banned (Resnik et al. 1999, Baylis and Robert 2004).

And, of course, opponents of enhancement have critiqued these arguments in favor of enhancement.

We do not attempt to resolve the debate about genetic enhancement, but we make a few observations that we believe are relevant for evaluating the positions (for and against). First, the distinction between enhancement and therapy is tenuous because it rests on the imprecise concept of disease (Resnik et al. 1999, Buchanan et al. 2001, Shamoo and Cole 2004). An intervention is regarded as therapeutic if helps to treat or prevent a disease. Enhancements, then, are interventions designed to do something other than treat or prevent a disease. According to an influential definition of disease, a disease is a deviation from normal functioning that causes a person to suffer or experience ill effects (Buchanan et al. 2001). For example, congestive heart failure (CHF) is disease in which the human heart's ability to pump blood is compromised due to weakness of the cardiac muscle. As a result, people with CHF experience shortness of breath, fatigue, edema, and other symptoms that interfere with life. Most people can agree that CHF is a disease, and there is little controversy concerning its diagnosis.

However, there are many other diseases where the definition is not so clear and the diagnosis is disputed because social, economic, or other factors play a large role in determining what counts as normal functioning. For example, attention-deficit/hyperactivity disorder (ADHD) is defined by reference to a cluster of behaviors exhibited by a person with this condition, such as inattentiveness and difficulty sitting still. Children who are diagnosed with this condition are often treated with drugs, such as Ritalin, which can help improve school performance. But what is the difference between a child with mild or moderate ADHD and a healthy child who simply doesn't like school or is a troublemaker? This is not an easy question to answer. Whether a child is diagnosed with ADHD often depends on social and economic factors. For example, parents or teachers may take steps to have children classified as ADHD so they will be easier to manage. Some commentators believe that ADHD is overdiagnosed (Baughman 2006, Conrad 2007). Indeed, in countries that have very different educational systems (or none at all), such as many developing nations, ADHD is not a concern at all. Other diseases in which diagnosis and treatment depend on social, economic, or cultural factors include depression, anxiety, short stature, chronic fatigue syndrome, premenstrual syndrome, erectile dysfunction, male menopause, baldness, and chronic pain (Shamoo and Cole 2004, Conrad 2007). The general point here is that the difference between disease and health is often not a clear and distinct line drawn by human biology and that there is a strong tendency to medicalize social problems (Conrad 2007). We also note that it was not very long ago that homosexuality was regarded as a disease.

If the difference between disease and health often depends on social, economic, political, and cultural factors, then the difference between therapy and enhancement does also. So, the definition of genetic enhancement (or therapy) is not as clear as one might think. If this is the case, then those who argue against enhancement should recognize that enhancement may not be easy to ban or even regulate, because government control of an activity is possible only if one can define it. If the government attempts to control genetic enhancement, what counts as "enhancement" may become a highly contested social and political issue. Moreover, the difference between therapy and enhancement may be a moving target, as technologies that were created for therapeutic purposes are used for other purposes (Rothman and Rothman 2004). Rather than banning or regulating enhancement technologies per se, the best strategy may be to evaluate each new technology on its own terms, examining its benefits and risks. This is what the U.S. Food and Drug Administration (FDA) currently does with new drugs, biologics, and medical devices. Although many drugs can be used for nonmedical (enhancement) purposes, the FDA does not ban this type of use. Instead, it approves products for specific uses.

Second, the idea of genetic determinism—the notion that our genes causally determine our physical and behavioral traits—plays a large role in arguments against and for enhancement. Opponents of genetic enhancement are concerned about the negative consequences of using genetics and reproductive technology to control human traits, while, proponents of genetic enhancement are excited about the possibility of improving human traits. This is especially true in mental disorders, such as in schizophrenia and bipolar disorders, where there is a 30–80% rate of homozygosity. Genes play a critical role in the manifestation of the disease, so gene therapy will be extremely beneficial (Shamoo and Cole 2004). Nevertheless, in general, both of these views—one pessimistic, the other optimistic—vastly underestimate the difficulty of using genetics and reproductive technology to manipulate, control, or engineer human traits. All human traits are produced by many different genetic and environmental factors. Some traits, such as eye color, are strongly genetically determined, while others, such as intelligence, are not. (A trait is strongly genetically determined if 95% of the organisms with the gene associated with the trait will develop the trait.) Furthermore, many human traits do not have a single genetic cause. For example, there is no single gene for intelligence, height, or musical or athletic ability. Opponents and proponents of genetic enhancement often assume that enhancing a human being would be as easy to do. But this is not likely to be the case for the foreseeable future, if ever, due to complexity of the problem and our scientific and technical limitations. Recognizing this fact should take some of rhetorical steam out of arguments against and for genetic enhancement, because negative consequences and the desired result may not be as easy to bring about as has been assumed. With even tremendous advances in science, it may never be possible to intentionally design the next Mozart or Einstein or select the traits of one's child as one would select features on a new automobile (Resnik and Vorhaus 2006).

GENETIC TESTING

The discovery of the structure of DNA by James Watson and Francis Crick in 1953 marked the beginning of a new era in science: molecular biology (Mayr 1982). Although scientists had been studying proteins, nucleic acids, amino acids, hormones, and many other important biological molecules prior to Watson and Crick's discovery, their model of the structure of DNA provided the key insight to understanding inheritance, protein synthesis, embryonic development, cell regulation and differentiation, and many other processes at the molecular level. One of the first fruits of molecular biology was developing a better understanding of genetic diseases as well as tests for genetic diseases. In 1963, Massachusetts developed a genetic screening program for phenylketonuria (PKU). PKU is a rare but treatable disorder in which the body lacks an enzyme for metabolizing the amino acid phenylalanine. If children have this condition and ingest phenylalanine, they can develop severe mental retardation. The treatment for PKU is fairly simple—eliminate phenylalanine from the diet. Because PKU is a treatable genetic disorder and the PKU test is reliable, effective, and inexpensive, all 50 states now require PKU screening for infants (Clayton 2001).

Physicians had known that sickle cell anemia (SCA) was an inherited blood disorder in which red blood cells have a sickle shape and do not transport oxygen normally. They were soon able to understand that the disease results from malformed hemoglobin molecules caused by a genetic mutation. The disease follows a heterozygous recessive pattern: People with only one copy of the mutated allele (heterozygotes) do not develop the disease, but people with two copies (homozygotes) do (Mayr 1982). The disease is most common in people with North African ancestry. By the late 1960s, researchers had developed a test for SCA. In the early 1970s, 16 states passed SCA screening legislation, which required that African-American newborns, pregnant women, preschool children, and couples applying for a marriage license be tested for SCA. The rationale for SCA screening was to identify people with the disease so that they could be treated, and to provide prospective parents with information they could use in making reproductive choices. The Air Force adopted an SCA screening program, as well: A person with SCA would be excluded from training to become a jet pilot (Andrews 2002).

Critics raised numerous objections to the SCA screening programs, arguing that they were discriminatory, coercive, and often improperly administered. Many states did not distinguish carefully between SCA carriers (heterozygotes) and people with SCA (homozygotes). States then changed their SCA screening programs to include it as part of the routine screening of newborns for a variety of genetic diseases, including PKU, cystic fibrosis, thalassemia, and galactosemia (Andrews 2002).

Since the 1970s, researchers have developed tests for many different types of diseases, including Huntington's disease, Down's syndrome (mentioned above), fragile-X syndrome, early-onset Alzheimer's dementia (AD), some types of breast and ovarian cancer, and some types of heart disease. Many of these tests can be performed at all stages of the human life cycle, including at conception, in utero, at birth, during childhood, or during adulthood. The completion of the Human Genome Project (HGP) in 2002, as well as other discoveries and innovations in genetics, genomics, and biotechnology, has accelerated the development of genetic tests (Andrews 2002). DNA chip technology has made it possible to test for thousands of genetic variants using a single assay. New fields of medicine, such as pharmacogenomics, may make extensive use of genetic tests to tailor treatments to an individual's genetic constitution (Service 2005). The speed of gene sequencing has continued to increase, while the costs of sequencing have decreased. Researchers are hoping one day to sequence an individual's entire genome for less than $1,000 (Service 2006).

In thinking about the ethical, legal, and social issues (ELSI) pertaining to genetic testing, it is important to distinguish between testing for clinical purposes and testing for research purposes. In clinical genetic testing, the goal is to acquire information about an individual that is likely to be useful in diagnosing, preventing, or treating a disease in that individual. In research genetic testing, the goal is to acquire information that is likely to be useful in developing generalizable knowledge about human health and disease. Genetic information gathered as part of a research study may or may not have any use in diagnosis, prevention, or treatment. (This distinction corresponds to the distinction between therapy and research discussed in chapter 12.)

Clinical Genetic Testing

Clinical genetic testing raises a number of issues, which we briefly mention here. The first is validity: Is the test accurate? Reliable? Is it conducted by a laboratory that has appropriate quality assurance procedures and regulatory approvals? What is the test's false-positive rate? What is its false-negative rate? False-positive and false-negative test results both pose problems for patients: A false-positive can cause someone needless worry, whereas a false-negative can cause someone to fail to take action to deal with a treatable problem. These and other issues relating to the validity of the test must be addressed before using the test results to make clinical decisions (Human Genome Project Information 2007).

Another issue is clinical significance: What do the test results mean for the patient? Is the patient likely to develop a disease? What is the progression of the disease? Only a few genetic tests indicate that a person will, without doubt, develop a disease; most tests indicate only that a person has an increased risk of developing a disease. For example, a woman with *BRCA1* mutations has a 90% chance of developing breast cancer by age 80 (Robson and Offit 2007). This risk is more than seven times the risk of a woman without this mutation. The *APOE-e4* gene is a risk factor for early-onset AD. People with one copy of the gene have an increased risk of developing the disease, and people with two copies have an even greater risk. For example, 9.3% of 65-year-old women without the *APOE-e4* gene will develop AD before they die, but 23% of 65-year-old women with one copy of the gene will develop AD, and 53% with two copies will develop it. Thus, a 65-year-old woman who is homozygous for the *APOE-e4* gene has five times the risk of developing AD compared with a woman without the gene (Liddell et al. 2001). If one's relative risk of developing a disease, given a positive test result, is not very high (e.g., less than 1.2 times the risk of a person without the gene), then it may not even be worth performing the test.

Another major clinical issue is whether anything can be done to prevent or treat the disease. A woman who has *BRCA1* mutations may take some measures to reduce her risk of developing breast cancer, such as altering her diet or lifestyle, having frequent mamograms, having all suspicious breast tissue biopsied, or the more radical option of having her breasts surgically removed (Robson and Offit 2007). However, a person who has the gene for Huntington's disease has little to do but wait for the disease to emerge, since there is currently no effective treatment or prevention for the disease. People with the gene for this disease can use that information in deciding whether to reproduce, but they cannot use this information to protect their own health. Many people would rather not know whether they have a gene for a disease, if there is no way of preventing or treating the disease. Also, it may be unfair to test children for such a disease, since this could burden the child with the implications of the test results without reasonable benefits in return. Some of these implications might include despair or hopelessness if the child learns about the results, and discrimination if other parties learn about the results. There is an ethical controversy about whether to test children for adult-onset genetic diseases if there is no way to treat or prevent these diseases (Rhodes 2006).

As mentioned above, an additional ethical concern related to the clinical significance of genetics tests would be if the tests are performed on fetuses or embryos, and the results could inform a decision to abort a fetus or discard an embryo. This raises the issue of whether it would be appropriate to make the decisions for reasons other than to prevent diseases, for example, to prevent the birth of a child of a particular sex or sexual orientation.

Another issue raised by genetic testing is discrimination. One of the major concerns with genetic testing is whether outside parties, such as insurers or employers, might be able to gain access to test results and use them to discriminate against the patient. If a health insurer pays for a genetic test, then it has a legal right to know the results. If the patient pays for the test out of his or her own pocket, then the insurer does not have a right to learn about those results, unless the results would qualify as a preexisting condition that could disqualify the patient from insurance coverage (Actuarial Foundation 1998). Thus, it is possible that patients could have trouble obtaining or keeping health insurance, in the United States, if their health insurers learn the results of their genetic tests. (The possibility of losing health insurance is not problem in countries that already guarantee health care coverage for all citizens, e.g., the United Kingdom, France, or Canada.) Employers might be able to use the results of genetic tests to make decisions adverse to the person's interests. For example, a company could fire, refuse to hire, or reassign all people who test positive for a particular gene. As noted above, the Air Force used the results of SCA tests to exclude cadets from its jet pilot training program (Kitcher 1997). However, this type of discrimination is now illegal in the United States: In 2008 the United States passed the Genetic Information Nondiscrimination Act (GINA), which protects Americans from both health insurance and employment discrimination based on their genetic information (see Hudson et al. 2008).

Surveys have shown that the public is very concerned about genetic discrimination. However, thus far actual cases of genetic discrimination still appear to be quite rare (Hall and Rich 2000a,b). Genetic discrimination is not common for several reasons. First, many states have adopted laws banning various forms of genetic discrimination. The federal government also has laws that would prohibit various types of genetic discrimination, such as laws against discrimination based on race, ethnicity, or disability (National Human Genome Research Institute 2007), in addition to GINA. Many of these laws have not been tested, so it is not clear how much protection they give patients. Even so, insurers and employers may not be willing to test them. Second, insurers and employers may find other types of health information more valuable than genetic information, such as information about one's health history or the health history of one's parents or siblings (Andrews 2002). Knowing that a person's parents both died of heart disease before age 60 could be more valuable to a life insurer than performing a genetic test on that person that might indicate an increased risk of a heart attack. Third, genetic testing is still not very cost-effective for insurers and employers. Although costs continue to go down, many tests still cost from several hundred to several thousand dollars (Andrews 2002). These circumstances may change in the future, of course, but at this juncture genetic testing does not appear to be a very useful decision tool for insurers or employers.

Genetic research has the potential to lead to not only discrimination against specific individuals, but also stigmatization of specific racial or ethnic groups, due to their association with genetic conditions or diseases (National Institutes of Health 2007d). For example, SCA is much more common in African Americans than in the rest of the U.S. population. Cystic fibrosis, Tay-Sachs disease, and certain types of hereditary breast cancer (see below) are more common among Ashkenazi Jews. What is ironic about this situation is that researchers would not have discovered these facts if these populations had refused to participate in genetic studies. Thus, participation in genetic research can lead to benefit for a population as a result of genetic knowledge, as well as harm related to discrimination or stigmatization (Clayton 2001).

Genetic testing also raises the issue of privacy and confidentiality: Even though the risk of genetic discrimination is still quite low, it is significant enough that it is important to protect the confidentiality of genetic information to protect patients from discrimination or the fear of discrimination.

Additionally, it is important to protect the confidentiality of genetic information so that patients can avoid embarrassment or shame, and also control who has access to their personal health information (PHI). Surveys have shown that patients undergoing genetic testing are very concerned about confidentiality and privacy (Hall and Rich 2000a,b). Many U.S. states have laws protecting the confidentiality of genetic information. In addition, federal and state laws pertaining to medical confidentiality also provide some protection for genetic information (Hall and Rich 2000a). The Privacy Rule of the Health Insurance Portability and Accountability Act (HIPAA) protects patients from unauthorized uses and disclosures of confidential PHI gathered by hospitals, medical centers, medical insurers, and other covered entities. HIPAA applies to all types of medical information collected and kept by health providers, including genetic information. Except in specific circumstances (e.g., public health reporting or criminal investigations), PHI cannot be disclosed to outside parties without consent of the patient (Lo et al. 2005).

A final issue we address regarding clinical genetic testing relates to family members: Because genes are the basic units of heredity, family members have common genetic characteristics. One can often make inferences about the genetic constitution of members of a family from information about their close relatives. For example, if a child tests positive for SCA, and we know that neither of his parents have the disease, we can infer that they both are SCA carriers. Conversely, if a couple both are SCA carriers, we can infer that they have a 25% chance of having a child with SCA.

Information about family members can be very valuable in understanding a person's genetic constitution and in learning about the inheritance of genetic diseases. For this reason, when performing a genetic test on an individual, it is often useful to ask family members to submit to genetic testing (Kitcher 1997). However, ethical questions can arise pertaining to family genetic testing. First, some family members may not want to be tested. If it is necessary to test a family member in order to help a patient with a suspected genetic disease, this could create a conflict between the right of the family member to avoid testing, and the health of the patient. Second, sometimes testing one family member may reveal the genetic constitution of another family member, who may not want to know this information or have this information known by others. For example, suppose a young adult has a positive test result for the gene for Huntington's disease. He tells his parents, who have not been tested. His father wants to take the test, but his mother does not. His father has a negative test result. So, it can be inferred that his mother has the gene. (Huntington's follows a heterozygous dominant pattern.) In this case, her right not to know her genetic constitution would conflict with the father's right to know. Third, other important information, such as paternity, may be discovered during the course of genetic testing. This can create a dilemma about whether to share this information and possibly disrupt family dynamics. Since genetic testing within families can strain ordinary understandings of patient confidentiality and autonomy, many commentators recommend that families reach agreements about the sharing of genetic information within the family before undergoing genetic testing (Doukas 2003).

Research Genetic Testing

Research genetic testing raises issues similar to those raised by clinical genetic testing, but in a slightly different way. For example, the same issues concerning validity mentioned above, such as accuracy, reliability, and quality, also arise in a research setting. However, validity can be more problematic in a research setting, because very often one of the main reasons for performing the test is to learn more about it. The validity of a genetic test may not be well understood when genetic testing is conducted as part of a research project. A good example of this problem would be a situation where researchers are analyzing an entire genome and testing for thousands of different genetic variants to determine whether they increase the risk of a particular disease. It will not be known, at the beginning of the study, whether any of these variants reliably indicate an increased risk for a disease (Renegar et al. 2006).

Another issue raised by clinical genetic testing is clinical significance: In research genetic testing, the purpose of conducting the tests is to develop generalizable knowledge, not to obtain information that can be used in making a clinical decision about a particular research subject (or patient). Thus, researchers often convey this information to subjects and also tell them that they will not receive any results from genetic tests. However, sometimes genetic information is likely to be useful in diagnosis, treatment, or prevention, and researchers must determine whether (and how) they will share this information with subjects/patients. Researchers must decide whether they will give subjects the option to receive this information and, if they do, how they will share this information. Will they tell the subject's doctor? Will they counsel the subject? Researchers also must decide a threshold of clinical significance for sharing test results. Should they share information because a subject wants to receive it? Should they share information only if it is absolutely vital that the subject receive this information to avoid drastic health outcomes? When should they share information? Immediately? Researchers and ethicists continue to debate these and other questions (Weir et al. 2004, Renegar et al. 2006).

Privacy and confidentiality are important in genetic research for the same reasons that they are important in clinical genetics: to protect people from potential discrimination, unwanted disclosures of private information, and so forth. There are different strategies for protecting confidentiality of data pertaining to research subjects, such as restricting access to the data, keeping the data in a secure place, and removing personal identifiers, such as name, social security number, and address, from the data. Some researchers remove all personal identifiers from the data to protect confidentiality and to facilitate research review by IRBs, since research on anonymous samples may be exempt from research review, or reviewed on an expedited basis, depending on the situation, under the federal research regulations. However, sometimes researchers may need to be able to identify research subjects for safety monitoring, follow-up, future studies, or to inform them about test results. When this is the case, researchers cannot use anonymous data. Instead, particular research subjects can be identified by a number (or other symbol) and linked to that number by a code. Access to the code should be restricted. In some cases, for example, when subjects must be frequently identified for testing or monitoring, the best practice may be to include the subjects' names with the data (Weir et al. 2004). The best practice for identifying subjects will vary according to the nature of the research being conducted. Investigators should anonymize data, whenever this is practical, to provide maximum confidentiality protection. Some commentators have argued that even anonymization may not offer 100% confidentiality protection, because when data are shared with the public, people may be able to access to various databases and powerful computer search tools to link genetic data to particular research subjects (McGuire and Gibbs 2006). While we recognize that this is a remote possibility in some situations, in most cases genetic research data anonymization will help to ensure confidentiality.

One final tool for protecting confidentiality (in the United States) is a Certificate of Confidentiality, which is a document issued by the Department of Health and Human Services that can prevent researchers from being compelled to disclose confidential information as part of a legal proceeding (Weir et al. 2004). However, we recognize that once the genetic makeup of an individual is known and stored in an accessible database, it will provide a signature reference to that individual. Thus, privacy could become breached with the proper technology. This risk is recognized by Lowrance and Collins (2007) as one of the three risks of identifying individuals from their gene sequence. The authors claim that by regulatory means, we can protect the identity of the individuals.

Yet another issue worth considering involves the use of genetic samples: Researchers should inform subjects if they plan to use the samples collected for genetic testing for future research, whether they plan to use the samples for commercial purposes, whom they will share the samples with, whether subjects can withdraw their samples, and whether the researchers will destroy the samples. There has been some debate about whether subjects can consent to the general use of their samples, for example, including the phrase "Your samples will be used for genetic research" on the informed consent form, or whether they should consent to more specific uses, for example, "Your samples will be used for research on disease X." From an investigator's perspective, a general consent is preferable to a specific one, because it gives the investigator flexibility to use the samples in future studies that cannot be specified in advance. From the subject's point of view, a specific consent gives the subject greater control over his own samples (Weir et al. 2004, Renegar et al. 2006). There has also been some debate about whether subjects should share in the benefits when their samples are used for commercial purposes. Currently, subjects are usually told that they will not receive any financial compensation, such as royalties, from commercial products derived from their samples. Some commentators have argued that this type of arrangement is unfair and that benefits should be shared with the population, if they are not shared with individual research subjects (Weir et al. 2004, Sheremeta and Knoppers 2003). There is also a debate about whether genes, DNA, or genetic tests should be commercialized at all, mentioned in our discussion of intellectual property in chapter 9.

Issues concerning family members also arise in genetic testing for research purposes: Many research studies today test several members in a family to identify genes that are implicated in disease. For example, to determine whether a type of cancer has a genetic basis, it may be useful to test patients with that cancer as well as their parents, siblings (if any), and other relatives. Genes for Huntington's disease, SCA, and some types of breast cancer (e.g., breast cancer resulting from *BRCA1* mutations) were discovered by testing family units (Kitcher 1997). Family members may become inadvertently involved in genetic studies when researchers collect information about a research subject's family history. For example, if a research subject informs investigators that her father has prostate cancer, she would be sharing private health information that he might not want to have disclosed. Researchers should be aware of the familial implications of genetic studies and take appropriate steps to protect the rights and welfare of family members, even when the family is not being studied as a unit (Botkin 2001).

GENETIC ENGINEERING AND GENE THERAPY

In the early 1970s, molecular biologists developed techniques for transferring genes from one organism to another. These techniques, known as recombinant DNA, take advantage of the fact that bacteria naturally transfer DNA sequences among each other via small vesicles known as plasmids. When the plasmid attaches to the surface of the bacteria, the DNA it contains enters the cell and is incorporated into its genome. The additional DNA can be inherited and can also function in the cell to make a protein. Bacteria can transfer many useful genes to their neighbors this way, such as genes for antibiotic resistance. In 1973, Stanley Cohen and Herbert Boyer developed a method for splicing DNA into plasmids, which could be used to transfer that DNA into bacteria. Ananda M. Chakrabarty's oil-spill bacterium (see chapter 9) was produced by transferring a gene into bacteria that coded for a protein that allowed it to metabolize crude oil. Since the 1970s, scientists have developed other methods of transferring DNA into cells, such as using viruses as vectors, and injecting DNA directly into the cell. Scientists also have developed artificial chromosomes, which can be inserted into cells. Artificial chromosomes can carry many different genes. Gene transfer is useful in developing genetically modified (GM) organisms, because genes that are transferred to germ cells will be present in the adult and can be inherited. Gene transfer techniques have been used to transfer genes to plants, such as tomatoes, corn, and rice, as well as sheep, mice, and rhesus monkeys. Genetically modified mouse models are playing an increasingly important role in biomedical research.

Gene transfer experiments on microbes, plants, animals have caused a great deal of excitement as well as concern among both scientists and the general public. In the early 1970s, many people became concerned about genetic accidents that might occur if genetically engineered "superbugs" escaped from a laboratory. Michael Crichton's book *The Andromeda Strain* (1971) and Kit Pedler and Gerry Davis's book *Mutant 59: The Plastic Eaters* (1972) depict such an event. Other science fiction writers have speculated about biological warfare and the future evolution of the human race. In February 1975, more than 140 concerned scientists from all over the world held a historic meeting at Asilomar, California, to discuss the benefits and risks of recombinant DNA and genetic engineering. They recommend that scientists proceed with the experiments with appropriate regulations and safeguards to minimize the risks from biohazards. Their recommendations led to the formation of the Recombinant DNA Advisory Committee, which oversees gene transfer protocols falling under regulations of the National Institutes of Health (NIH) or FDA.

Ever since gene transfer experiments began, they have faced some outspoken critics, such as Jeremy Rifkin (1998). For example, during the 1990s, a controversy erupted in Europe over GM foods and crops, which led to some countries banning these products. People have expressed concerns about the risks to their health and safety posed by GM foods and crops. Others are concerned about environmental hazards or about commercial control over agriculture. Critics have raised similar objections to GM animals. To the extent that these concerns have a legitimate basis, they should be based on sound scientific evidence. In a sense, GM plants and animals are no different from the plants and animals that have been produced through selective breeding for thousands of years. So one might argue that the risks are minimal, or at least no greater than risks we already accept. Critics argue, however, that gene transfer experiments can produce novel combinations of genes, such as tomatoes with fish genes. Critics also point out the GM plants and animals let loose in the wild, like a nonnative species, can upset the ecological balance. In any case, since these issues concern empirical questions relating to risks and hazards, they require further study (Rollin 1995, Reiss and Straughan 1996).

As researchers continued to make progress in gene transfer in plants and animals, they began to consider the possibility of gene transfer in humans. Critics of human gene transfer have raised a number of different objections to this procedure (discussed above in the section on eugenics). In the 1970s, theologians Paul Ramsey and Hans Jonas warned of the grave moral implications of genetic engineering of humans. In 1982, the President's Commission for the Study of Ethical Problems in Medicine and Biomedical and Behavioral Research issued the report *Splicing Life: A Report on the Social and Ethical Issues of Genetic Engineering with Human Beings*. This report raised some important questions and warned about the possibility of enhancing human beings. It also called for continued dialogue on the issues and federal oversight.

In 1986, philosopher Leroy Walters published an important paper on the ethics of gene therapy that made two influential distinctions. The first is between somatic gene therapy (SGT) and germline gene therapy (GLGT). The goal of SGT is to transfer genes to somatic cells to treat or prevent disease; the goal of GLGT is to transfer genes to germ cells, such as gonadal tissue, gametes, or embryos, to prevent a disease in their child. Because therapies designed to affect somatic cells may inadvertently affect germ cells, SGT poses a small risk of accidental GLGT (Walters 1989, Holtug 1997, Walters and Palmer 1997). The second distinction made by Walters was between genetic therapy and genetic enhancement, as discussed above: Genetic therapy aims to treat or prevent diseases in human beings, whereas enhancement aims to change, improve, or enhance human beings.

W. French Anderson (1989), who pioneered SGT, argued that this procedure is ethically acceptable but that GLGT and genetic enhancement (somatic or germline) are ethically questionable. The main ethical argument for proceeding with SGT was that it was conceptually similar to existing forms of therapy, such as chemotherapy, radiation therapy, and surgery. The rationale for all of these procedures is the same: The benefits of therapy outweigh the risks for an individual (existing) patient. Many therapies, such as cancer treatments, have a high degree of risk, but these risks can be justified if they offer important benefits, such as the only hope for a cure for a terminal disease. If SGT passes the risk/benefit test, then it can be used in human beings. (This argument assumes, of course, that we can distinguish between "therapy" and "enhancement," discussed above.)

Anderson was able to convince regulatory agencies in the United States to approve SGT clinical trials. The first SGT experiment took place in 1990 to treat a child with adenosine deaminase deficiency, which causes severe combined immunodeficiency (SCID). The goal of the experiment was to transfer a gene into the child's liver cells, so that they could start producing a necessary liver enzyme. People lacking the adenosine deaminase enzyme cannot break down dATP, which is toxic to immature immune system cells when it accumulates in the body. Anderson's experiment was successful. Since then, a variety of other SGT procedures have been used to treat conditions such as glycogen storage disease, antitrypsin deficiency, familial hypercholesterolemia, cystic fibrosis, Duchene muscular dystrophy, Huntington's disease, and breast cancer (Walters and Palmer 1997, Beaudet et al. 2001).

The general idea behind SGT is to use a vector, such as a plasmid or a virus, to deliver a gene to a cell that lacks a normal copy of the gene. For this procedure to be effective, one needs to deliver the gene to the cell and to ensure that the gene functions properly in the cell (i.e., it is expressed as a protein in the right way, at the right time, in the right amount). Although SGT has had some successes, it still suffers from technical difficulties. The body often develops a powerful immune response to the vectors used in gene therapy as well as to foreign genes, so the immune system often destroys foreign transgenes (genes that are transferred from one organism to another). Even when transgenes penetrate the body's defenses, they may not integrate into the genome or function in the cell. SGT has suffered some setbacks, such as deaths in clinical trials (Gelsinger and Mohr) (Gelsinger and Shamoo 2008) noted in chapter 1 and 12, and it has yet to deliver on the promises made in the early 1990s. However, it is a young science that will take some time to develop (Resnik 2001a).

Most of the controversy surrounding SGT has related to GLGT and genetic enhancement, as discussed above. Jesse Gelsinger's tragic death, which we have discussed several times in this book, illustrates most of the other areas of ethical concern with SGT: safety and risk, conflicts of interest, and overestimation of the benefits of therapy in the informed consent process, also known as the therapeutic misconception.

REGULATION OF ASSISTED REPRODUCTION

We have already mentioned some assisted reproduction technologies (ART) in our discussion of eugenics, but we mention them again here in the context of regulation. On July 25, 1978, Louise Brown became the first child conceived via in vitro fertilization (IVF). In this procedure, doctors give women hormones to cause ovulation, and then they use a needle to extract eggs from their ovaries. The eggs can be fertilized in a petri dish and implanted in the womb or stored for future use. Because couples almost never use all of the embryos created in IVF, thousands of unused embryos are destroyed each year, which itself raises ethical issues for people who regard embryos as having moral value or status. In some instances, divorcing couples have fought battles for "ownership" or "custody" of the embryos. At the time that Louise Brown was born, the risks and benefits of IVF were not well documented. Although the procedure had been used in animals, it had not been tried in humans. The procedure was performed under the guise of "therapy" and therefore was able to sidestep the questions, delays, and red tape that would have been involved in a research protocol. Fortunately, Louise Brown was born healthy, and more than 150,000 "test-tube babies" have been born since then. The procedure is now considered "routine," although at one time people were concerned that it could cause defects. The procedures developed for IVF also have made it possible for women to donate their oocytes (or eggs) for use in reproduction. These procedures involve some discomfort and risks, and egg donors often receive a fee for their services. Many commentators are concerned about the emerging market for human eggs and question whether human reproductive materials should be treated as commodities (McGee 1997, Andrews 2000).

Surrogate pregnancy, which emerged in the 1960s, became more widespread in the 1980s. In surrogate pregnancy, a woman agrees to become pregnant and give the child up for adoption to the couple. The woman may be artificially inseminated with the man's sperm, or she may carry an IVF embryo. In the first instance, she would be biologically related to the child; in the second, she need not be. Although, some surrogates act for purely altruistic reasons, many act as surrogates for a fee. Surrogate motherhood is another procedure that was developed as a form of therapy, not as an experiment. It has emerged, in the United States, with very little regulation or oversight. Some states have adopted laws defining "biological" and "gestational," and some have refused to recognize surrogacy contracts (Andrews 2000).

Unlike many other areas of medical practice or research, ART in the United States developed with very little regulation or oversight. Some commentators argue that additional regulations are necessary to protect children, parents, and society (Andrews 2000, President's Council on Bioethics 2004). One troubling aspect of ART procedures is that they are often performed as "therapy" rather than "research," which avoids the regulation and oversight that accompanies medical research. As a result, there continue to be concerns about the health and safety of patients and future children. One health and safety issue that has become a sticking point is the high incidence of multiple births, such as sextuplets, septuplets, and higher (ISLAT Working Group 1998). To ensure a successful pregnancy, ART practitioners usually transfer several embryos to the womb. Also, some women who take fertility drugs for ART do not heed warnings from their physicians and become pregnant with four or more fetuses. In these cases, many obstetricians recommend selective abortion to reduce the number of fetuses in the womb, to improve the health of the other unborn children and the mother. But many women refuse selective abortion on moral grounds.

Many critics of ART are also concerned about its impacts on the traditional family structure (Kass 1985, Callahan 1998). Lesbian couples and single mothers can use artificial insemination to have children. Male homosexuals can use surrogate mothers to have children. A woman's sister, mother, daughter, or cousin could be the surrogate mother for her child. A family could have children with a variety of biological, social, and gestational parents, and other variations are possible. Some claim that children who are born into these nontraditional families may suffer irreparable psychosocial harms. According to the critics, ART will harm society as well as undermine the traditional family. Proponents of ART, on the other hand, point out that these procedures benefit parents and the children who would not have been born without ART (Robertson 1994). A life in a nontraditional family is far superior to no life at all, one might argue. Besides, it is futile to insist that all families conform to some "traditional" norm in this era of working mothers, divorce rates as high as 50%, adoptions, grandparents raising children, gay couples, single-parent households, and so on. A family, one might argue, exists as long as there is love among human beings living together. Moreover, reproduction is an important human right that should not be restricted unless it causes harms to other people. As long as ART procedures are safe and effective, couples should have access to them. Apart from the safety issues, ART could result in separating the human activity of sex from reproduction. This may have some adverse outcomes, especially if society does not provide a period of dialogue and transition.

Some critics object to the commercial aspects of ART (Andrews and Nelkin 2001), for example, selling reproductive products, such as sperm or eggs, or reproductive services, such as serving as a surrogate mother or an adoption agent. Critics argue that many of the financial arrangements in ART can lead to exploitation of women (or parents) and may be equivalent to "baby selling."

THE ABORTION DEBATE

Before discussing the final two topics of this chapter, cloning and embryonic stem cell research, we briefly discuss the abortion issue, because it has had tremendous influence on many of the debates concerning human reproduction and genetics. This issue has sharply divided U.S. citizens ever since the monumental 1973 *Roe v. Wade* decision by the Supreme Court. This decision overturned anti-abortion laws adopted by various states and granted women a constitutional right to abortion based on the right to privacy. The right to privacy ensures that U.S. citizens have dominion over a sphere of private decisions, such as reproduction, family life, privacy in the home, child rearing, and refusal of medical treatment. The right to privacy is not mentioned in the Constitution, but judges and legal scholars have found a basis for privacy rights in the first, fourth, ninth, and fourteenth amendments. These rights are not absolute, however, because states can restrict privacy rights in order to prevent harm to people or to protect state interests.

In the Court's ruling in *Roe v. Wade*, the majority held that a woman's right to an abortion could only be restricted in order to protect her health or to promote the state's interests in human life. The Court did not state that the fetus has a legal right to life, but it carved out the "trimester" system, effectively giving the fetus some legal status in the third trimester. According to its ruling, abortions should be available to women on demand in the first trimester of pregnancy, with few restrictions. In the second trimester, the state can impose additional restrictions on abortion in order to safeguard the health of the mother. In the third trimester, the state's interests in promoting human life become paramount, and abortions should be allowed only in exceptional circumstances. The third trimester represented an important legal demarcation, according to the Court, because fetuses are considered to be viable by this time. One problem with this position is that medical technology continues to change the time of viability—the ability to live outside the mother's womb. Currently, in the U.S. fetuses born at 22–23 weeks of gestation are considered viable. In theory, abortions are legal at any time during pregnancy, even though the state can impose various restrictions.

Ever since this important ruling, various states have enacted laws that attempt to whittle away abortion rights, such as requiring parental consent for minors to have abortions, a 48-hour waiting period for all abortions, and mandatory counseling about the psychological risks of abortion. The Supreme Court has not overturned *Roe v. Wade*, but it has moved away from the analysis in *Roe*. Recent Supreme Court decisions have followed the "undue burden" doctrine articulated in *Webster v. Reproductive Health Services* (1989). According to the doctrine, government restrictions on abortion are constitutional as long as they do not place an undue burden on the woman. A restriction would be an undue burden if the goal of the restriction is to prevent women from having abortions.

Leaving aside medical and legal issues for a moment, the key moral issue in abortion is the status of the fetus. Abortion opponents (the "pro-life" view) regard the fetus as a human being from the moment of conception, with at least some moral rights, including the right not to be killed. Abortion, according to this view, is equivalent to murder. Abortion proponents (the "pro-choice" view) regard the fetus (especially the early embryo) as a mere clump of cells with only the "potential" to become human being. Since the fetus is not an actual human being with moral rights, abortion may be irresponsible or wasteful, but it is not murder. There are also positions in between these two extreme views. According to a third view, the fetus acquires moral value as it develops. As an early embryo, the fetus has only minimal moral value, but by the time it is ready to be born, it has the moral value equivalent to an infant. Restrictions on abortion, therefore, should follow fetal development and increase as the fetus develops (Green 2001). A fourth view holds that embryos that have been implanted in the womb are morally different from embryos created by IVF and have not been implanted. An embryo in the womb will continue to develop and become an infant, if nothing interferes with this natural process. An IVF embryo, however, is in a state of "limbo" and must be implanted in order to develop (Agar 2007).

In the United States, the abortion debate affects political elections and key decisions in government. Pro-life and pro-choice advocates often stage protests, and some resort to intimidation and violence. We do not take a stand on the moral status of the fetus in this book, but we do suggest that this issue is similar to questions one might raise about the rights of animals (see chapter 11). A coherent moral position would be one that comes to terms with the moral status of various entities, such as embryos, fetuses, animals, plants, and ecosystems.

The abortion issue also has far-reaching effects on debates about genetics and human reproduction. For example, the main reason for conducting prenatal genetic testing is to obtain knowledge that could lead to an abortion. Those who oppose abortion therefore have problems with prenatal genetic testing. In IVF, unused frozen embryos are destroyed. But those who view the fetus (or embryo) as a human being may oppose IVF because it involves the creation and destruction of embryos. Abortion opponents also oppose the use of fetal tissue in research because they believe that this may encourage people to have abortions and because using fetal tissue shows a lack of respect for the fetus. As discussed below, the abortion issue has also affected debates about cloning and stem cell research.

HUMAN CLONING

A clone, in biotechnology, is understood to be an exact copy of a molecule, such as DNA or a protein, or an exact copy of a cell or an organism. For many years, botanists and agronomists have been able to clone plants from cuttings. For example, a jade plant grown from a leaf will have almost exactly the same genome as its parent (there may be some slight genetic variations due to mutations). For years scientists have also been able to clone mammals, such as cows (Rollin 1995). It is possible to split an early embryo or take a cell from it without damaging the offspring. This procedure, known as "twinning" or "embryo splitting," has been used successfully for many years to produce cows and other livestock. Indeed, identical human twins are natural clones.

Before 1996, however, scientists had not produced a mammalian clone from an adult cell. Philosophers and scientists had discussed this possibility, and science fiction writers had written about it. For example, Aldous Huxley's book *Brave New World* (1932) depicts a fascist state that uses cloning and other technologies to produce distinct castes; the movie *The Boys from Brazil* (1978) is about scientists who clone Hitler, and the movie *Jurassic Park* (1993) depicts the cloning of dinosaurs. However, cloning of this sort did not become a reality until the birth of Dolly, the world's most famous sheep.

On February 23, 1997, Ian Wilmut and his colleagues at the Roslin Institute in Scotland announced that they had produced Dolly, who was actually born on July 5, 1996. The story soon made headlines around the world as the media rushed to report the event. On the following day, U.S. president Clinton asked the National Bioethics Advisory Commission (NBAC) to review the legal and ethical implications of this new development. A few days later, President Clinton banned the use of federal funds for cloning a human being (National Bioethics Advisory Commission 1997), and several European countries have since followed suit. Congress has considered a ban on human cloning several times but has failed to pass any legislation. The commission appointed by President George W. Bush, the President's Council on Bioethics, has recommended a complete ban on reproductive cloning, that is, cloning for the purposes of producing progeny, and a temporary moratorium on cloning for other purposes, such as research (President's Council on Bioethics 2002).

Today both types of cloning are legal in the United States, although a number of states have banned it. Nineteen European countries have banned reproductive cloning, and several have banned all types of cloning. In 2005, the United Nations adopted a treaty banning reproductive cloning and requiring respect for human dignity and human life. So far, the biomedical community is observing a voluntary moratorium on reproductive cloning, although a few individuals and groups, such as the Raelian religious cult, have said they would attempt to clone human beings.

The procedure that Wilmut and colleagues used to create Dolly involved taking a nucleus from a mammary cell from a sheep and transferring it into an egg that had been stripped of its nucleus. They then stimulated the egg with an electric charge to induce normal embryonic development, and transferred the developing embryo into the womb of a sheep. Thus, Dolly is an identical twin of the sheep that donated the nucleus. It took 277 failed attempts to produce Dolly. Most of these failures occurred before implantation, but many others occurred after implantation; the embryos either failed to implant or were aborted (Pence 1998). This was a new milestone in reproductive biology because the enucleated egg was "reprogrammed" with the full genetic complement from an adult sheep to go through the life cycle again. Prior to this feat, scientists were able to speculate that cloning would be feasible from embryonic cells but not from adult cells because adult (differentiated) cells have lost their ability to develop into a new organism.

In 1993, two reproductive biologists were able to produce embryonic clones by splitting human embryos. This event also prompted a great deal of media attention and public debate and prompted the National Advisory Board on Ethics in Reproduction to hold a workshop to hear from embryologists, ethicists, reproductive biologists, public policy experts, legal scholars, journalists, and government officials (Cohen 1994). In 1993 as in 1997, reasonable people disagreed about the ethics of human cloning (Annas 1994, Macklin 1994).

The arguments for reproductive cloning are similar to the arguments for developing other reproductive technologies: Cloning could be justified on the grounds that (a) it benefits children who would not exist without the procedure, (b) it benefits infertile couples or other people who want to reproduce, (c) it respects the right to procreation, and (d) it benefits society if it is used to reduce the economic and social burden of genetic disease. The strongest case for reproductive cloning would be for a couple who lack fertile gametes because of exposure to radiation or chemicals. Others have argued that cloning allows homosexual couples to have genetically related children or that cloning might be useful in preventing genetic diseases (Pence 1998). Some people have suggested that it might be appropriate to clone a child with leukemia in order to produce a bone marrow donor, or to clone a "replacement" for a dead child.

However, many people have raised objections to reproductive cloning. The strongest objection raised so far is that cloning is too risky. This procedure has not been tested on human beings, and we do not know whether it is safe or effective. Although scientists have produced many mammalian clones, these offspring have had some physical problems. Cloning may also require the creation and destruction of hundreds of embryos to produce a single child. Others have pointed out the possible psychological harms to the clone. Would clones regard themselves as freaks? As objects produced or manufactured to fulfill their parents' goals? Although the psychological risks are real, it is important to note that most children born via ART techniques have faced these potential psychological problems and most have not been harmed significantly. However, the physiological risks of cloning are far greater than the risks of IVF or surrogate pregnancy, because these procedures do not involve nuclear transfer. In a sense, cloning can be regarded as a type of germline therapy because it involves manipulation of the germline (American Association for the Advancement of Science 2000, Resnik and Langer 2001).

Many of the other objections to reproductive cloning are similar to objections to other ART procedures:

1. Cloning will have adverse impacts on the family structure, birth, death, and marriage (Shamoo 1997b).
2. Cloning threatens social or religious values, such as respect for human life.
3. Cloning is "playing God."
4. Cloning threatens the human gene pool.
5. Cloning leads to the commodification or commercialization of people.
6. Cloning will lead to various abuses by dictators, millionaires, and others.
7. Cloning will threaten the individuality, uniqueness, or dignity of the clone.

We do not explore all of these arguments in detail, but refer the reader to Pence (1998) and reports by the NBAC (National Bioethics Advisory Commission 1997) and the President's Council on Bioethics (2002). We also note, regarding objection 7, that the idea that a clone would not be a unique person is a common, though ill-informed, argument against cloning. In popular accounts of cloning, some people have assumed that the clones will be exactly alike in every way: the same hair, the same personality, and so on. But this idea goes against what we know about human development. All human traits are influenced by genetics and the environment. People who are genetically identical may nevertheless be very different, due to the influence of environmental factors. Identical twins are natural clones and have different personalities, likes and dislikes, talents, and so on. Twins also have anatomical differences. For example, twins do not have the same fingerprints, and they may have different birthmarks. We also know that twins have no trouble forming their own identities, and they do not consider themselves "copies." We should expect that a clone would also develop his or her own personality and sense of self (McGee 1997, Pence 1998).

Although we think that most of the arguments against reproductive cloning are unconvincing, we do not think that reproductive cloning would be a good idea for the foreseeable future, due to the safety concerns mentioned above. However, we do think there are some good reasons to move forward with cloning for research or therapy, which we discuss below.

Cloning for research has more public support than cloning for reproduction. Many countries that have banned reproductive cloning have refrained from banning research cloning. The U.N. treaty deals with research cloning in a similar way. Some U.S. states, such as California, have passed laws that specifically state that research cloning is legal, to undercut any federal legislation banning cloning (President's Council on Bioethics 2004). The main argument for research cloning is that experiments involving this procedure could improve our general understanding of human health and disease and may one day lead to new therapies. The main argument against research cloning is that it could open the door to cloning for other purposes by developing the very techniques that would be used in reproductive cloning (President's Council on Bioethics 2002). We do not find this argument against research cloning to be very convincing, because regulations, laws, and professional standards can deter people from engaging in reproductive cloning. Additionally, developments in animal cloning also help to pave the way for human cloning, but stopping animal cloning would greatly interfere with biomedical research.

STEM CELL RESEARCH

Research on human embryonic stem (HES) cells has also generated a great deal of moral and political controversy in the United States and around the world. Stem cells are cells that can differentiate into different cell types, and they are found in many different tissues in the body. Adult stem cells, which are found in differentiated tissues, such as bone marrow, fat, muscle, or nervous tissue, are multipotent, which means that they can become more than one cell type, but they cannot become just any type of cell in the body. For example, bone marrow stem cells can become red blood cells or several different types of white blood cells. On the other hand, stem cells found in the umbilical cord and placenta are pluripotent, which means they can differentiate into any tissue type in the body. And stem cells taken from the inner cell mass of a 4- to 5-day-old embryo are totipotent, which means that they can become any type of tissue in the body or can develop into a separate individual (National Bioethics Advisory Commission 1999).

In 1998, James Thompson and colleagues at the University of Wisconsin discovered how to grow HES cells in a culture medium. This discovery led to several patents, which are held by the Wisconsin Alumni Research Foundation (WARF). The significance of this research is that it may be a potential source for replacement tissues for defective or damaged tissues. Many human diseases, such as type I diabetes, CHF, cirrhosis of the liver, kidney disease, paralysis, and AD, are caused by defective or damaged tissue. Although adult, umbilical cord, and placental stem cells may also serve as a source for replacement tissues, HES cells may be more useful than these other types of cells, because they can differentiate into any type of tissue, and tissues derived from HES cells may overcome tissue rejection problems, because they lack the immunological factors found in more mature tissues. Some researchers are planning to use HES cells to develop replacement organs and body parts. To date, embryonic stem cell therapies have been successful in some animal studies. Researchers have used embryonic stem cells to successfully treat rodents with spinal cord injuries. As of the writing of this book, HES cell clinical trials have not yet begun. Geron Corporation has announced plans to treat spinal cord injuries with HES cells. It is worth noting that one type of adult stem cell therapy has a long history of success. For many decades, doctors have used donated bone marrow stem cells to treat leukemia (Pucéat and Ballis 2007).

Since the 1980s, there has been a ban on the use of federal funds for research that involves creating human embryos for research or destroying human embryos. President Clinton interpreted this ban as applying only to research on embryos, not research on tissues derived from embryos. Thus, under the Clinton administration, federal funds were available for HES cell research. Federally funded scientists used HES cells that were developed by private companies (Green 2001). Seven months after he took office in 2001, President George W. Bush announced a more restrictive HES funding policy. No additional federal funds could be used to purchase HES cells from private companies. Federally funded scientists could use the cell lines that they were already using, but no new cell lines could be used. This policy severely limited the amount of HES cell research that the NIH could sponsor. President Bush also appointed a committee, the President's Council on Bioethics, to provide him with advice on stem cell research policy and other bioethics issues. Congress has passed bills that loosen funding restrictions, but Bush has vetoed this legislation (Stout 2006).

In the absence of significant federal funding, private companies, such as Geron and Advanced Cell Therapeutics, and states, such as California, Illinois, Maryland, and New Jersey, have stepped forward. Some of these companies have appointed their own ethics boards (Lebacqz 1999). California has established a $3 billion fund for HES cell research. Foreign governments, such as Singapore, the United Kingdom, Australia, South Korea, and Israel, have also become major players in supporting HES cell research. Many stem cell scientists have migrated from the United States to the United Kingdom in search of funding and a research-friendly environment (Gruen and Grabel 2006). Thus, the Bush administration's decision to severely restrict funding for HES cell research has not stopped HES cell science from going forward, but it has drastically reduced the federal government's contribution to this new science. This is a very unusual and troubling development, since the U.S. government has been in the forefront of all areas of scientific innovation and discovery since World War II.

HES cells are morally controversial because current methods of deriving them require the destruction of the embryo. Researchers have been trying to develop methods of deriving or creating embryonic stem cells that do not require the destruction of the embryo. As of the writing of this book, none of these have been perfected, although some, such as deriving pluripotent stem cells from adult skin cells, show great promise (Weissman 2006, Green 2007). Even if alternative methods of acquiring pluripotent stem cells are developed, it is likely that researchers will want to continue pursuing research on HES cells, because of the important insights this research can yield for our understanding of development, cell differentiation, the properties of stem cells, cancer, and so on. It may also turn out that pluripotent stem cells made from embryos are clinically more useful than those made from adult cells. Only time will tell.

As we discussed above, many people regard human embryos as having moral value equivalent to that of a human being. For people who regard the embryo in such high esteem, destroying an embryo, even for a good cause, would be immoral—it would be equivalent to murder (Green 2001, 2007). Some people who highly value embryos argue that HES cell research can be justified as long as the embryos are obtained from couples seeking infertility treatment who donate their leftover embryos. Couples may create several hundred embryos when they attempt to conceive a child by IVF. The unused embryos remain frozen for some time, but they are usually destroyed. Each year, thousands of leftover embryos are destroyed. Since the leftover embryos are fated to die anyway, why not make good use of them in research? This is the position of many people who support HES cell research (Green 2007). Although this position makes a great deal of sense, someone who values embryos highly might still object on the grounds that destroying an embryo that is fated to die is still murder. We would not consider it morally acceptable to kill a terminally ill infant and use its body for research; likewise, we should not use human embryos in this way.

There is another method of deriving HES cells that is even more controversial than deriving them from leftover embryos, known as therapeutic or research cloning. The goal of this method is to create embryos from the tissues of patient so that the cells derived from those embryos will be nearly genetically identical to the patient. This procedure would dramatically reduce the risk of tissue rejection since the immunological signature proteins of the cells and patient would be almost identical. This procedure would involve removing the nucleus from one of the patient's somatic cells and transferring it into an embryo that has had its nucleus removed (Solter and Gearhart 1999, Weissman 2006). This technique has been performed in animals but not yet in human beings. In the Woo Suk Hwang case discussed in chapter 1, Hwang and colleagues fraudulently claimed to have successfully performed this procedure to derive patient-specific HES cells. This method is more controversial than using leftover embryos because it involves cloning, and because it creates embryos in order to use them (and eventually destroy them) for research purposes (Green 2007). Embryos created for IVF at least have a chance of being implanted in the womb and developing into human beings.

One of the unfortunate consequences of the Bush administration's decision to severely restrict federal funding for HES cell research is that federal agencies, such as the NIH, have not found it necessary to develop regulations or guidelines for this research. As a result, research has gone forward with little guidance from the federal government, other than what can be inferred from existing regulations and policies pertaining to research with human subjects. In the absence of explicit guidance from the federal government, various organizations have proposed some guidelines for HES cell research, including the National Academy of Sciences, the California Institute of Regenerative Medicine, and the International Society for Stem Cell Research (ISSCR). According to the ISSCR guidelines, there should be rigorous review and oversight of HES cell research through existing institutional bodies, such as institutional review boards (IRBs). Institutions may decide the best way to conduct this review. There is no need, according to the ISSCR, to form a new committee to review HES cell research. The ISSCR also recommends that no experiments be conducted that raise strong ethical concerns, such as human reproductive cloning, in vitro culture of human embryos beyond 14 days old, and creation of human–animal chimeras likely to harbor human gametes. The 14-day limit for research is justified on the grounds that experiments on these very early embryos are distinct from experiments on older embryos, because early embryos have not developed to the point where they are unique individuals. Prior to the 14-day limit, an embryo can spontaneously split into two twins, or merge with another embryo to form a chimera. The ISSCR guidelines also recommend that informed consent be obtained from all people who participate in stem cell research, including gamete donors and somatic cell donors. The ISSCR guidelines also express concern about the potential for undue inducement that exists when egg donors are offered large amounts of money (e.g., $2,500 or more), but the guidelines do not set any limits on the amount of money that may be offered to egg donors. The guidelines encourage oversight committees to carefully evaluate financial incentives offered to egg donors in HES cell research (International Society for Stem Cell Research 2007).

QUESTIONS FOR DISCUSSION

1. Should we try to "improve" the human race? Why or why not?
2. Couples often seek to produce children that are free from genetic diseases, and some seek to produce children with enhanced traits. Is there a moral difference between these "eugenics" decisions made by couples and those made by the state?
3. Some people have developed sperm banks of famous male scientists. In theory, one could also develop egg banks for famous female scientists. Would you contribute to such banks? Would you use them?
4. Are there limits to the future of genetic engineering technology? Should society impose limits? If so, what should they be?
5. If you were married and you and your spouse were expecting a child that you knew would survive with an IQ of 10, how would you feel about selective abortion? Would you use it?
6. What if in the scenario in question 5, you were offered prenatal gene therapy to have a normal child? Would you use it? Why? What about somatic gene enhancement to ensure a high IQ for the baby? Why or why not?
7. Your sister had genetic testing for a dozen diseases. She is a writer and is writing a popular book to talk about her results. What do you think she should do regarding your rights, if anything? Would you interfere with her book publication? Why or why not?
8. Three major events in our lives are interwoven into our social fabric: birth, marriage, and death. Do you think cloning could change the social context of these three events? Why? How?

CASES FOR DISCUSSION

Case 1

The Human Genome Project (HGP) was one of the most significant and expensive biomedical research projects of the twentieth century. The project was conceived in the mid-1980s, begun in 1990, and completed in 2002, several years ahead of schedule. The goal of the HGP was to map and sequence the entire genome. Scientists expected that accomplishing this goal would have tremendous payoffs for our understanding of human genetics and human health and would lead to new and innovative therapies. The price tag for the project was $15 billion. The HGP was funded by the Department of Energy in collaboration with the NIH and several genome centers. Collectively, the centers were known as the Human Genome Sequencing Consortium. The first director of the HGP was Nobel Prize winner James Watson. In 1992 he was replaced by National Human Genome Research Institute director Francis Collins (Human Genome Project 2003).

The HGP generated controversy from the moment it was conceived. Biomedical researchers were concerned that it would draw funding away from smaller science projects and that the investment in a large project would not be worthwhile. Some researchers argued it would be better to sequence individual genes on an as-needed basis, rather than sequence the entire DNA in the genome, which might contain redundant and meaningless sequences. Since the HGP would sequence the genome from five individuals of European ancestry, scientists also wondered whether the project would fail to capture important human genetic diversity. Ethicists, politicians, and religious leaders were concerned about the potential ethical, political, and social implications and social consequences of the HGP, such as eugenics, discrimination, loss of privacy, and commercialization of biomedical research. In response to these ethical concerns, the HGP leaders agreed to devote 3–5% of the project's funds to research and educational programs on HGP ELSI (Kitcher 1997).

In May 1998, the scientists working on the HGP found themselves in competition with a private effort to sequence the human genome. J. Craig Venter helped found Celera Genomics, a private firm with the goal of sequencing the entire human genome in three years. Venter, who had worked at the NIH, left the agency in 1991 after becoming frustrated with government bureaucracy and epistemological conservatism. Venter had proposed using a novel "shotgun" approach for genome sequencing instead of the clone-by-clone method used by the NIH and other researchers. Under the shotgun approach, researchers take apart the entire genome, sequence all the different pieces using automated sequencing machines, and then use supercomputers and statistics algorithms to reassemble the pieces. Venter collaborated with Michael Hunkapillar, head of scientific equipment at PE Biosystems, to assemble a large collection of automated sequencing machines. NIH researchers harbored serious doubts about the "shotgun" approach, because they thought it would not be as accurate or reliable as the clone-by-clone approach. In February 2000, the "shotgun" approach yielded its first victory when Celera published a high-quality genome sequence of the fruit fly (*Drosophila melanogaster*) (Resnik 2003d).

From the outset, the relationship between the consortium and Celera was strained. The consortium was a public effort committed to free and rapid release of the genome sequence data prior to publication. Researchers working in the consortium deposited their data in an Internet database known as GenBank (Collins et al. 2003). As a private company, Celera hoped to profit from research on human genetics and planned to patent some human genes and share its data with customers for a fee. Researchers working for Celera deposited genome sequence data on the company's secure servers, not on a public site like GenBank. Despite having very different approaches, researchers for both groups cooperated with each other, sharing ideas, tools, technologies, techniques, data, and results (Marshall 2001).

The competition between the consortium and Celera, as well as the infusion of millions of dollars in private money, helped to accelerate the pace of research, and both groups finished ahead of schedule. The two groups reached an agreement to publish their draft versions of the human genome on February 21, 2001. Celera published its result in *Science;* the consortium published in *Nature. Science* allowed the Celera researchers to publish their results without requiring them to make their data available on a public Web site. Celera agreed to allow academic researchers to view the data for free, provided that they sign an agreement to refrain from commercializing the data. The consortium and Celera published final versions of the human genome in April 2003.

As predicted, the HGP has had a tremendous impact on biology and medicine. Molecular geneticists have a better understanding of gene regulation and interaction, protein synthesis, mutation, the relationship between DNA and RNA, and the importance of noncoding regions of the genome. Anthropologists and evolutionary biologists have studied the human genome sequence to learn more about phylogenetic relationships among different human populations. Microbiologists and epidemiologists have a better understanding of how viruses and other microbes have affected the human gene pool. Many different fields of medical research and practice have benefited from genomic data, including gene therapy, genetic testing and counseling, oncology, endocrinology, pharmacology, reproductive medicine, and pathology. New fields have grown up alongside the HGP, including genomics, proteomics, pharmacogenomics, and bioinformatics (Collins and McKusick 2001).

As also predicted, the HGP has also brought significant issues to light, as well as political implications. The availability of so much genomic data from the HGP has increased public concerns about genetic discrimination and privacy, DNA fingerprinting, the commercialization of genomic research, eugenics, human cloning, genetic determinism, and personal responsibility. The decision to set aside a significant amount of funding for ELSI educational and research projects has proven to be a wise move, as ELSI grant recipients have helped to advance the public understanding of human genetics and have made key contributions to scholarly disciplines that study ELSI, such as bioethics, political science, law, and religion.

- Do you think ELSI helped or hindered the effort to sequence the human genome?
- What are some of the advantages of private genetic research?
- What are some of the disadvantages?

Case 2

You are conducting a study on genetic factors in breast cancer. You are collecting blood samples and cheek scrapings from a cohort of 5,000 women who have had breast cancer and a similar cohort of 5,000 women who have not had the disease (i.e., controls). You are using gene-chip assays to test for thousands of genetic variants, and statistical algorithms to search for variants (or combinations of variants) associated with breast cancer. You are also collecting information about environmental factors (e.g., exposures to pollutants and pesticides), lifestyle factors (e.g., diet and smoking), and health history. You are coding the data so that you have the ability to recontact the women for follow-up studies. After enrolling more than 3,000 cases and 3,000 controls, you discover a genetic variant that doubles a women's risk of developing breast cancer. More than 200 women in the control group have this variant. In the informed consent document, you told the subjects that they would not receive the results of their genetic tests, but now you think that may have been a mistake. You are considering contacting subjects with the genetic variant and telling them that they have an increased risk of developing breast cancer.

- Should you contact subjects with the variant?
- What factors should you consider in making this decision?
- If you contact the subjects, what should you say to them?

Case 3

You are planning a study on the genetics of asthma. You will collect genetic samples from 5,000 patients with asthma and 5,000 controls. You will store the samples indefinitely and would like to share them with other researchers. You would develop an asthma DNA bank for use in future studies. The informed consent document states, "Your DNA samples will be kept for an indefinite time and may be shared with other researchers. Your samples will be used to study asthma and other diseases. You will not receive any financial benefits from commercial products developed from your samples." You submit your protocol, consent forms, and other documents to the IRB. The IRB gives your study conditional approval, provided that you specify clearly the other diseases that you plan to study with the samples. The IRB states that the phrase "Your samples will be used to study asthma and other diseases" is too general and needs to be more specific. Do you agree with the IRB's decision? Why or why not?

Case 4

You are conducting research on the genetics of testicular cancer. You are recruiting 500 subjects from different clinics around the country. You will collect genetic samples, conduct various blood tests, and obtain information about the subjects' medical and family history. You plan to ask the subjects about any relatives who have had testicular cancer. You will also ask subjects to provide you with information about their relatives' general health, and so on.

- Since you will be collecting private information about the subjects' relatives, do you also need to contact the relatives to ask them for permission for this information?
- What steps should you take to protect rights and welfare of the relatives?

Case 5

You are applying for an NIH grant on the genetics of homosexuality. You plan to collect DNA samples from 1,000 male homosexuals and 1,000 male heterosexuals, controlling for education, income, race, ethnicity, and other environmental factors. Your goal is to identify genetic variants associated with homosexuality.

- What are the potential social and ethical implications of discovering genetic variants associated with homosexuality?
- How should you deal with those implications?
- Should the NIH consider these implications when it decides whether to fund this study?
- Should an IRB consider them when it decides whether to approve it?

Case 6

A group of researchers, consisting of anthropologists, geneticists, and physicians, identified a rare genetic disease in an indigenous population of Native Americans (the Beewoks) living on coast of western Canada. People who have the disease (known Wajan's disease) have 12 fingers and 12 toes, white hair, blue eyes, and enlarged hearts. People with Wajan's disease usually die before age 25 as a result of heart failure. The Beewoks do not regard this condition as a disease. They believe that people who are born with this condition are blessed by the gods and have special gifts of empathy, prophecy, and spiritual healing. They also believe that people with this condition have been a part of the population since the beginning of time. The Beewoks also believe that they have always lived on the place where they are currently living.

After studying Wajan's disease, its genetics, and the population, the researchers publish an article describing and explaining the condition and its history in the local population. They refer to the condition as a "disease" and an "illness." They say that the people with Wajan's disease have "weak hearts." They also claim that the Beewoks share many unique genetic variants with Eskimos in northern Alaska, and that it is likely that the Beewoks migrated from northern Alaska less than 500 years ago. A local newspaper prints a story about the scientific article. Some of the leaders of the Beewok population learn about the results of the research from the local paper, and they are very upset about how people with the disease are portrayed and how the population and its history are described. They demand that the researchers return their blood samples, and they do not want the researchers to publish any more articles about them. They also say that they will never allow any scientists to study them again.

- How could the researchers have avoided this unfortunate situation?
- What steps should they have taken to protect the Beewoks and promote their interests?

Case 7

A pharmaceutical company is developing a gene therapy protocol for male pattern baldness. The product is a cream that contains a retrovirus similar to the viruses that cause the common cold. The virus contains a gene that promotes hair growth. To be effective, the cream needs to be rubbed on the bald skin and not removed (by washing) for at least three hours. In animal studies, the cream has successfully grown hair on bald rodents with few side effects, other than some respiratory symptoms similar to the common cold. The company would like to begin phase I trials on healthy, adult men with male pattern baldness.

- Do you see any potential ethical and social problems with this gene therapy study?
- If you were an IRB member, would you approve it?
- What else would you like to know about this research before you make a decision?

Case 8

Investigators at University X are planning to recruit college students to donate biomaterials for an embryonic stem cell research protocol. The investigators plan to experiment with different methods deriving HES cells. They plan to create some embryos by IVF and some by nuclear transfer from somatic tissue. The investigators will offer the women $5,000 per egg donation. Each woman can donate up to four times in a year. Sperm and tissue donors will be paid $50 per donation and can donate once a week. Do you have any ethical concerns about this plan?

Case 9

It is the year 2010 and scientists have discovered a gene that confers an increased risk (25%) for ADHD. A group of educators are proposing that the state enact a law requiring mandatory screening for this gene, because studies have shown that early intervention for ADHD can yield excellent results. If a child tests positive for the gene, parents can be educated about the condition and what steps to take to provide an appropriate environment for their child, including possible treatment with medication. The educators argue that this screening program will be very cost-effective, because about 10% of all children have ADHD, the test is inexpensive, and the state can save a great deal of money on special education programs and medication through early intervention. How would you react to this for your children?

Case 10

H.T. is a 44-year-old white male hospitalized for impaction. He is mentally retarded and has physical deformities, which have been present since birth. He has one "normal/functional" limb, his right arm, and three abnormal ones (short, clubbed, stiff, drawn in). His family lives far away from town but wants to transfer his care to the outpatient service; they like the care he has received while a patient at the hospital. After examining the patient and seeing his niece, the attending physician begins to wonder whether H.T. has a genetic syndrome. A resident also comments that he has seen other members of the family (nieces, nephews, and siblings) who have a very similar appearance (face, skin, hair, body size, and build). H.T. is currently living with his niece.

The plan of care is to stabilize his medications and send him home with a prescription for a laxative and recommend some physical therapy to help him with constipation. The team will also schedule a follow-up appointment at the outpatient clinic. The attending physician comments that this would be an interesting case to publish as a Case Report, and that it might be worth testing the family to identify genes or genetic variants responsible for the condition.

- Should the team bring up the issue of genetic tests for H.T. or possibly family members?
- What are the possible benefits and risks of these tests?
- Should the family be tested under the guise of therapy or under a research protocol?

Case 11

Investigators are developing a protocol to create different human–animal chimeras with embryonic stem cells from different species. Human embryonic stem cells will be injected to animal embryos, so that the different lineages will develop and grow together. The investigators hope to learn more about basic mechanisms of development, cell differentiation, and cell regulation. Nonhuman species that will be used include rodents, cats, and rhesus monkeys. The chimeras will be implanted in the wombs of these species and aborted prior to birth. Do you have any ethical concerns with this protocol?

15

International Research

This chapter provides an overview of ethical, social, and policy issues related to international research with human subjects, with a special emphasis on research in the developing world. The chapter covers such topics as international codes of ethics, relativism in international research, exploitation, benefit sharing, commercialization, informed consent, and the use of placebos in research.

International research is a new and emerging topic in human research ethics. Pharmaceutical and biotechnology companies, as well as government agencies, have become increasingly involved in sponsoring or conducting international research. The conduct of research with subjects in developing countries has even entered the popular culture: The 2005 Oscar-winning movie *The Constant Gardener* depicts a widower determined to find out how his wife died after she discovered serious abuses of human subjects in a clinical trial.

There are several reasons for this increased attention to international research. First, the globalization of commerce, trade, industry, and travel means that diseases can spread easily across the globe. HIV, for example, spread from Africa to North America and the rest of the world. Each year, new strains of the influenza virus emerge in Southeast Asia and spread throughout the globe. Because diseases have become international in origin, etiology, and impact, biomedical research must also become international. Second, globalization has had an impact on science. Scientists from different countries now routinely collaborate with indigenous scientists on research projects. Conferences usually have attendees from all over the world, and editorial boards often include members from many different countries. Third, some private companies view international research as a profitable business enterprise. Companies can hire cheap labor (e.g., scientists and research subjects) to test their products. Companies can also take advantage of lax regulations in developing nations (Shamoo 2005, Shah 2006). International research is funded by a variety of sources, such as industry, the U.S. government (e.g., National Institutes of Health [NIH], Centers for Disease Control and Prevention [CDC]), and private foundations such the Bill and Melinda Gates Foundation and William J. Clinton Foundation.

INTERNATIONAL GUIDELINES

Most countries in the developed world have their own laws and regulations concerning research with human subjects. Chapter 12 focuses on the laws and regulations in the United States. We do not discuss the laws and regulations in other countries in this chapter, but refer the reader to some useful resources (Office of Human Research Protections 2007). Most of these laws and regulations, like U.S. laws and regulations, emphasize the key principles of human research discussed in chapter 12, such as informed consent, risk minimization, reasonable benefit/risk ratio, and confidentiality. The U.S. regulations are based on the three Belmont Report principles: respect for person, beneficence, and justice. We think these principles are applicable globally.

Instead of focusing on the laws and regulations of any particular country, we explore some influential ethics codes and guidelines, some of which are mentioned in chapter 12, such as the Nuremberg Code and the Helsinki Declaration. As noted there, the main significance of the Nuremberg Code is that it was the first international code of ethics for research with human subjects. Another important aspect of the code is that it emphasizes the importance of informed consent in research, which had not been discussed much before the code was adopted: "The voluntary consent of the human subject is absolutely essential. This means that the person involved should have legal capacity to give consent; should be so situated as to be able to exercise free power of choice, without the intervention of any element of force, fraud, deceit, duress, over-reaching, or other ulterior form of constraint or coercion; and should have sufficient knowledge and comprehension of the elements of the subject matter involved as to enable him to make an understanding and enlightened decision" (Nuremberg Code 1949).

One problem with the code's statement about informed consent is that it leaves no room for enrolling subjects who cannot provide consent in research, since the code requires that the person involved in research should have the legal capacity to give consent. However, as discussed in chapters 12 and 13, it is ethically acceptable to enroll subjects who do not have the capacity to consent in research studies, provided that proper protections are in place. The Helsinki Declaration was adopted to provide additional guidance for researchers beyond what was included in the Nuremberg Code. It also explicitly states guidelines for including people who cannot give consent in research projects:

> For a research subject who is legally incompetent, physically or mentally incapable of giving consent or is a legally incompetent minor, the investigator must obtain informed consent from the legally authorized representative in accordance with applicable law. These groups should not be included in research unless the research is necessary to promote the health of the population represented and this research cannot instead be performed on legally competent persons. When a subject deemed legally incompetent, such as a minor child, is able to give assent to decisions about participation in research, the investigator must obtain that assent in addition to the consent of the legally authorized representative. Research on individuals from whom it is not possible to obtain consent, including proxy or advance consent, should be done only if the physical/mental condition that prevents obtaining informed consent is a necessary characteristic of the research population. (World Medical Association 2004)

So, the Helsinki Declaration allows people who cannot give consent to be enrolled in research, provided that there is appropriate proxy consent, the opportunity for assent, and protections from unnecessary research risks.

The Nuremberg Code also includes several provisions that deal with minimizing risks to subjects:

> The experiment should be so conducted as to avoid all unnecessary physical and mental suffering and injury. No experiment should be conducted where there is an a priori reason to believe that death or disabling injury will occur; except, perhaps, in those experiments where the experimental physicians also serve as subjects.... Proper preparations should be made and adequate facilities provided to protect the experimental subject against even remote possibilities of injury, disability, or death.... During the course of the experiment the scientist in charge must be prepared to terminate the experiment at any stage, if he has probable cause to believe, in the exercise of the good faith, superior skill and careful judgment required of him that a continuation of the experiment is likely to result in injury, disability, or death to the experimental subject. (Nuremberg Code 1949)

The Helsinki Declaration includes similar provisions for minimizing risks: "It is the duty of the physician in medical research to protect the life, health, privacy, and dignity of the human subject.... Physicians should abstain from engaging in research projects involving human subjects unless they are confident that the risks involved have been adequately assessed and can be satisfactorily managed. Every precaution should be taken to...minimize the impact of the study on the subject's physical and mental integrity and on the personality of the subject" (World Medical Association 2004). Finally, the Nuremberg Code requires that the risks to the research subjects are justified in terms of the benefits to science and society: "The experiment should be such as to yield fruitful results for the good of society, unprocurable by other methods or means of study, and not random and unnecessary in nature.... The degree of risk to be taken should never exceed that determined by the humanitarian importance of the problem to be solved by the experiment" (Nuremberg Code 1949). The Helsinki Declaration has similar provisions: "Every medical research project involving human subjects should be preceded by careful assessment of predictable risks and burdens in comparison with foreseeable benefits to the subject or to others.... Physicians should cease any investigation if the risks are found to outweigh the potential benefits or if there is conclusive proof of positive and beneficial results" (World Medical Association 2004).

The Helsinki Declaration, which is much longer than the Nuremberg Code, covers a broad assortment of topics not specifically addressed by the code, such as privacy and confidentiality, research oversight, protocol development, protection of vulnerable subjects, publication, scientific design, the use of placebos, and access to treatments (World Medical Association 2004). The guidelines developed by the Council for International Organizations of Medical Sciences (CIOMS), which are longer than the Helsinki Declaration, address the topics covered by the Helsinki Declaration and other topics in considerable detail. The CIOMS guidelines also discuss ethical review committees, documentation of informed consent, inducement in research, research with specific populations such as children and pregnant women, compensation for research injury, and research capacity building (Council for International Organizations of Medical Sciences 2002). The International Conference on Harmonization's Good Clinical Practice Guidelines are an influential source of guidance for industry-sponsored clinical trials. The guidelines include ethical rules as well as recommendations for clinical trial design, execution, and management (International Conference on Harmonization 1996). We discuss specific provisions of these different guidelines as we consider some ethical controversies in international research.

EXPLOITATION IN BIOMEDICAL RESEARCH

The most fundamental issues in international research concern questions of exploitation and justice. The history of the relationship between developed nations and developing nations is marked by slavery, colonization, military domination, proselytizing, unfair trading practices, murder, and theft. Western, developed nations have taken natural resources, slaves, and land from developing nations, offering little in return. Although the worst episodes of exploitation occurred from about 1500 to 1900, social, political, and economic problems still occur today. Although the globalization of trade, industry, science, and technology has clearly helped many nations to escape from poverty and despair, it also produced some undesirable consequences, such as the destruction of cultures, the disruption of economies, political strife, racism, and unfair trading practices (Shah 2006). While it is not the purpose of this book to pass moral judgment on particular leaders, countries, religions, or political or economic philosophies, we think that it is important to consider the history of exploitation between developed and developing nations when thinking about the ethics of international research, because many of the effects of exploitation still remain, such as institutionalized racism, distrust, resentment, and poverty.

Exploitation has been a major theme in many of the ethical controversies concerning research in the developing world, including the use of placebos in clinical trials, access to medications, and commercialization of research (Resnik 2003a, Ballantyne 2005). Exploitation is a relationship (or transaction) in which a person (or persons) takes unfair advantage of another person (or persons). For example, someone with a tree removal company who engages in price gouging following a hurricane would be taking unfair advantage of its customers. Customers would be willing to pay way above the fair market value for the service, due to their desperate circumstances. There are three basic elements of exploitation: (a) injustice, (b) disrespect, and (c) harm. For a relationship or transaction to be exploitative, at least one of these elements must be present. In some cases, all three elements are present (Wertheimer 1999). For example, slavery usually involves injustice because the benefits and burdens of the relationship are not distributed fairly; disrespect, because slaves have no free choice concerning different tasks they are supposed to perform and they are treated like property; and harm, because slaves may be physically or psychologically abused or neglected.

A relationship can be exploitative even though both parties consent to it and benefit from it. If the relationship involves injustice, it could still be exploitative. For example, consider the price gouging example again. Suppose that a customer freely chooses to give her money to the tree removal company, and that she benefits from having her trees removed. Most people would still consider this relationship to be exploitative because the company benefits unfairly from the situation by charging a price way above the fair market value, that is, the price it could charge if there had not been a hurricane. How does one know whether a relationship is unjust? Wertheimer (1999) argues that one way to determine this is to consider whether benefits and burdens are distributed fairly in relationship. If one party reaps almost all of the benefits with minimal burdens, while the other party bears most of the burdens with few benefits, then the relationship would be unjust. To use another example, consider someone operating a diamond mine. Suppose that laborers earn $20 per day but they collect about $5,000 worth of raw diamonds per day. To determine whether this arrangement would be exploitative, one would need to consider all of the different costs the owners of the mine must bear, such as paying for equipment, electricity, water, health care, and food, as well as the burdens that laborers must bear, such as pain, fatigue, and health risks. We could consider the relationship exploitative if, in the final tally, the miners do not receive a fair share of the benefits, given the burdens they bear.

It is also important to note that labeling a transaction or relationship as exploitative does not automatically imply that it should be banned or avoided, since some types of exploitation may be ethically acceptable because they promote other values, such as social utility (Wertheimer 1999). When faced with the choice between promoting social utility and preventing exploitation, the best choice may sometimes be to promote social utility. For example, suppose that a textile factory relocates its operations in a developing nation to save money on labor. The factory offers workers what could be considered an unfair wage, because it is much less than the value of the labor, so it exploits the cheap labor available in the country. Nevertheless, one might argue that this situation, while exploitative, is still ethical, because it can benefit the country by providing jobs and economic development. Some types of exploitation may be morally justified on the grounds that they respect individual rights. For example, in the United States, people are free to enter unfair contracts, such as selling a painting worth $4,000 for only $4. Although this transaction might be exploitative to the seller, we would still allow it because it respects his right to sell the painting.

In this analysis of exploitation we have focused on exploitation in transactions or relationships, ignoring broader social, economic, cultural, and historical conditions that can affect these transactions or relationship. It is important to also consider these other conditions because they can affect a party's ability to voluntarily engage in transactions or relationships (Conference on Ethical

Aspects of Research in Developing Countries 2004). If there is an imbalance of power (or social status) between the parties, the stronger party will be able to control the terms of the transaction or relationship. The two parties will not be able to negotiate a fair deal. Also, people with limited choices may be willing to accept unfair deals, in order to receive the benefits. For example, someone who faces extreme poverty may have few options but to work for an employer who does not offer a fair wage. A person who has no access to medical care may enroll in a clinical trial to receive some medical care. Under such conditions, it becomes difficult to avoid exploitation, and one should plan ahead to avoid it (Benatar 2000).

Having discussed exploitation in general, we can apply this analysis to exploitation in international research. Two types of situations might be regarded as exploitative: (a) exploitation of research subjects and (b) exploitation of populations, communities, or nations (Resnik 2001b, 2003a). Exploitation of a research subject could occur if the subject does not receive a fair share of the benefits of research, the subject is harmed, the subject does not consent to the research, or all three. For example, suppose that a drug company conducts a phase I study of a new hypertension drug in a developing nation and pays the research subjects a wage far below the average wage in that country. Suppose that most of the subjects become seriously ill due to drug toxicity, and that some suffer permanent damage. Suppose, too, that the drug, if developed and approved, is likely to be very expensive and not available to most of the people in the developing nation. The company also has no plans to help provide any other benefits to the country, such as enhancing its research or health care infrastructure or providing education or free medicine to the people. This study could be regarded as exploiting the research subjects and exploiting the population, community, or nation. Although this example is hypothetical, it is based in reality, because pharmaceutical companies have been outsourcing clinical trials to the developing world. Some developing nations, such as India, have courted pharmaceutical companies (Nundy and Gulhati 2005, Shah 2006). We return to the topic of exploitation below when examine particular controversies relating to the ethics of international research.

THE USE OF PLACEBOS IN INTERNATIONAL RESEARCH

To understand some of the controversies concerning the use of placebos in international research, it is useful to provide some background to these disputes. Placebos are used in randomized controlled trials to control for the placebo effect. Studies have shown that patients who are given placebos often have significant health improvements, compared with patients who receive no therapy. The placebo effect is most pronounced in medical conditions with a strong psychological component, such as depression or chronic pain. Scientists are not certain why placebos can have a positive effect on health, but it is thought that placebos exert their effects through the patient's belief system, which, in turn, may influence the immune system, the cardiovascular system, and other parts of the body (Wampold et al. 2005). In drug studies that include a placebo control group, the investigators and subjects are usually not aware of the subjects' group assignments, so they do not know who is receiving a placebo and who is receiving a pharmacologically active therapy. When a new drug is compared with a placebo in a clinical trial, the aim of the trial is to demonstrate that the drug is more effective than the placebo (Wampold et al. 2005).

The use of placebos in clinical trials has been a controversial issue in the ethics of research with human subjects for many years and remains so today (Emanuel and Miller 2003). The problem is that the patients/subjects in the placebo control group might not receive optimal medical care. Most researchers and ethicists agree that it is unethical to give placebos to research subjects in clinical trials when the subjects have a serious illness and there is an effective therapy. If a therapy has been proven to be effective, then all subjects should receive that therapy, or perhaps an experimental therapy thought to be more effective than the accepted therapy. For example, a phase II study of a new blood pressure medication should provide all of the subjects some type of medication to control their blood pressure, because uncontrolled blood pressure can cause heart damage, a stroke, or other serious medical problems. Some subjects could receive a standard therapy, while others could receive an experimental one, but no subjects should receive a placebo. Placebos can only be given to patients with a serious illness when there is no known effective therapy for the illness.

There is some disagreement about why giving placebos in medical research is sometimes unethical. According to the standard view, it is unethical to administer placebos when patients have a serious illness because doctors have an ethical duty to provide optimal medical care for their subjects/patients, and giving a placebo instead of an effective treatment violates that duty (London 2001). Placebos can be given only when clinical investigators are uncertain whether there is an effective treatment for a condition (Freedman 1987, London 2001), because then research subjects with the condition who receive placebos are not being denied medical care that they might otherwise receive. According to a more recent view, it is unethical to give placebos in clinical research not because of the obligation to provide optimal medical care, but because giving placebos instead of an effective treatment is exploitative, since subjects/patients do not receive a fair share of the benefits of research and may be harmed, and clinical researchers have a duty not to exploit their subjects (Miller and Brody 2002).

The use of placebos in clinical trials raises many questions that we do not address in this book, such as, What is clinical uncertainty? What is a serious medical condition? Do placebos provide some medical benefits to patients/subjects? We encourage the reader to consult other sources (Miller and Brody 2002, Emanuel and Miller 2003, Wampold et al. 2005, Chiong 2006, Veatch 2007).

Returning to the topic of international research, a controversy concerning the use of placebos in clinical trials emerged in 1997, when two members of the Public Citizen's Health Research Group, Peter Lurie and Sidney Wolfe (1997), published an article in the *New England Journal of Medicine* (*NEJM*) in which they argued that 15 clinical trials taking place in sub-Saharan Africa and other developing nations were unethical. The editor of the *NEJM*, Marcia Angell (1997a), also argued that the clinical trials were unethical. She compared the trials to the infamous Tuskegee Syphilis Study and also accused the researchers of holding a double-standard, one for the developed world and one for the developing world. NIH director Harold Varmus and CDC director David Satcher (1997) published a response to the allegations by Lurie and Wolfe in the next issue of the *NEJM*, and an international debate ensued.

The controversial studies were attempting to determine whether perinatal (mother-to-child) transmission of HIV could be effectively prevented using a method that was much less expensive than the method currently being used to prevent perinatal HIV transmission in developed nations. The standard of care in developed nations, known as the 076 protocol, involved the administration of $800 worth of azidothymidine (zidovudine; AZT) to the mother during pregnancy and labor, and to the child following birth. Breast-feeding mothers would also receive AZT. This method was shown to reduce the rate of perinatal HIV transmission from 25% to 8%. The controversial studies were attempting to determine whether perinatal HIV transmission could be reduced using about $80 worth of AZT and fewer health care services. The drug would be administered less frequently than it would be under the 076 protocol. None of the nations where the studies took place could afford the medications needed to administer the 076 protocol on a large scale. The countries also did not have sufficient health care infrastructure to execute the 076 protocol. The trials were approved by the local leaders and authorities, by the World Health Organization (WHO) and the U.N. Joint Programme on HIV/AIDS, and by the CDC and NIH, which helped to sponsor the trials (Resnik 1998b). Local researchers helped to design and implement the trials and recruit subjects. Less than a year after the controversy began, the investigators were able to show that a 10% dose of AZT given at the end of pregnancy can reduce rate of transmission of HIV by 50% (De Cock et al. 2000).

Most of the ethical controversy concerning these trials focused on the research design, since the trials included control groups of subjects who only received placebos. The reason for including placebo groups was to prove that the lower dose of AZT was more effective than a placebo. It was already known that the higher dose was effective, but it was not known whether the lower dose would be. The reason for attempting to determine whether the lower dose would be effective is that few people in developing nations can afford the higher dose. The researchers wanted to test a cheaper method of preventing perinatal HIV transmission.

Lurie, Wolfe, and others objected to this research design on the grounds that it denied subjects in the control group a proven, effective therapy. They argued that since AZT has already been shown to prevent perinatal HIV transmission, all of the subjects should receive the drug. Giving placebos instead of an effective therapy was unethical and exploitative, they argued. The investigators were sacrificing some research subjects for scientific or public health goals. Lurie and Wolfe argued that the studies should have used active controls rather than placebo controls. (An active control group is a control group where subjects receive an effective treatment, e.g., some clinical trials compare different drugs with respect to health outcomes.)

Varmus, Satcher, and other defenders of the trials argued that an active control design would lack the scientific rigor of a placebo control design. An active control design would also require a much larger sample size to ensure that the studies had sufficient statistical power. The sample would need to be much larger because a study that used active controls would be attempting to detect a very small difference between treatment groups. It would probably also take a much longer time to complete active control trials. Placebo control trials would take less time, cost less money, and yield clearer, more rigorous results. Defenders of the controversial studies also argued that the subjects who were receiving placebos were not being exploited or mistreated, because they did not have access to AZT anyway. Participation in the study did not make the subjects any worse off and could have benefited them by giving them access to medical care (other than AZT therapy). Critics of the studies argued that it did not matter whether subjects lacked access to the treatments needed to prevent the perinatal transmission of HIV, since the treatment had been proven effective and was available in developed countries. The medical standard of care should be universal, not local. The studies were exploitative because they were taking advantage of the fact that subjects did not have access to AZT (Resnik 1998b, London 2001).

This controversy concerning the use of placebos to develop cost-effective treatments for developing nations has still not been resolved. The Helsinki Declaration and CIOMS guidelines have been revised since the controversy erupted. These revisions provide some useful guidance but do not entirely resolve the issues. According to the Helsinki Declaration: "The benefits, risks, burdens and effectiveness of a new method should be tested against those of the best current prophylactic, diagnostic, and therapeutic methods. This does not exclude the use of placebo, or no treatment, in studies where no proven prophylactic, diagnostic or therapeutic method exists" (World Medical Association 2004).

Lurie, Wolfe, and others argued that the disputed studies violated this provision of the Helsinki Declaration, since there was an effective therapy in existence, and the studies did not test the new method against that therapy. The declaration was revised in 2002. A footnote to the 2002 revision states:

> Extreme care must be taken in making use of a placebo-controlled trial and that in general this methodology should only be used in the absence of existing proven therapy. However, a placebo-controlled trial may be ethically acceptable, even if proven therapy is available, under the following circumstances:
>
> • Where for compelling and scientifically sound methodological reasons its use is necessary to determine the efficacy or safety of a prophylactic, diagnostic or therapeutic method; or

- Where a prophylactic, diagnostic or therapeutic method is being investigated for a minor condition and the patients who receive placebo will not be subject to any additional risk of serious or irreversible harm. (World Medical Association 2004)

It is not entirely clear whether the 2002 revision Helsinki Declaration would forbid the type of studies criticized by Lurie, Wolfe, and others. Although the declaration states that placebos should be used only if no effective therapy exists, it allows an exception when there is a compelling and scientifically sound reason to use a placebo to determine the safety or efficacy of a treatment. Defenders of the controversial studies could argue that there was a compelling and scientifically sound reason to use placebos to determine the efficacy of AZT at a lower dose.

The CIOMS guidelines have also been revised since the controversy took place. According to the CIOMS guidelines:

As a general rule, research subjects in the control group of a trial of a diagnostic, therapeutic, or preventive intervention should receive an established effective intervention. In some circumstances it may be ethically acceptable to use an alternative comparator, such as placebo or "no treatment." Placebo may be used:

- When there is no established effective intervention;
- When withholding an established effective intervention would expose subjects to, at most, temporary discomfort or delay in relief of symptoms;
- When use of an established effective intervention as comparator would not yield scientifically reliable results and use of placebo would not add any risk of serious or irreversible harm to the subjects. (Council for International Organizations of Medical Sciences 2002, guideline 11)

This part of the guidelines would seem to prohibit the studies criticized by Lurie, Wolfe, and others, since those studies do not appear to meet any of the three exceptions noted in the guidelines. However, a comment on this guideline appears to make an exception for the controversial placebo studies:

Exceptional Use of a Comparator Other than an Established Effective Intervention
An exception to the general rule is applicable in some studies designed to develop a therapeutic, preventive or diagnostic intervention for use in a country or community in which an established effective intervention is not available and unlikely in the foreseeable future to become available, usually for economic or logistic reasons. The purpose of such a study is to make available to the population of the country or community an effective alternative to an established effective intervention that is locally unavailable. Accordingly, the proposed investigational intervention must be responsive to the health needs of the population from which the research subjects are recruited and there must be assurance that, if it proves to be safe and effective, it will be made reasonably available to that population. Also, the scientific and ethical review committees must be satisfied that the established effective intervention cannot be used as comparator because its use would not yield scientifically reliable results that would be relevant to the health needs of the study population. In these circumstances an ethical review committee can approve a clinical trial in which the comparator is other than an established effective intervention, such as placebo or no treatment or a local remedy.

(Council for International Organizations of Medical Sciences 2002, guideline 11, commentary)

This passage makes it clear that a placebo may be used when an established intervention is not available in a country for economic or logistic reasons, which was the case with the controversial HIV studies. The studies were conducted because the standard treatment was not available in those countries, due to its high cost and the complexity of administering it. However, the passage also makes it clear that the purpose of such research is to make an effective alternative available to the population. The treatment should be responsive to the health needs and that population and should be made reasonably available when the research is completed. The overall ethical rationale seems to be that the studies may be justified because they will benefit the societies in which they are conducted. The ideas of "reasonable availability" and "responsiveness to community needs" bring us to another important issue in the ethics of international research.

SOCIETAL BENEFITS IN INTERNATIONAL RESEARCH

Critics of the perinatal HIV studies also charged that the research would not be reasonably available to members of the community, because the therapy would still not be affordable for most of the people who need it. Even $80 worth of AZT would be well beyond the budgets of governments that spend $100 or less per person on health care each year (Resnik 1998b). Defenders of the research countered that the therapy would still be reasonably available, since it would be marketed to the countries and a large number of people would have access to it. This aspect of the debate raised two important questions: What does it mean for a treatment to be reasonably available? Why is it important for a treatment to be reasonably available?

We have already considered the answer to the second question in this chapter: A treatment should be reasonably available so that the research does not exploit the population or community. By making a treatment reasonably available, investigators and sponsors help to ensure that the population or community will derive some benefits from the research, so that the relationship with the population or community will be fair (Conference on Ethical Aspects of Research in Developing Countries 2004). Although they may disagree about how to define the concept of exploitation, most commentators agree that avoiding exploitation is the principal reason for ensuring that populations or communities benefit.

The more difficult question is to define what it means for a therapy to be "reasonably available." Is a therapy reasonably available only if everyone who needs it has access to it, or is it enough that 50% or only 25% have access? The Helsinki Declaration does not offer much guidance on how to address this issue. It says only that "medical research is only justified if there is a reasonable likelihood that the populations in which the research is carried out stand to benefit from the results of the research" (World Medical Association 2004). The CIOMS guidelines do not define "reasonable likelihood" or explain how a community or population might benefit from research, but they do mention that researchers have an obligation to strengthen capacity for ethical and scientific review:

> Many countries lack the capacity to assess or ensure the scientific quality or ethical acceptability of biomedical research proposed or carried out in their jurisdictions. In externally sponsored collaborative research, sponsors and investigators have an ethical obligation to ensure that biomedical research projects for which they are responsible in such countries contribute effectively to national or local capacity to design and conduct biomedical research, and to provide scientific and ethical review and monitoring of such research. Capacity-building may include, but is not limited to, the following activities:
>
> • Establishing and strengthening independent and competent ethical review processes/committees
> • Strengthening research capacity
> • Developing technologies appropriate to health-care and biomedical research
> • Training of research and health-care staff
> • Educating the community from which research subjects will be drawn (Council for International Organizations of Medical Sciences 2002, guideline 20)

There seem to be at least two justifications for requiring researchers and sponsors to help strengthen a country's capacity for ethical and scientific review. First, this is necessary so that the host country can conduct ethical and scientific review of research that is to be conducted in it. Second, this can provide an important benefit for the host country (White 2007).

The participants in the 2001 Conference on Ethical Aspects of Research in Developing Countries argue that the reasonable availability standard is not sufficient for avoiding exploitation of populations, communities, or countries in research. They also argue that the reasonable availability standard is too vague in that it does not define "reasonable availability," does not say to whom duties are owed, and does not specify the strength of researchers' or sponsors' obligations (Conference on Ethical Aspects of Research in Developing Countries 2004). The participants argue that the "fair benefits" standard is a better one to use. The fair benefits standard holds that "there should be a comprehensive delineation of tangible benefits to the research participations and the population from both the conduct and the results of research" (Conference on Ethical Aspects of Research in Developing Countries 2004, p. 22). Some of these benefits include:

• Improved health of research subjects
• Posttrial access to medications for research subjects
• Health services and public health measures made available to the population
• Employment and economic activity
• Availability of the intervention when research is complete
• Improvements to the health care infrastructure and research capacity
• Long-term research collaboration
• Sharing of financial rewards, including intellectual property

The fair benefits approach would appear to be effective at preventing exploitation at the level of the individual research subject and the level of the population, community, or nation, and it offers clearer guidance than the Helsinki Declaration or CIOMS guidelines. However, some commentators argue that someone could justify any research with human subjects as long as other benefits, not related to the research itself, can be used to justify unnecessary research. All agree that research on male pattern baldness and cosmetic surgery, at the moment, should not receive high priority, but it might be justified under the fair benefits approach (Shamoo 2005). Others contend that the fair benefits approach does not go far enough to address issues of exploitation or benefit the population. Alex John London (2005) criticizes the fair benefits approach on the grounds that its notions of exploitation and justice are too limited. He argues that to understand exploitation and justice one must look beyond particular transactions or relationships and consider the larger social, economic, cultural, and political context. The fair benefits approach is based on the idea, according to London, of a fair agreement between researchers/ sponsors and a host population, community, or country. The agreement is fair if both parties consent to it and benefits are distributed fairly. The problem with this idea, according to London, is that it ignores the larger context in which the agreement is made, such as extreme poverty, famine, and disease in the host country, or the history of relationship between the two countries (e.g., racism, slavery, theft, or exploitation). An agreement cannot be truly fair unless one also addresses this larger context. To do this, researchers and sponsors must do more than provide fair benefits: They must rectify past injustices and promote social, economic, and political development in the host nation. London calls this idea the human development approach (London 2005).

While London's approach is admirable in some ways, one might argue that it is also too idealistic and unrealistic (Shamoo 2005). It asks researchers and sponsors to do much more for the host countries than they can possibly be expected to do. If implemented, London's approach would make biomedical research in the developing world prohibitively expensive and complicated. Sponsors would choose to forgo research in the developing world to avoid paying the high costs of nation building (White 2007). Although promoting economic, social, and political development in developing nations is a worthwhile goal, it is a job best left to the United Nations, the International Monetary Fund, the World Bank, and other organizations whose main goal is development. The main goal of research is...research. Researchers and sponsors should provide meaningful and fair benefits to the populations they work with, but they need not do more.

POSTTRIAL ACCESS TO MEDICATIONS

One issue related to fair benefits to human subjects in research is the availability of medications after the study is completed. The availability of medications is usually not an issue in the United States and other developed countries because health insurers or government agencies, such as Medicaid or Medicare, often will pay for medications after the subject completes a clinical trial. Someone in the United States who enrolls in a clinical trial of a new hypertension drug usually does not have to worry about having access to a hypertension medication after the trial is over. In developing nations, however, research subjects may not have access to medications once a trial is completed. Indeed, their only opportunity to receive a medication may be to participate in a study. When the study is completed, they would lose access. The problem of lack of posttrial access to medications raises a couple of concerns. First, one might consider it to be unfair and exploitative not to provide subjects with access to medications when the study is complete: It would be like saving someone from drowning so they can work on your boat, then tossing them back in the water when the work is done. Second, stopping medications when a study is complete can often be dangerous or life-threatening. With these concerns in mind, the Helsinki Declaration specifically addresses posttrial access to treatments: "At the conclusion of the study, every patient entered into the study should be assured of access to the best proven prophylactic, diagnostic and therapeutic methods identified by the study" (World Medical Association 2004). The CIOMS guidelines state that "sponsors are ethically obliged to ensure the availability of: a beneficial intervention or product developed as a result of the research *reasonably available* to the population or community concerned" (guideline 21, emphasis added). In the commentary to this guideline, it adds that research protocols include a provision for "continuing access of subjects to the investigational treatment after the study, indicating its modalities, the individual or organization responsible for paying for it, and for how long it will continue."

While posttrial access to medications is a laudable goal for researchers and sponsors, it may be financially difficult to pursue. Treating a single HIV patient with antiretroviral drugs can cost thousands of dollars a year for many years. There is some evidence that researchers and sponsors often fail to meet the goal of posttrial access to medications, even when they include posttrial access as part of the protocol. The posttrial access requirement is another added cost that may prevent sponsors from pursuing research projects that can benefit people in developing nations (Ananworanich et al. 2004). The NIH has struggled with funding research under the posttrial access requirement (Hellemans 2003). Although researchers and sponsors should make a concerted effort to guarantee posttrial access to treatment for all research subjects, this commitment should be tempered by realism and an understanding of the purpose of research. Researchers' and sponsors' primary responsibility is to protect the rights and welfare of subjects during the study. Once a study is complete, they should still try to protect subjects from harm, but the posttrial obligations need not be as strong as those that occur during the trial. Other stakeholders with an interest in the health of populations in developing nations, such as the United Nations, WHO, Doctors Without Borders, and various governments from developing and developed nations should help researchers and sponsors provide posttrial access to medications (Merritt and Grady 2006).

INFORMED CONSENT

Critics of the perinatal HIV trials also charged that the subjects did not provide adequate informed consent (Resnik 1998b). Although all the major international research codes and guidelines require that the research subject (or the subject's representative) provide informed consent, there are some significant challenges to implementing this requirement in developing nations. First, there are linguistic barriers to effective consent. Interpreters must often be used to converse with subjects in their native language, and translators must be used to translate consent documents and other materials, such as brochures or questionnaires. Some words may not translate easily into different languages. Also, some populations may have no written language, so use of a consent form or other document can be problematic. Second, there can be conceptual or cultural obstacles to effective consent. People from non-Western, nonindustrialized nations may have little comprehension of Western concepts such as disease, cause and effect, genetics/DNA, virus, bacteria, and so on. While it may be possible to explain these concepts to research subjects, some ideas may be so alien to people from non-Western, nonindustrialized nations that they may find them difficult to comprehend. Third, many African nations have tribal governance. The leaders of the tribe may need to give permission before any member of the tribe can be recruited into a study. Members of the tribe may not believe that they have the right to decide whether to participate, or may not even comprehend the notion of individual decision making. They may believe that they must do whatever the tribal leaders say to do (Préziosi et al. 1997, Weijer and Emanuel 2000, Fitzgerald et al. 2002, Doumbo 2005). Attaining and documenting a truly valid informed consent should remain the goal, however (Bhutta 2004).

The Helsinki Declaration does not give any recognition to the consent problems that may arise in research in developing nations. The model of consent outlined in the declaration is Western and individualistic (World Medical Association 2004). The CIOMS guidelines, however, include some commentaries that address these consent issues:

Cultural considerations. In some cultures an investigator may enter a community to conduct research or approach prospective subjects for their individual consent only after obtaining permission from a community leader, a council of elders, or another designated authority. Such customs must be respected. In no case, however, may the permission of a community leader or other authority substitute for individual informed consent. In some populations the use of a number of local languages may complicate the communication of information to potential subjects and the ability of an investigator to ensure that they truly understand it. Many people in all cultures are unfamiliar with, or do not readily understand, scientific concepts such as those of placebo or randomization. Sponsors and investigators should develop culturally appropriate ways to communicate information that is necessary for adherence to the standard required in the informed consent process. Also, they should describe and justify in the research protocol the procedure they plan to use in communicating information to subjects. For collaborative research in developing countries the research project should, if necessary, include the provision of resources to ensure that informed consent can indeed be obtained legitimately within different linguistic and cultural settings. (Council for International Organizations of Medical Sciences 2002, guideline 4, commentary)

Nothing in this guidance suggests that the individual model of informed consent should be abandoned in some settings: All research subjects should still have the right to consent (or refuse to consent) to participation in a study. However, this guidance suggests that individual decision making may need to be supplemented with community consultation in some cases. In some situations, the community leaders may need to give permission before the study could begin (Weijer and Emanuel 2000). The guidance also makes the important point that investigators may need to draw on the expertise of people who are familiar with the language or customs of the local population to ensure that the consent process is valid and culturally appropriate.

It is worth noting that the federal research regulations allow institutional review boards (IRB) to waive the requirement that consent be documented if it finds "(1) That the only record linking the subject and the research would be the consent document and the principal risk would be potential harm resulting from a breach of confidentiality. Each subject will be asked whether the subject wants documentation linking the subject with the research, and the subject's wishes will govern; or (2) That the research presents no more than minimal risk of harm to subjects and involves no procedures for which written consent is normally required outside of the research context" (45 CFR 46.117(c)). These provisions would allow researchers who are operating under the Common Rule (45 CFR 46) to refrain from documenting the consent process in a situation where the population has no written language and risks of the research are minimal or involve at most the potential loss of confidentiality. If the risks are more than minimal, researchers operating under the Common Rule would need to document consent, even if the population has no written language. This might be handled, for example, by explaining the consent document to the subject in the subject's language and asking the subject to make a mark on a document indicated that he/she consents. The U.S. Food and Drug Administration (FDA) regulations require that consent always be documented (21 CFR 50.27). Hence, international studies on medical products subject to FDA approval would need to provide documentation consistent with the FDA's regulations.

COMMERCIALIZATION

One of the key issues in international research is the commercialization of the products of research, such as drugs, biologics, and genetic tests. Many commentators and activists have argued against intellectual property rights in biomedicine on the grounds that intellectual property rights can deny people access to care by making medical products and services prohibitively expensive. To promote access to care, biomedical patenting should banned or restricted. HIV/AIDS activists have held the pharmaceutical industry accountable for the HIV pandemic in sub-Saharan Africa, because of the high costs of patented drugs. Activists have placed political pressure on drug companies to make their medications more accessible to patients in the developing world, and companies have responded to this pressure by lowering their prices in developing nations and giving away free medications. Although pharmaceutical companies obviously did not cause the HIV/AIDS pandemic, they have a moral obligation to help deal with this crisis (Resnik 2001a). Companies should work with local authorities, nongovernmental organizations, and international coalitions to promote access to medications in the developing world.

Much of the controversy concerning commercialization in international research has focused on the Trade-Related Aspects of Intellectual Property Rights (TRIPS) agreement. TRIPS is an agreement signed by members of the World Trade Organization in 1994 for the protection of patents, copyrights, and trademarks. It has been renegotiated several times. Although signatory nations retain their own sovereignty with respect to intellectual property, they agree to enact laws consistent with the TRIPS agreement (World Trade Organization 2007). TRIPS sets standards for copyrights, patents, and trademarks. Patents, for example, should last 20 years. TRIPS requires countries to cooperate with each other concerning the enforcement of intellectual property rights and includes provisions for countries to settle disputes about intellectual property. The agreement also includes a provision that allows countries to override patent rights to deal with national emergencies:

> Where the law of a Member allows for other use of the subject matter of a patent without the authorization of the right holder, including use by the government or third parties authorized by the government, the following provisions shall be respected:
>
> (a) authorization of such use shall be considered on its individual merits;
> (b) such use may only be permitted if, prior to such use, the proposed user has made efforts to obtain authorization from the right holder on reasonable commercial terms and conditions and that such efforts have not been successful within a reasonable period of time. This requirement may be waived by a Member in the case of a national emergency or other circumstances of extreme urgency or in cases of public non-commercial use. In situations of national emergency or other circumstances of extreme urgency, the right holder shall, nevertheless, be notified as soon as reasonably practicable. In the case of public non-commercial use, where the government or contractor, without making a patent search, knows or has demonstrable grounds to know that a valid patent is or will be used by or for the government, the right holder shall be informed promptly. (World Trade Organization 2007)

This provision of the agreement gives member nations the right to authorize a third party to use a patented invention without permission of the patent in holder in the case of a national emergency. For example, if a patented drug is urgently needed to deal with a national emergency, such as the HIV/AIDS crisis, a country could allow companies other than the patent holder to manufacture and market the drug. Although the TRIPS agreement clearly gives member nations the right to override patents, few nations have chosen this option as a way of dealing with the HIV/AIDS crisis because pharmaceutical companies and developed nations urged countries to not exercise these options. Pharmaceutical companies sued the government of South Africa for importing cheap drugs from countries that do not honor the TRIPS agreement. They later dropped this lawsuit in response to political pressure from advocacy groups (Sterckx 2004).

Although there are strong humanitarian reasons for countries to take advantage of their options under TRIPS to deal with public health emergencies, such as HIV/AIDS, there are also some reasons for exercising some discretion and restraint. The best option is for a country to try to reach an agreement with a patent holder before taking the extreme step of overriding the patent. It may be possible to negotiate a deal with the company so that it can significantly lower its prices, license other companies to make its drugs, or institute drug giveaway programs (Resnik and DeVille 2002). If an agreement cannot be reached, countries should take steps to protect the interests of the company that will be affected. For example, the country should make sure that the drugs will only be used domestically and will not be exported to other countries, to prevent its decision to override patents from having an impact on the global market for the drug. The country should also carefully define "national emergency" to avoid overriding patents for trivial reasons. The decision to override patents should not be a step down a slippery slope toward erosion of intellectual property rights (Resnik and DeVille 2002).

As noted in chapter 9, it is important to protect intellectual property rights to promote scientific innovation and private investment in research and development (R&D). If companies cannot trust that their intellectual property rights will be honored when they allocate R&D money for HIV/AIDS medications, they may stop funding that type of research. One explanation of why 90% of the world's R&D funds are allocated toward only 10% of the world's disease burden is that private companies, which sponsor more than 60% of biomedical R&D, find it more profitable to invest money in diseases that affect people who live in developed nations than in diseases that affect people who live primarily in developing nations. It is more profitable to invest money in diseases of the developed world for numerous reasons— the market is larger and stronger, for example—but one of those reasons is that intellectual property protection is more certain in the developed world. Some would argue that developing nations should strengthen, not weaken, their intellectual property protections to encourage private investment in R&D (Resnik 2004c).

RELATIVISM AND DOUBLE STANDARDS

The final issue we consider in this chapter is whether there should be a single ethical standard for research conducting in different countries, or whether ethical standards can vary from one country to the next. This was an important issue in the disputed HIV studies mentioned above. Angell

(1997a) argued that the studies embodied a double standard—one standard for developed countries and another for developing ones—but that the same standard should apply across the globe. If it would be unethical to use a placebo control group for a study conducted in the United States, then it would also be ethical if the study were conducted in Uganda, Mali, Ethiopia, or any other country. Angell's criticisms of the study raise an important meta-ethical question mentioned briefly in chapter 2: Are standards of research ethics absolute or relative? Can there be some variance in standards among different nations or societies?

Arguments in favor of ethical relativism have traditionally appealed to facts about transnational variations in social behaviors and customs (Macklin 1999a,b). There is considerable evidence for variation in many different types of social behaviors and customs across different nationalities, including marriage practices, sexual behavior, religious rituals, education, sports, and so on. There is also some evidence that behaviors and customs related to the conduct of research, such as individual versus group decision making and human rights, vary across different countries. As noted above, consent practices in some tribal communities are very different from the individual model of consent embodied in the U.S. regulations and international ethics codes. Human rights abuses have occurred in many different countries around the globe, such as China, Burma, Cuba, Russia, Colombia, North Korea, Pakistan, Darfur, and Sudan. Some countries, such as Saudi Arabia, Iran, and Kuwait, accord men and women different rights. From this evidence of behavioral variation, the relativist makes the inference that ethical standards ought to vary—when in Rome, do as the Romans do (Macklin 1999a,b).

This type of inference—inferring a normative claim from a descriptive (or factual) claim—is logically flawed. From the fact that "more than 90% of drivers on interstate highways drive above the posted speed limit," one cannot logically infer that "more than 90% of drivers on interstate highways *should* drive above the posted speed limit." Likewise, from the fact that "there are no free elections in Burma," one cannot logically infer that "there *should* be no free elections in Burma." The problem is simply that there is a logical gap between normative statements and descriptive ones. One of the problems with this argument for relativism is that it attempts to bridge an unbridgeable chasm. Normative claims must be based on other normative claims, or on claims that combine normative and factual components. For example, one can derive the normative claim "John ought to try to earn a college degree" from the normative claim "John wants to become a doctor" and the factual claim "to become a doctor one must earn a college degree." Although the claim "Johns wants to become a doctor" appears to be factual, it is not, because it assumes that John's wants are morally acceptable. For example, one cannot infer the claim "John should shoot Harold" from the claims "John wants to kill Harold" and "Shooting is an effective way of killing a human being," because wanting to kill Harold is not morally justified.

Relativism concerning rules or standards makes sense when one can show that the rules or standards are mere conventions, with little or no import for basic human values. Matters of mere convention can and should vary across different countries. For example, consider motor vehicle laws. Many of these laws are mere conventions. In some countries, driving on the right side of the road is the law; in other countries driving on the left is the law. In some countries, you cannot receive a driver's license until you are 18; in others you can receive a license before age 18. Other laws that show some variation include speed limits, making right turns at red lights, and so on.

Not all driving laws are mere conventions, however. We know of no country where is legal to use a motor vehicle as a weapon to kill someone. We know of no country where it is legal to steal automobiles. We no of know country where it is legal to drive recklessly, endangering human life. None of these behaviors are legal because of their implications for human life and human rights. If any of these behaviors were legal in some country, we would say that they ought to be illegal. If a country made it legal to use a motor vehicle as a weapon, we would say that the law ought to be changed. The point of this digression into motor vehicle laws is that some social behaviors are mere conventions that can and should vary across different countries, but that others not. Social behaviors with implications for basic moral values or principles should not vary across different countries.

To bring the discussion back to research ethics, some of the rules and customs pertaining to research are mere conventions, which can and should vary across nations, but others are not. Some standards apply to all countries. For example, we consider informed consent to be a universal requirement of research with human subjects. However, different countries may interpret this principle in different ways. The exact manner in which consent is obtained may vary across different countries, or even populations or communities. It makes little difference, from an ethical point of view, whether a researcher explains a study in English, Spanish, or Chinese, as long as the subject understands the language that is spoken. It also does not make any difference whether the researcher discusses the risks of the study before or after discussing the benefits, or whether the subject asks questions while the study is being explained or afterward. Documentation of consent is also mostly a matter of convention. It makes no difference, from an ethical point of view, whether consent is documented using a paper or electronic record, whether the typeface is Times Roman or Arial, whether there are 200 or 400 words on a page, and so on. Other universal principles that may be interpreted differently in different countries include protection of confidentiality and privacy, minimization of risks, and extra protection for vulnerable groups.

Another potential source of variation across nations occurs when researchers must decide how to resolve conflicts among different ethical principles when they apply them to particular cases. As noted in chapter 2, resolving conflicts among ethical principles is an important part of ethical decision making. Since there is often more than one acceptable way to resolve a conflict, researchers in different countries might resolve conflicts in different ways. For example, conflicts sometimes occur between scientific rigor and other ethical principles, such as informed consent. In some types of behavioral experiments, fully informing the subjects may compromise the scientific rigor of the study. For example, in an experiment that tests a person's willingness to obey an authority, informing the subject that their willingness to obey authority is being studied may cause the subject to change her behavior. To avoid this type of bias, researchers may need to deceive research subjects about the true nature of the experiment until it is completed. Many commentators have argued that it is ethical to deceive human subjects in behavioral research, provided that the deception is necessary to conduct the research, deception does not pose a significant risk to the subjects, and the investigators have made adequate plans for debriefing after the experiment is complete. However, others might settle the conflict in a different way by prohibiting deceptive research in social science. Researchers who want to study human behavior must design experiments that do not involve deception, such as role playing (Wendler 1996). We think that these two different approaches to deception in social science research are both morally defensible, and that it would be acceptable for different countries to choose different ways of settling this conflict: One country could allow deceptive research under certain conditions, while another might ban it. This situation illustrates how ethical variations could arise among different nations in a perfectly legitimate way. Ethical principles would still be universal, but conflicts would be resolved differently.

Let us reconsider Angell's critique of the perinatal HIV studies. She argued that the studies were unethical because they constituted a double standard when there should be a single standard for all research. While we agree with Angell that there should be a set of ethical principles that apply across the globe, we recognize that there can be differences in how these principles are interpreted and applied. The use of placebo controls in clinical research has always represented a conflict between scientific rigor, on the one hand, and protecting the welfare of human subjects, on the other. In the controversial HIV studies, the conflict also involved questions concerning social benefit versus exploitation. The controversy, in essence, was about different ways of resolving conflicts among these different ethical principles (Resnik 1998b). Defenders of the studies emphasized the importance of maintaining scientific rigor and producing social benefits, while critics emphasized the importance of protecting human subjects and avoiding exploitation. No one on either side of the argument claimed that any of these four moral principles should be abandoned: Defenders of the studies did not claim that protecting the welfare of human subjects is unimportant, nor did critics of the studies claim that scientific rigor is unimportant. The two sides disagreed about how these different principles should be balanced when they conflict. We think that reasonable people could disagree about how to apply ethical principles in this case, and that different countries could resolve these dilemmas in different ways. Thus, we think it would be perfectly reasonable for people in different countries to disagree about the studies. Authorities in Germany could disapprove a protocol with a placebo control group, but authorities in

Uganda could approve it. Local social and economic circumstances can play a key role in the application of ethical principles. We close this discussion with a quote from Edward Mbidde, director of the Uganda Cancer Institute at Makerere University in Uganda:

> The ethics of the design of clinical trials to prevent transmission of HIV-1 from mother to child in developing countries have been criticized. However, a discussion of ethical principles in biomedical research that ignores the socioeconomic heterogeneity is not worth holding. Policies regarding health management differ within and between industrialized and developing countries because of their different economic capabilities. Whereas it is established policy that all HIV-positive pregnant women in the United States and other developed countries are offered azidothymidine (AZT), this is not achievable in many developing countries because the costs of the drug and logistical support are prohibitive...to conduct research in Uganda based on such a regimen would produce results applicable only in a research setting. (Mbidde 1998)

What Mbidde is saying, in effect, is that the clinical research design should reflect local social and economic circumstances. Research studies should be expected to produce results that will be useful to the people in the local population and host country. Indeed, one might argue that refusing to acknowledge the importance of local circumstances in the ethics of human research is to engage in a kind of ethical imperialism (Resnik 1998b, Macklin 1999a,b).

QUESTIONS FOR DISCUSSION

1. Should the ethical principles be universal? How would you compare these to universal human rights?
2. Is requiring the application of universal ethical principles a cultural imperialism? Why?
3. Should the use of placebo be allowed in developing countries? Why?
4. Should the FDA require that clinical trials being conducted in developing countries follow the same standards as those conducted in the United States? Should the FDA then require prior approval of an investigational new drug, as it does in the United States?
5. How would you deal with a corrupt government in a developing country in order to conduct a clinical trial?
6. Is informed consent essential in developing countries?

CASES FOR DISCUSSION

Case 1

You are planning to conduct a study on the women's attitudes toward birth control in a developing nation. The population you are planning to work with has a tribal governance structure: All research projects must be approved by the tribal leaders. You present your proposed research project to members of the tribe, describing the protocol and consent process. You plan to conduct a medical history and physical exam on each woman. You will also conduct a thirty-minute interview in which you ask them questions about their sexual behavior and attitudes toward sex. Each woman will also receive tests for HIV, chlamydia, herpes, and other sexually transmitted diseases. The tribal leaders give their tentative approval to the research with the understanding that if the research subject is married, her husband must be present for the consent process and interview. If she is not married, the leading male member of her family, such as her father or grandfather, must be present. The husband/leading male family member must also give his approval for the woman's participation in the study. You have some serious concerns about these conditions, because you think that complying with them could bias your data and compromise the ethical integrity of the project.

- Would it be ethical to conduct the research under the conditions set forth by the tribal leaders?
- How should you discuss these issues with the tribal leaders?

Case 2

A contract research organization (CRO) is planning to move its operations to a developing nation in Southeast Asia. The CRO specializes in conducting phase I trials (i.e., dosing studies) on healthy subjects for pharmaceutical companies. It is moving its operations to this country to save money on labor costs and to take advantage of a less burdensome regulatory system. The research subjects will be paid $5 per day for participating in research, which is about the average daily pay in that country. The CRO has no plans to compensate research subjects for injuries. Health care in the Southeast Asian nation is very limited. Most of the drugs tested by the company are likely to be used to treat chronic diseases that affect people in developed nations, such as hypertension, congestive heart failure, arthritis, and depression. Most of the drugs will be too expensive for patients in the country to afford, at least until their patents expire. The CRO is helping to establish an IRB to review the research at a local hospital. The CRO will conduct some educational and training activities for the IRB members and pay them the same wage as the subjects. Local leaders are delighted that the CRO has moved its operations to their vicinity, since this will create jobs and stimulate economic activity. They have reached a deal with the CRO that includes tax breaks.

- Do you see any ethical problems with the CRO's plans?
- Is the local population receiving fair benefits from this deal?
- Is this situation exploitative? Unethical?

Case 3

Investigators sponsored by the NIH and private foundation are conducting an HIV vaccine trial in a sub-Saharan African nation. In the trial, HIV-negative adolescents will be randomized to receive either an experimental vaccine or a placebo. They will be instructed on ways to prevent HIV infection and provided with condoms. They will be followed for five years to see if either group has a higher HIV infection rate. They will be tested for HIV on a regular basis and interviewed concerning their sexual activities and other risky behaviors. If they develop HIV, they will be treated according to the standard of care in the sub-Saharan nation, which includes access to only some of the HIV medications available in the developing world. Subjects who develop HIV will receive free medication for a maximum of one year. There are no plans to compensate subjects for research-related injuries. Interviews with some of the subjects participating in the trial revealed a profound lack of understanding of the purpose of the study and its procedures. Many participants thought they were receiving treatment for HIV rather than an HIV vaccine. A data monitoring committee found that people who received the vaccine had an increased risk for arthritis and diabetes.

- What are some of the potential ethical problems with this study?
- Should an HIV vaccine be tested on adolescents?

Case 4

A U.S. researcher is collaborating with researchers in a developing nation in South America studying a hereditary disorder found only in an isolated population living deep in the jungle. The South American researchers are planning to send genetic samples to the U.S. researcher for analysis. They hope to identify genes associated with this disorder. The South American researchers have a protocol and consent form, which have been approved by a local IRB. They have submitted an amendment to their local IRB to include the U.S. researcher as a co-investigator. In order for the U.S. researcher to collaborate with the South American researchers on this project, the IRB at his institution must also approve the study. The U.S. researcher submits a proposal to his IRB. The South American consent form includes all the legally required elements (by U.S. law) for informed consent, such as a description of the study, the procedures, benefits, risks, and confidentiality protections. However, the consent form is very sketchy and short on details. It is far below U.S. standards for consent documents. It is only about one single-spaced, typed page (400 words). The U.S. researcher presents his proposal at a meeting of the IRB. IRB members are not receptive to the proposal. The U.S. researcher acknowledges that the consent document is substandard by U.S. standards, but he urges the IRB to approve the document on the grounds that the document is acceptable by the local standards and his role in the research can be very beneficial to the local population.

- Should the IRB approve this study?
- What are some of the potential ethical or legal problems with approving this study?

Case 5

For as long as anyone in the local population can remember, members of a South Pacific island community have used a native plant known as pongo-pongo to cure fevers. Dr. Langston and his collaborators have learned about the plant while studying the population. They have collected and analyzed leaves from the plant and found that they contain powerful, broad-spectrum antibiotics. A compound isolated from the pongo-pongo plant even kills bacteria in cultures that are resistant to most antibiotics. Dr. Langston is planning to patent the compound isolated from the pongo-pongo plant, conduct animal studies, and if these are successful, conduct clinical trials. He hopes that one day the compound will play an important role in fighting bacterial infections, especially infections resistant to most antibiotics.

- Should Dr. Langston share any financial gains with the local population?
- Would it be unfair or exploitative if he does not share?
- Should he be able to patent this compound?

Case 6

Malaria is a devastating disease in Africa and in many developing countries. Nearly one million Africans die annually from the disease. *Plasmodium falciparum* is a protozoan parasite, one of the species of *Plasmodium* that cause malaria in humans. It is carried by *Anopheles* mosquitoes. *P. falciparum* malaria has the highest rates of mortality. Acute malaria in pregnancy is associated with increased mortality, spontaneous abortion, and deformities. The current treatment for malaria is chloroquin and sulfadoxine-pyrimethamine. The use of these two drugs has resulted in the development of *P. falciparum* that is more than 50% resistant to these drugs. Moreover, the two drugs lower the development of acquired immunity to malaria. Therefore, there is a great deal of need to develop a new drug for pregnant women. Another drug, malarone, composed of a fixed dose of atovaquone and proguanil, is licensed for treatment of malaria and the prevention of malaria for travelers. Clinical trials have indicated that it is fairly safe when used in adults and children. However, there has been no determination of safety of malarone in pregnant women, although there are some studies indicating effectiveness.

The proposed trial is to test the drug in pregnant women. The study was approved by the local research ethics committee (REC), a group that is similar to an IRB. The drug is manufactured by GlaxoSmithKline (GSK). Many organizations have approved the study, such as the United Nations, WHO, and UNICEF. The opponents of the study claim that malarone is expensive to the community ($50 per preventive regiment), GSK may be producing it for expatriates, and GSK should guarantee the accessibility of the drug post-trial to the local population. The proponents of the study point out that there is a need for an effective and safe treatment for malaria in pregnant women. It is a high-priority public health problem, and cost can be accommodated later with new sources of funding.

- Should the study go forward?
- If you were a member of the IRB, how would you vote?
- What other conditions, if any, you would require?

16

Conclusion

> This final chapter recommends some steps researchers, institutional officials, government agencies, and scientific organizations can take to promote ethical conduct in scientific research, such as ethical leadership, education and training in responsible conduct of research, policy development, compliance activities, and international cooperation.

In this book, we have discussed many different aspects of the responsible conduct of research (RCR), ranging from foundational topics like ethical theories and decision making to such applied topics as conflicts of interest and research misconduct. We hope that students, trainees, investigators, and administrators find our book to be a valuable source of information and guidance. In this final chapter, we make some suggestions for putting our proposals into practice. We discuss some steps that researchers, institutional officials, government agencies, and scientific organizations can take to promote ethical conduct in scientific research. We group these recommendations into five areas: (a) leadership, (b) education and training, (c) policy development, (d) compliance activities, and (e) international cooperation.

LEADERSHIP

There's a saying in business that "it all begins at the top." This means that for a business to be successful, it needs ethical leadership. One need not look very far for examples of how bad corporate leadership leads to poor decisions, unethical and illegal conduct, and business failures. The Enron scandal is a perfect illustration of the consequences of bad leadership for a corporation's employees and stockholders and the general public. Enron was an energy-trading corporation led by CEO Kenneth Lay. In 2000, the company had more than 22,000 employees and claimed more than $100 billion in revenues. In 2001, the company came crashing down when its financial success was revealed to be an illusion sustained by systematic and creatively planned accounting fraud (McLean and Elkind 2004). Even before the Enron scandal broke, successful businesses understood the importance of a commitment to ethics and integrity, and many had adopted corporate ethics statements and instituted corporate ethics programs (Murphy 1998). The scandal provided additional proof of the importance of promoting an ethical corporate culture in business.

Just as ethical leadership is important for success in business, it is also important for success in scientific research. Research institutions are like businesses in many ways: A typical research university is a large organization with thousands of employees, a complex bureaucracy, and an annual budget of hundreds of millions of dollars. Unethical leadership at research institutions can lead to corruption and abuse, mismanagement of funds, loss of income, scandals, legal liability, inefficient operations, a bad reputation, and low employee morale. Without ethical leadership, a research institution will not only fail to produce the tangible products of research, such as publications, patents, contracts and grants, but it may also fail to live up to standards of quality, morality, and integrity.

What is ethical leadership at research institution? There are many different people who hold leadership positions at research institutions. At the lowest level of organization, a leader may be a laboratory or program director; at a higher level, a department or center director; and on up the ladder to deans, vice presidents, presidents, and CEOs. Ethical leaders are people whose words and deeds embody a commitment to ethics. An ethical leader sets an example for other people in the organization through his or her conduct, demeanor, attitudes, and actions. Ethical leaders stress the importance of ethics in their speeches and public statements and in their budgets and priorities. They do good work, and they inspire others to do good work. Ethical leaders ensure that the institution makes strong commitments of financial and human resources to ethics. The decision to commit a significant portion of the Human Genome Project's budget to ethical, legal, and social issues, mentioned in chapter 14, is an example of ethical leadership. The National Institutes of Health (NIH) decision in 2000 to require all researchers and trainees in its intramural program to have education in research ethics, mentioned in chapter 1, is another example of ethical leadership.

The importance of leadership illustrates how research ethics is an institutional or organizational concern, not just an individual one (Shamoo and Davis 1989, Berger and Gert 1997, Shamoo and Dunigan 2000). Although most ethical theories, principles, and concepts focus on the ethical duties of individuals, institutions and organizations also have ethical and legal responsibilities (De George 1995). For example, a research institution has a duty to ensure that the various laws, regulations, and policies that govern the institutional are obeyed. A research institution also has a duty to do good work for society and to avoid harming individuals, society, or the environment. While all members of the institution are responsible for helping to ensure that it lives up to its obligations and duties, institutional leaders play a key role in this endeavor, since they have a greater ability to affect institutional decisions and policies. Leaders can help institutions to honor their ethical obligations by promoting activities that encourage ethical conduct, including ethics education and training, ethics policy development, and compliance activities.

EDUCATION AND TRAINING

Most organizations with an interest in research ethics agree that ethics education and training are very important for establishing an ethical culture at a research institution and promoting ethical behaviors (Sigma Xi 1986, 1999,

National Academy of Sciences 1992, 1994, 1997, 2002). The main rational for RCR education is that it is a form of preventative ethics: It is better to prevent ethical problems in research than to deal with them when they arise (Vasgird 2007). Although it seems obvious to many that education and training can have a positive impact on ethical behavior, there has not been a great deal of research so far on this topic. Little research has been done on the effect of ethics education and training in science in part because research ethics is still a very young field, and it takes time to establish empirical research programs. Some barriers to progress in this field include disagreement about the basic concepts, principles, measurements, and methodologies. Other barriers include a lack of standardization of education and training activities at different universities, and the difficulty of following subjects after they have completed their studies in research ethics (Steneck and Bulger 2007).

Despite these problems, a few studies of the effects of education and training have been published. RCR education may affect students by increasing knowledge of ethics, improving awareness of ethical issues, enhancing decision-making skills, altering attitudes, or changing behavior (Pimple 1995). Powell et al. (2007) surveyed students before and after a short course in RCR. They measured the impact of the course on knowledge of RCR, decision-making skills, attitudes toward RCR, and discussions of RCR topics outside of class. They found that the only statistically significant effect of the course was an improvement in RCR knowledge. A survey of students' perceptions of the effectiveness of RCR courses by Plemmons et al. (2006) also found similar results. Although the students reported that courses had a variety of positive impacts, the impact of the course on knowledge of RCR was much greater than the impact on RCR attitudes or skills (e.g., decision making). Finally, a survey by Anderson et al. (2007) found that courses in RCR did not reduce a variety of unethical behaviors but that mentoring in RCR did promote ethical behavior.

These studies cast some doubt on the effectiveness of RCR courses. Although these courses may increase knowledge, there is no evidence, so far, that they significantly change attitudes, behaviors, and skills. These results should come as no surprise to those who understand the nature of ethics. Being ethical involves much more than knowing some principles, concepts, or rules. Ethics is a skill or an art, not just a body of knowledge. To become ethical, one must practice ethical thinking, decision making, and behavior. Taking courses in biochemistry is only part of what it takes to be a good biochemist; one must also practice the skills, techniques, methods, and thought processes of biochemistry. Graduate education in biochemistry involves much more than coursework; it also includes mentoring, independent research, and informal instruction. Likewise, education in RCR must include much more than a course in research ethics; it should also include mentoring, integrated learning of ethics, and experiential learning of ethics (National Academy of Sciences 1997, 2002). Chapter 4 discusses mentoring, which is an effective method for learning ethics because it allows students to learn by example. Students can see how an established researcher deals with ethical problems and dilemmas. Integrated learning of ethics occurs when students have an opportunity to discuss ethical issues when they naturally arise outside of a formal course in ethics, such as a discussion in the laboratory or classroom. A course in research methods, for example, might include some ethics topics. Experiential learning would include experiences related to dealing with actual ethical dilemmas in research, such as observing meetings of institutional review boards (IRBs), institutional animal care and use committees (IACUCs), or conflict of interest committees. All of these different forms of education are different from classroom instruction in that they involve ethics in context. A discussion of authorship means much more to student, for example, when that student is having a disagreement with a coauthor on the order of authorship on a paper, than it does when the topic is brought up in a class on ethics.

To promote ethics education and training, research organizations must provide financial support for these activities and make room for them in the curriculum and work schedule. Although instruction in RCR can begin at the undergraduate level, it is probably more effective at the graduate level, when students are actively involved in research. Such instruction may include refresher courses in RCR and periodic updates on specialized topics in RCR. It should include the topics covered in this book—data management, conflicts of interest, authorship and so on—as well as specialized topics tailored to an institution's particular needs, such as ethical aspects of bioweapons research or nanotechnology. As noted in chapter 4, institutions must provide incentives for researchers to engage in mentoring activities; otherwise, they may decide that they cannot spare any time or energy for mentoring. To help support integrated learning of ethics, institutions can encourage researchers to discuss ethics topics in their courses. Institutions can also support informal gathering for ethics discussions, such as brown-bag lunch meetings, focus groups, or workshops, and they can support ethics consultations by professional ethicists or researchers with ethics knowledge. To encourage experiential learning, IRBs, IACUCs, and other committees that deal with ethical issues should invite students and researchers to observe their meetings, provided that confidential information is adequately protected.

Although we have focused on educational activities of universities, other institutions or organizations involved in research, such as pharmaceutical companies, government laboratories, journals, nonprofit research foundations, contract research organizations, and independent (for-profit) IRBs, should also support educational and training activities to promote an ethical climate.

POLICY DEVELOPMENT

To help support education and training activities, and to establish standards of conduct and behavior, institutions should develop policies and procedures pertaining to RCR. Topics for policy development should include the following:

- Data management (acquisition, storage, analysis, sharing, etc.)
- Research misconduct (definition, procedures for investigation and adjudication)
- Conflict of interests
- Intellectual property
- Technology transfer
- Authorship and publication
- Peer review
- Mentoring
- Collaboration
- Harassment
- Human subjects
- Animal subjects
- Radiation safety, biological safety, and laboratory safety

Many university policies on, for example, research misconduct and conflict of interest, are already mandated by government funding agencies. For policy development to be meaningful and useful, it should include input from all of the key stakeholders in research (i.e., junior and senior investigators and staff), not just input from research administrators and attorneys. Researchers should perceive policies as serving important needs, not as useless red tape catering to bureaucratic whims. All policies that are developed should also be well publicized. Copies of policies should be distributed to faculty and staff and should be available on institutional Web sites. Policies should be discussed and reviewed during educational or training sessions. Additionally, knowledgeable people should be available to answer questions about the interpretation and application of these policies. For example, an IRB chair or administrator could answer questions about human subjects policies, and a technology transfer officer could answer questions about technology transfer. A person who specializes in research ethics could answer questions about a variety of different policies, including data management, authorship, conflict of interest, and research misconduct (de Melo-Martín et al. 2007). For an example of a thoughtful, comprehensive, and well-ordered set of university policies, see Duke University (2007).

It is also important to mention that other research organizations, such as scientific journals, granting agencies, and government laboratories, should also develop policies. Many journals have created policies for a variety of topics, including authorship, duplicate publication, disclosure of financial interests, data sharing, human subjects protections, clinical trial registration, and peer review (Krimsky and Rothenberg 2001, Cooper et al. 2006). The ICMJE (International Committee of Medical Journal Editors 2007) and COPE (Committee on Publication Ethics 2007) have developed ethical guidelines for authors, reviewers, and editors. More than 650 biomedical journals follow the ICMJE guidelines, and nearly 400 belong to COPE.

The NIH has a number of different research ethics policies pertaining to extramural and intramural research. The NIH's *Guidelines for the Conduct of Research in the Intramural Research Program at NIH* (National Institutes of Health 2007d) covers all of the core RCR topics.

COMPLIANCE ACTIVITIES

Compliance activities must also be a part of any institutional ethics program. Compliance activities include various efforts to enforce ethical and legal rules and polices at the research institution. Whereas education focuses on the prevention of ethical and legal transgressions, compliance focuses on detecting and investigating misbehaviors and implementing punishments. Compliance is like police work. Some important compliance activities include the following:

- Establishing institutional mechanisms for reporting, investigating, and adjudicating unethical or illegal behavior, such as research misconduct and violations of human or animal research rules
- Establishing and staffing committees to deal with ethics issues in the institution, such conflicts of interest, human subjects research, animal research, and biosafety
- Developing mechanisms for quality assurance and quality improvement in research
- Developing mechanisms for auditing research records, including data, time and effort sheets, informed consent documents, research protocols, and standard operating procedures (Shamoo 1988, 1989)
- Developing mechanisms for auditing other aspects of research, such as the informed consent process, occupational and laboratory safety, and laboratory animal welfare
- Providing counseling and advice to researchers concerning compliance with various policies, rules, regulations, and so on

It is also important to mention that other types of research organizations (journals, granting agencies, etc.) should also engage in compliance activities, where appropriate. The Korean stem cell scandal discussed in chapter 1 has prompted many journals to consider how they should deal with misconduct and other problems (Check and Cyranoski 2005, Resnik et al. 2006). Many journals are now using computers to audit articles that are submitted. Computer programs can indicate whether a digital image has been inappropriately manipulated and can detect some types of plagiarism (Butler 2007).

INTERNATIONAL COOPERATION

Many different research projects today involve international cooperation and collaboration. Articles, books, and monographs often have coauthors from different countries, journal editorial boards usually have members from around the globe, and most scientific conferences have attendees from many different nationalities. Because science has become increasingly global in scope, international cooperation on ethical and policy issues is crucial for promoting scientific progress and ensuring the public's trust of research. The Korean stem cell scandal illustrates the need for international cooperation regarding research ethics (Resnik et al. 2006). This scandal involved researchers in two different countries and had global implications. However, the two main countries involved in the scandal, South Korea and the United States, do not have common ethical rules or guidelines pertaining to some of the important question concerning the research, such as manipulation of digital images and authorship on scientific papers.

Because different countries may have their own laws and policies pertaining to scientific research, the most important step toward international cooperation is the development of international codes of ethics similar to the Helsinki Declaration or Council for International Organizations of Medical Sciences (CIOMS) guidelines, discussed in chapter 15. While the Helsinki Declaration and CIOMS guidelines are useful and important, they apply only to ethical issues concerning research with human subjects and do not cover the full range of ethical issues in research, such as data management, misconduct, authorship, and conflicts of interest. An international code of research ethics should have the full range of issues that arise in research. A set of international guidelines could establish a model code of ethics for different nations. If a nation has no laws or policies pertaining to scientific research, then the code could provide some guidance in the absence of any local rules.

A key practical question in developing an international code of ethics is who would sponsor the code. For the code to be successful, it needs to be sponsored by a well-known and well-respected organization with considerable influence over scientific research. Some potential candidate organizations include the U.N. Educational, Scientific and Cultural Organization (UNESCO), ICMJE, and COPE. Alternatively, a group of professional organizations from around the globe, such as the American Association for the Advancement of Science, the British Association for the Advancement of Science, the European Council of Applied Sciences and Engineering, the European Academy of Sciences and Arts, and the Third World Academy of Sciences could work together to develop an international code of ethics. If no organization steps forward to undertake this task, it may be necessary for a group of researchers from around the world to take the first steps toward developing an international code of ethics. A grassroots movement of this sort could draw the attention, support, and endorsement of large organizations.

Appendix 1

Sample Policy and Procedures for Responding to Allegations of Research Misconduct

This sample policy and procedures is intended to assist those institutions with limited resources or experience in addressing research misconduct to develop research misconduct policies and procedures consistent with 42 CFR Part 93. However, it does not create a standard or expectation for those policies and procedures. This sample policy and procedures is non-binding and is not intended to create any right or benefit, substantive or procedural, enforceable at law by a party against the United States, its agents, officers, or employees.

Where the requirements of the regulation are reiterated in this sample, the term "must" or "shall" is used and a citation to the pertinent section or sections of the regulation is included. Where the sample policy and procedures sets forth a way of complying with the regulation that is one of several ways to achieve the result required by the regulation, the term "should" or "will" is used and other options may be suggested. In case of any conflict between the sample and 42 CFR Part 93, the regulation shall prevail.

I. INTRODUCTION

A. General Policy

[*NOTE:* Institutions may wish to insert in this section general statements about their philosophy and that of the scientific community related to research integrity. These might include institutional values related to the responsible conduct of research, a statement of principles, and the institution's position on preventing misconduct in research and protecting the positions and reputations of good faith complainants, witnesses and committee members.]

B. Scope

This statement of policy and procedures is intended to carry out this institution's responsibilities under the Public Health Service (PHS) Policies on Research Misconduct, 42 CFR Part 93. This document applies to allegations of research misconduct (fabrication, falsification, or plagiarism in proposing, performing, or reviewing research, or in reporting research results) involving:

- A person who, at the time of the alleged research misconduct, was employed by, was an agent of, or was affiliated by contract or agreement with this institution;[1] and
- (1) PHS support biomedical or behavioral research, research training or activities related to that research or research training, such as the operation of tissue and data banks and the dissemination of research information, (2) applications or proposals for PHS support for biomedical or behavioral research, research training or activities related to that research or research training, or (3) plagiarism of research records produced in the course of PHS supported research, research training or activities related to that research or research

training. This includes any research proposed, performed, reviewed, or reported, or any research record generated from that research, regardless of whether an application or proposal for PHS funds resulted in a grant, contract, cooperative agreement, or other form of PHS support.[2]

This statement of policy and procedures does not apply to authorship or collaboration disputes and applies only to allegations of research misconduct that occurred within six years of the date the institution or HHS received the allegation, subject to the subsequent use, health or safety of the public, and grandfather exceptions in 42 CFR § 93.105(b).

[*Option:* Under 42 CFR § 93.319 institutions may adopt additional standards of conduct that go beyond the PHS standards in 42 CFR Part 93. These additional standards may be included in the same document as the PHS standards, but if an institution chooses that option, it should make certain that the two sets of standards and actions under them are easily distinguishable. These additional institutional standards will apply only to the internal decisions at the institution. They must not apply to decisions made under 42 CFR Part 93.]

II. DEFINITIONS

Terms used have the same meaning as given them in the Public Health Service Policies on Research Misconduct, 42 CFR Part 93.

[*Option:* This section may include definitions that do not appear in 42 CFR Part 93. For example, institutions may want to designate Research Integrity Officers (RIO) and Deciding Officials (DO) to carry out various institutional responsibilities under the regulation, even though such designations are not required by the regulation. In that case, this section might include definitions of those terms, as follows:

Deciding Official (DO) means the institutional official who makes final determinations on allegations of research misconduct and any institutional administrative actions. The Deciding Official will not be the same individual as the Research Integrity Officer and should have no direct prior involvement in the institution's inquiry, investigation, or allegation assessment. A DO's appointment of an individual to assess allegations of research misconduct, or to serve on an inquiry[3] or investigation committee, is not considered to be direct prior involvement.

Research Integrity Officer (RIO) means the institutional official responsible for: (1) assessing allegations of research misconduct to determine if they fall within the definition of research misconduct, are covered by 42 CFR Part 93, and warrant an inquiry on the basis that the allegation is sufficiently credible and specific so that potential evidence of research misconduct may be identified; (2) overseeing inquires and investigations; and (3) the other responsibilities described in this policy. For completeness, this sample policy and procedures sets forth duties and responsibilities that might be appropriate for DOs and RIOs in subsequent sections.

III. RIGHTS AND RESPONSIBILITIES

A. Research Integrity Officer

The [designated institutional official] will appoint [*Option:* will serve as] the RIO who will have primary responsibility for implementation of the institution's policies and procedures on research misconduct. A detailed listing of the responsibilities of the RIO is set forth in Appendix A. These responsibilities include the following duties related to research misconduct proceedings:

- Consult confidentially with persons uncertain about whether to submit an allegation of research misconduct;
- Receive allegations of research misconduct;
- Assess each allegation of research misconduct in accordance with Section V.A of this policy to determine whether it falls within the definition of research misconduct and warrants an inquiry;
- As necessary, take interim action and notify ORI of special circumstances, in accordance with Section IV.F. of this policy;
- Sequester research data and evidence pertinent to the allegation of research misconduct in accordance with Section V.C of this policy and maintain it securely in accordance with this policy and applicable law and regulation;
- Provide confidentiality to those involved in the research misconduct proceeding as required by 42 CFR § 93.108, other applicable law, and institutional policy;
- Notify the respondent and provide opportunities for him/her to review/comment/respond to allegations, evidence, and committee reports in accordance with Section III.C of this policy;
- Inform respondents, complainants, and witnesses of the procedural steps in the research misconduct proceeding;
- Appoint the chair and members of the inquiry and investigation committees, ensure that those committees are properly staffed and that there is expertise appropriate to carry out a thorough and authoritative evaluation of the evidence;
- Determine whether each person involved in handling an allegation of research misconduct has an unresolved personal, professional, or financial conflict of interest and take appropriate action, including recusal, to ensure that no person with such conflict is involved in the research misconduct proceeding;
- In cooperation with other institutional officials, take all reasonable and practical steps to protect or restore the positions and reputations of good faith complainants, witnesses, and committee members and counter potential or actual retaliation against them by respondents or other institutional members;

- Keep the Deciding Official and others who need to know apprised of the progress of the review of the allegation of research misconduct;
- Notify and make reports to ORI as required by 42 CFR Part 93;
- Ensure that administrative actions taken by the institution and ORI are enforced and take appropriate action to notify other involved parties, such as sponsors, law enforcement agencies, professional societies, and licensing boards of those actions; and
- Maintain records of the research misconduct proceeding and make them available to ORI in accordance with Section VIII.F. of this policy.

B. Complainant

The complainant is responsible for making allegations in good faith, maintaining confidentiality, and cooperating with the inquiry and investigation. As a matter of good practice, the complainant should be interviewed at the inquiry stage and given the transcript or recording of the interview for correction. The complainant must be interviewed during an investigation, and be given the transcript or recording of the interview for correction. [*Option:* As a matter of policy or on the basis of case-by-case determinations, the institution may provide to the complainant for comment: (1) relevant portions of the inquiry report (within a timeframe that permits the inquiry to be completed within 60 days of its initiation); and (2) the draft investigation report or relevant portions of it. The institution must require that comments on the draft investigation report be submitted within 30 days of the date on which the complainant received the draft report. The institution must consider any comments made by the complainant on the draft investigation report and include those comments in the final investigation report.]

C. Respondent

The respondent is responsible for maintaining confidentiality and cooperating with the conduct of an inquiry and investigation. The respondent is entitled to:

- A good faith effort from the RIO to notify the respondent in writing at the time of or before beginning an inquiry;[4]
- An opportunity to comment on the inquiry report and have his/her comments attached to the report;[5]
- Be notified of the outcome of the inquiry, and receive a copy of the inquiry report that includes a copy of, or refers to 42 CFR Part 93 and the institution's policies and procedures on research misconduct;[6]
- Be notified in writing of the allegations to be investigated within a reasonable time after the determination that an investigation is warranted, but before the investigation begins (within 30 days after the institution decides to begin an investigation), and be notified in writing of any new allegations, not addressed in the inquiry or in

the initial notice of investigation, within a reasonable time after the determination to pursue those allegations;[7]
- Be interviewed during the investigation, have the opportunity to correct the recording or transcript, and have the corrected recording or transcript included in the record of the investigation;[8]
- Have interviewed during the investigation any witness who has been reasonably identified by the respondent as having information on relevant aspects of the investigation, have the recording or transcript provided to the witness for correction, and have the corrected recording or transcript included in the record of investigation;[9] and
- Receive a copy of the draft investigation report and, concurrently, a copy of, or supervised access to the evidence on which the report is based, and be notified that any comments must be submitted within 30 days of the date on which the copy was received and that the comments will be considered by the institution and addressed in the final report.[10] The respondent should be given the opportunity to admit that research misconduct occurred and that he/she committed the research misconduct. With the advice of the RIO and/or other institutional officials, the Deciding Official may terminate the institution's review of an allegation that has been admitted, if the institution's acceptance of the admission and any proposed settlement is approved by ORI.[11]

[*Optional Addition:* As provided in 42 CFR § 93.314(a), the respondent will have the opportunity to request an institutional appeal if the institution's procedures provide for an appeal.]

D. Deciding Official

The DO will receive the inquiry report and after consulting with the RIO and/or other institutional officials, decide whether an investigation is warranted under the criteria in 42 CFR § 93.307(d). Any finding that an investigation is warranted must be made in writing by the DO and must be provided to ORI, together with a copy of the inquiry report meeting the requirements of 42 CFR § 93.309, within 30 days of the finding. If it is found that an investigation is not warranted, the DO and the RIO will ensure that detailed documentation of the inquiry is retained for at least 7 years after termination of the inquiry, so that ORI may assess the reasons why the institution decided not to conduct an investigation.[12] The DO will receive the investigation report and, after consulting with the RIO and/or other institutional officials, decide the extent to which this institution accepts the findings of the investigation and, if research misconduct is found, decide what, if any, institutional administrative actions are appropriate. The DO shall ensure that the final investigation report, the findings of the DO and a description of any pending or completed administrative actions are provided to ORI, as required by 42 CFR § 93.315.

IV. GENERAL POLICIES AND PRINCIPLES

A. Responsibility to Report Misconduct

All institutional members will report observed, suspected, or apparent research misconduct to the RIO [*Option:* also list other officials]. If an individual is unsure whether a suspected incident falls within the definition of research misconduct, he or she may meet with or contact the RIO at [contact information] to discuss the suspected research misconduct informally, which may include discussing it anonymously and/or hypothetically. If the circumstances described by the individual do not meet the definition of research misconduct, the RIO will refer the individual or allegation to other offices or officials with responsibility for resolving the problem. At any time, an institutional member may have confidential discussions and consultations about concerns of possible misconduct with the RIO [*Option:* list other officials] and will be counseled about appropriate procedures for reporting allegations.

B. Cooperation with Research Misconduct Proceedings

Institutional members will cooperate with the RIO and other institutional officials in the review of allegations and the conduct of inquiries and investigations. Institutional members, including respondents, have an obligation to provide evidence relevant to research misconduct allegations to the RIO or other institutional officials.

C. Confidentiality

The RIO shall, as required by 42 CFR § 93.108: (1) limit disclosure of the identity of respondents and complainants to those who need to know in order to carry out a thorough, competent, objective and fair research misconduct proceeding; and (2) except as otherwise prescribed by law, limit the disclosure of any records or evidence from which research subjects might be identified to those who need to know in order to carry out a research misconduct proceeding. The RIO should use written confidentiality agreements or other mechanisms to ensure that the recipient does not make any further disclosure of identifying information. [*Option:* The institution may want to provide confidentiality for witnesses when the circumstances indicate that the witnesses may be harassed or otherwise need protection.]

D. Protecting Complainants, Witnesses, and Committee Members

Institutional members may not retaliate in any way against complainants, witnesses, or committee members. Institutional members should immediately report any alleged or apparent retaliation against complainants, witnesses or committee members to the RIO, who shall review the matter and, as necessary, make all reasonable and practical efforts to counter any potential or actual retaliation and protect and restore the position and reputation of the person against whom the retaliation is directed.

E. Protecting the Respondent

As requested and as appropriate, the RIO and other institutional officials shall make all reasonable and practical efforts to protect or restore the reputation of persons alleged to have engaged in research misconduct, but against whom no finding of research misconduct is made.[13] During the research misconduct proceeding, the RIO is responsible for ensuring that respondents receive all the notices and opportunities provided for in 42 CFR Part 93 and the policies and procedures of the institution. Respondents may consult with legal counsel or a non-lawyer personal adviser (who is not a principal or witness in the case) to seek advice and may bring the counsel or personal adviser to interviews or meetings on the case. [*Option:* Some institutions may decide not to permit the presence of lawyers at interviews or meetings with institutional officials. Some institutions that permit lawyers to be present at interviews and meetings restrict the lawyer's role to advising (as opposed to representing) the respondent.]

F. Interim Administrative Actions and Notifying ORI of Special Circumstances

Throughout the research misconduct proceeding, the RIO will review the situation to determine if there is any threat of harm to public health, federal funds and equipment, or the integrity of the PHS supported research process. In the event of such a threat, the RIO will, in consultation with other institutional officials and ORI, take appropriate interim action to protect against any such threat.[14] Interim action might include additional monitoring of the research process and the handling of federal funds and equipment, reassignment of personnel or of the responsibility for the handling of federal funds and equipment, additional review of research data and results or delaying publication. The RIO shall, at any time during a research misconduct proceeding, notify ORI immediately if he/she has reason to believe that any of the following conditions exist:

- Health or safety of the public is at risk, including an immediate need to protect human or animal subjects;
- HHS resources or interests are threatened;
- Research activities should be suspended;
- There is a reasonable indication of possible violations of civil or criminal law;
- Federal action is required to protect the interests of those involved in the research misconduct proceeding;
- The research misconduct proceeding may be made public prematurely and HHS action may be necessary to safeguard evidence and protect the rights of those involved; or
- The research community or public should be informed.[15]

V. CONDUCTING THE ASSESSMENT AND INQUIRY

A. Assessment of Allegations

Upon receiving an allegation of research misconduct, the RIO will immediately assess the allegation to determine whether it is sufficiently credible and specific so that potential evidence of research misconduct may be identified, whether it is within the jurisdictional criteria of 42 CFR § 93.102(b), and whether the allegation falls within the definition of research misconduct in 42 CFR § 93.103.[16] An inquiry must be conducted if these criteria are met. The assessment period should be brief, preferably concluded within a week. In conducting the assessment, the RIO need not interview the complainant, respondent, or other witnesses, or gather data beyond any that may have been submitted with the allegation, except as necessary to determine whether the allegation is sufficiently credible and specific so that potential evidence of research misconduct may be identified. The RIO shall, on or before the date on which the respondent is notified of the allegation, obtain custody of, inventory, and sequester all research records and evidence needed to conduct the research misconduct proceeding, as provided in paragraph C. of this section.

B. Initiation and Purpose of the Inquiry

If the RIO determines that the criteria for an inquiry are met, he or she will immediately initiate the inquiry process. The purpose of the inquiry is to conduct an initial review of the available evidence to determine whether to conduct an investigation. An inquiry does not require a full review of all the evidence related to the allegation.[17]

C. Notice to Respondent; Sequestration of Research Records

At the time of or before beginning an inquiry, the RIO must make a good faith effort to notify the respondent in writing, if the respondent is known. If the inquiry subsequently identifies additional respondents, they must be notified in writing. On or before the date on which the respondent is notified, or the inquiry begins, whichever is earlier, the RIO must take all reasonable and practical steps to obtain custody of all the research records and evidence needed to conduct the research misconduct proceeding, inventory the records and evidence, and sequester them in a secure manner, except that where the research records or evidence encompass scientific instruments shared by a number of users, custody may be limited to copies of the data or evidence on such instruments, so long as those copies are substantially equivalent to the evidentiary value of the instruments.[18] The RIO may consult with ORI for advice and assistance in this regard.

D. Appointment of the Inquiry Committee

[Note: The regulation does not require that the inquiry be conducted by a committee, but many institutions have used such committees.] The RIO, in consultation with other institutional officials as appropriate, will appoint an inquiry committee and committee chair as soon after the initiation of the inquiry as is practical. The inquiry committee must consist of individuals who do not have unresolved personal, professional, or financial conflicts of interest with those involved with the inquiry and should include individuals with the appropriate scientific expertise to evaluate the evidence and issues related to the allegation, interview the principals and key witnesses, and conduct the inquiry.[19] [Option: As an alternative, the institution may appoint a standing committee authorized to add or reuse members and use experts when necessary to evaluate specific allegations.] [*Option:* An institution may establish a procedure for notifying the respondent of the proposed committee membership to give the respondent an opportunity to object to a proposed member based upon a personal, professional, or financial conflict of interest. If so, the institution should limit the period for submitting objections to no more than 10 calendar days. The institution would make the final determination of whether a conflict exists.]

E. Charge to the Committee and First Meeting

The RIO will prepare a charge for the inquiry committee that:

- Sets forth the time for completion of the inquiry;
- Describes the allegations and any related issues identified during the allegation assessment;
- States that the purpose of the inquiry is to conduct an initial review of the evidence, including the testimony of the respondent, complainant and key witnesses, to determine whether an investigation is warranted, not to determine whether research misconduct definitely occurred or who was responsible;
- States that an investigation is warranted if the committee determines: (1) there is a reasonable basis for concluding that the allegation falls within the definition of research misconduct and is within the jurisdictional criteria of 42 CFR § 93.102(b); and, (2) the allegation may have substance, based on the committee's review during the inquiry.
- Informs the inquiry committee that they are responsible for preparing or directing the preparation of a written report of the inquiry that meets the requirements of this policy and 42 CFR § 93.309(a).

At the committee's first meeting, the RIO will review the charge with the committee, discuss the allegations, any related issues, and the appropriate procedures for conducting the inquiry, assist the committee with organizing plans for the inquiry, and answer any questions raised by the committee. The RIO will be present or available throughout the inquiry to advise the committee as needed.

F. Inquiry Process

The inquiry committee will normally interview the complainant, the respondent, and key witnesses as well as examining relevant research records and materials. Then the inquiry committee will evaluate the evidence, including the testimony obtained during the inquiry. After consultation with the RIO, the committee members will decide whether an investigation is warranted based on the criteria in this policy and 42 CFR § 93.307(d). The scope of the inquiry is not required to and does not normally include deciding whether misconduct definitely occurred, determining definitely who committed the research misconduct or conducting exhaustive interviews and analyses.

However, if a legally sufficient admission of research misconduct is made by the respondent, misconduct may be determined at the inquiry stage if all relevant issues are resolved. In that case, the institution shall promptly consult with ORI to determine the next steps that should be taken. See Section IX.

G. Time for Completion

The inquiry, including preparation of the final inquiry report and the decision of the DO on whether an investigation is warranted, must be completed within 60 calendar days of initiation of the inquiry, unless the RIO determines that circumstances clearly warrant a longer period. If the RIO approves an extension, the inquiry record must include documentation of the reasons for exceeding the 60-day period.[20] [*Option:* The respondent will be notified of the extension.]

VI. THE INQUIRY REPORT

A. Elements of the Inquiry Report

A written inquiry report must be prepared that includes the following information: (1) the name and position of the respondent; (2) a description of the allegations of research misconduct; (3) the PHS support, including, for example, grant numbers, grant applications, contracts and publications listing PHS support; (4) the basis for recommending or not recommending that the allegations warrant an investigation; (5) any comments on the draft report by the respondent or complainant.[21] Institutional counsel should review the report for legal sufficiency. Modifications should be made as appropriate in consultation with the RIO and the inquiry committee. [*Option:* The inquiry report should include: the names and titles of the committee members and experts who conducted the inquiry; a summary of the inquiry process used; a list of the research records reviewed; summaries of any interviews; and whether any other actions should be taken if an investigation is not recommended.]

B. Notification to the Respondent and Opportunity to Comment

The RIO shall notify the respondent whether the inquiry found an investigation to be warranted, include a copy of the draft inquiry report for comment within [*suggested:* 10 days], and include a copy of or refer to 42 CFR Part 93 and the institution's policies and procedures on research misconduct.[22] [*Option:* The institution may notify the complainant whether the inquiry found an investigation to be warranted and provide relevant portions of the inquiry report to the complainant for comment [*suggested:* within 10 days]. A confidentiality agreement should be a condition for access to the report.] Any comments that are submitted by the respondent or complainant will be attached to the final inquiry report. Based on the comments, the inquiry committee may revise the draft report as appropriate and prepare it in final form. The committee will deliver the final report to the RIO.

C. Institutional Decision and Notification

1. *Decision by Deciding Official*   The RIO will transmit the final inquiry report and any comments to the DO, who will determine in writing whether an investigation is warranted. The inquiry is completed when the DO makes this determination.

2. *Notification to ORI*   Within 30 calendar days of the DO's decision that an investigation is warranted, the RIO will provide ORI with the DO's written decision and a copy of the inquiry report. The RIO will also notify those institutional officials who need to know of the DO's decision. The RIO must provide the following information to ORI upon request: (1) the institutional policies and procedures under which the inquiry was conducted; (2) the research records and evidence reviewed, transcripts or recordings of any interviews, and copies of all relevant documents; and (3) the charges to be considered in the investigation.[23]

3. *Documentation of Decision Not to Investigate*   If the DO decides that an investigation is not warranted, the RIO shall secure and maintain for 7 years after the termination of the inquiry sufficiently detailed documentation of the inquiry to permit a later assessment by ORI of the reasons why an investigation was not conducted. These documents must be provided to ORI or other authorized HHS personnel upon request.

VII. CONDUCTING THE INVESTIGATION

A. Initiation and Purpose

The investigation must begin within 30 calendar days after the determination by the DO that an investigation is warranted.[24] The purpose of the investigation is to develop a factual record by exploring the allegations in detail and examining the evidence in depth, leading to recommended findings on whether research misconduct has been committed, by whom, and to what extent. The investigation will also determine whether there are additional instances of possible research misconduct that would justify broadening the scope beyond the initial allegations. This is particularly important where the alleged research misconduct involves clinical trials or potential harm to human subjects or the general public or if it affects research that forms the basis for public policy, clinical practice, or public health practice. Under 42 CFR § 93.313 the findings of the investigation must be set forth in an investigation report.

B. Notifying ORI and Respondent; Sequestration of Research Records

On or before the date on which the investigation begins, the RIO must: (1) notify the ORI Director of the decision to begin the investigation and provide ORI a copy of the inquiry report; and (2) notify the respondent in writing of the allegations to be investigated. The RIO must also give the respondent written notice of any new allegations of research misconduct within a reasonable amount of time of deciding to pursue allegations not addressed during the inquiry or in the initial notice of the investigation.[25] The RIO will, prior to notifying respondent of the allegations, take all reasonable and practical steps to obtain custody of and sequester in a secure manner all research records and evidence needed to conduct the research misconduct proceeding that were not previously sequestered during the inquiry. The need for additional sequestration of records for the investigation may occur for any number of reasons, including the institution's decision to investigate additional allegations not considered during the inquiry stage or the identification of records during the inquiry process that had not been previously secured. The procedures to be followed for sequestration during the investigation are the same procedures that apply during the inquiry.[26]

C. Appointment of the Investigation Committee

[*Note:* The regulation does not require that the investigation be conducted by a committee, but many institutions have used such committees.] The RIO, in consultation with other institutional officials as appropriate, will appoint an investigation committee and the committee chair as soon after the beginning of the investigation as is practical. The investigation committee must consist of individuals who do not have unresolved personal, professional, or financial conflicts of interest with those involved with the investigation and should include individuals with the appropriate scientific expertise to evaluate the evidence and issues related to the allegation, interview the respondent and complainant and conduct the investigation. Individuals appointed to the investigation committee may also have served on the inquiry committee. [*Option:* When necessary to secure the necessary expertise or to avoid conflicts of interest, the RIO may select committee members from outside the institution.] [*Option:* As an alternative, the institution may appoint a standing committee authorized to add or reuse members or use consultants when necessary to evaluate specific allegations.] [*Option:* An institution may establish a procedure for notifying the respondent of the proposed committee membership to give the respondent an opportunity to object to a proposed member based upon a personal, professional, or financial conflict of interest. If so, the institution should limit the period for submitting objections to no more than 10 calendar days. The institution will make the final determination of whether a conflict exists.]

D. Charge to the Committee and the First Meeting

1. *Charge to the Committee*   The RIO will define the subject matter of the investigation in a written charge to the committee that:

- Describes the allegations and related issues identified during the inquiry;
- Identifies the respondent;
- Informs the committee that it must conduct the investigation as prescribed in paragraph E of this section;

- Defines research misconduct;
- Informs the committee that it must evaluate the evidence and testimony to determine whether, based on a preponderance of the evidence, research misconduct occurred and, if so, the type and extent of it and who was responsible;
- Informs the committee that in order to determine that the respondent committed research misconduct it must find that a preponderance of the evidence establishes that: (1) research misconduct, as defined in this policy, occurred (respondent has the burden of proving by a preponderance of the evidence any affirmative defenses raised, including honest error or a difference of opinion); (2) the research misconduct is a significant departure from accepted practices of the relevant research community; and (3) the respondent committed the research misconduct intentionally, knowingly, or recklessly; and
- Informs the committee that it must prepare or direct the preparation of a written investigation report that meets the requirements of this policy and 42 CFR § 93.313.

2. *First Meeting*   The RIO will convene the first meeting of the investigation committee to review the charge, the inquiry report, and the prescribed procedures and standards for the conduct of the investigation, including the necessity for confidentiality and for developing a specific investigation plan. The investigation committee will be provided with a copy of this statement of policy and procedures and 42 CFR Part 93. The RIO will be present or available throughout the investigation to advise the committee as needed.

E. Investigation Process

The investigation committee and the RIO must:

- Use diligent efforts to ensure that the investigation is thorough and sufficiently documented and includes examination of all research records and evidence relevant to reaching a decision on the merits of each allegation;[27]
- Take reasonable steps to ensure an impartial and unbiased investigation to the maximum extent practical;[28]
- Interview each respondent, complainant, and any other available person who has been reasonably identified as having information regarding any relevant aspects of the investigation, including witnesses identified by the respondent, and record or transcribe each interview, provide the recording or transcript to the interviewee for correction, and include the recording or transcript in the record of the investigation;[29] and
- Pursue diligently all significant issues and leads discovered that are determined relevant to the investigation, including any evidence of

any additional instances of possible research misconduct, and continue the investigation to completion.[30]

F. Time for Completion

The investigation is to be completed within 120 days of beginning it, including conducting the investigation, preparing the report of findings, providing the draft report for comment and sending the final report to ORI. However, if the RIO determines that the investigation will not be completed within this 120-day period, he/she will submit to ORI a written request for an extension, setting forth the reasons for the delay. The RIO will ensure that periodic progress reports are filed with ORI, if ORI grants the request for an extension and directs the filing of such reports.[31]

VIII. THE INVESTIGATION REPORT

A. Elements of the Investigation Report

The investigation committee and the RIO are responsible for preparing a written draft report of the investigation that:

- Describes the nature of the allegation of research misconduct, including identification of the respondent; [*Option:* The respondent's c.v. or resume may be included as part of the identification.]
- Describes and documents the PHS support, including, for example, the numbers of any grants that are involved, grant applications, contracts, and publications listing PHS support;
- Describes the specific allegations of research misconduct considered in the investigation;
- Includes the institutional policies and procedures under which the investigation was conducted, unless those policies and procedures were provided to ORI previously;
- Identifies and summarizes the research records and evidence reviewed and identifies any evidence taken into custody but not reviewed; and
- Includes a statement of findings for each allegation of research misconduct identified during the investigation.[32] Each statement of findings must: (1) identify whether the research misconduct was falsification, fabrication, or plagiarism, and whether it was committed intentionally, knowingly, or recklessly; (2) summarize the facts and the analysis that support the conclusion and consider the merits of any reasonable explanation by the respondent, including any effort by respondent to establish by a preponderance of the evidence that he or she did not engage in research misconduct because of honest error or a difference of opinion; (3) identify the specific PHS support; (4) identify whether any publications need correction or retraction; (5) identify the person(s) responsible for the misconduct; and (6) list any current support or known applications or proposals for support that the respondent has pending with non-PHS federal agencies.[33]

B. Comments on the Draft Report and Access to Evidence

1. *Respondent*   The RIO must give the respondent a copy of the draft investigation report for comment and, concurrently, a copy of, or supervised access to the evidence on which the report is based. The respondent will be allowed 30 days from the date he/she received the draft report to submit comments to the RIO. The respondent's comments must be included and considered in the final report.[34]

2. *Complainant*   [*Option:* As a policy applicable to all cases or on a case-by-case basis, the institution may provide the complainant a copy of the draft investigation report, or relevant portions of it, for comment. If the institution chooses this option, the complainant's comments must be submitted within 30 days of the date on which he/she received the draft report and the comments must be included and considered in the final report. See 42 CFR §§ 93.312(b) and 93.313(g).]

3. *Confidentiality*   In distributing the draft report, or portions thereof, to the respondent, [*Option:* and complainant] the RIO will inform the recipient of the confidentiality under which the draft report is made available and may establish reasonable conditions to ensure such confidentiality. For example, the RIO may require that the recipient sign a confidentiality agreement.

C. Decision by Deciding Official

The RIO will assist the investigation committee in finalizing the draft investigation report, including ensuring that the respondent's [*Option:* and complainant's] comments are included and considered, and transmit the final investigation report to the DO, who will determine in writing (1) whether the institution accepts the investigation report, its findings, and the recommended institutional actions; and (2) the appropriate institutional actions in response to the accepted findings of research misconduct. If this determination varies from the findings of the investigation committee, the DO will, as part of his/her written determination, explain in detail the basis for rendering a decision different from the findings of the investigation committee. Alternatively, the DO may return the report to the investigation committee with a request for further fact-finding or analysis.

When a final decision on the case has been reached, the RIO will normally notify both the respondent and the complainant in writing. After informing ORI, the DO will determine whether law enforcement agencies, professional societies, professional licensing boards, editors of journals in which falsified reports may have been published, collaborators of the respondent in the work, or other relevant parties should be notified of the outcome of the case. The RIO is responsible for ensuring compliance with all notification requirements of funding or sponsoring agencies.

D. Appeals

[*Option:* An institution's procedures may provide for an appeal by the respondent that could result in a reversal or modification of the institution's findings of research misconduct. If such an appeal is provided for, it must be completed within 120 days of its filing, unless ORI finds good cause for an extension, based upon the institution's written request for an extension that explains the need for the extension. If ORI grants an extension, it may direct the filing of periodic progress reports. 42 CFR § 93.314.]

E. Notice to ORI of Institutional Findings and Actions

Unless an extension has been granted, the RIO must, within the 120-day period for completing the investigation [*Option:* or the 120-day period for completion of any appeal], submit the following to ORI: (1) a copy of the final investigation report with all attachments [*Option:* and any appeal]; (2) a statement of whether the institution accepts the findings of the investigation report [*Option:* or the outcome of the appeal]; (3) a statement of whether the institution found misconduct and, if so, who committed the misconduct; and (4) a description of any pending or completed administrative actions against the respondent.[35]

F. Maintaining Records for Review by ORI

The RIO must maintain and provide to ORI upon request "records of research misconduct proceedings" as that term is defined by 42 CFR § 93.317. Unless custody has been transferred to HHS or ORI has advised in writing that the records no longer need to be retained, records of research misconduct proceedings must be maintained in a secure manner for 7 years after completion of the proceeding or the completion of any PHS proceeding involving the research misconduct allegation.[36] The RIO is also responsible for providing any information, documentation, research records, evidence, or clarification requested by ORI to carry out its review of an allegation of research misconduct or of the institution's handling of such an allegation.[37]

IX. COMPLETION OF CASES; REPORTING PREMATURE CLOSURES TO ORI

Generally, all inquiries and investigations will be carried through to completion and all significant issues will be pursued diligently. The RIO must notify ORI in advance if there are plans to close a case at the inquiry, investigation, or appeal stage on the basis that respondent has admitted guilt, a settlement with the respondent has been reached, or for any other reason, except: (1) closing of a case at the inquiry stage on the basis that an investigation is not warranted; or (2) a finding of no misconduct at the investigation stage, which must be reported to ORI, as prescribed in this policy and 42 CFR § 93.315.[38]

X. INSTITUTIONAL ADMINISTRATIVE ACTIONS [OPTIONAL]

If the DO determines that research misconduct is substantiated by the findings, he or she will decide on the appropriate actions to be taken, after consultation with the RIO. The administrative actions may include:

- Withdrawal or correction of all pending or published abstracts and papers emanating from the research where research misconduct was found;
- Removal of the responsible person from the particular project, letter of reprimand, special monitoring of future work, probation, suspension, salary reduction, or initiation of steps leading to possible rank reduction or termination of employment;
- Restitution of funds to the grantor agency as appropriate; and
- Other action appropriate to the research misconduct.

XI. OTHER CONSIDERATIONS

A. Termination or Resignation Prior to Completing Inquiry or Investigation

The termination of the respondent's institutional employment, by resignation or otherwise, before or after an allegation of possible research misconduct has been reported, will not preclude or terminate the research misconduct proceeding or otherwise limit any of the institution's responsibilities under 42 CFR Part 93. If the respondent, without admitting to the misconduct, elects to resign his or her position after the institution receives an allegation of research misconduct, the assessment of the allegation will proceed, as well as the inquiry and investigation, as appropriate based on the outcome of the preceding steps. If the respondent refuses to participate in the process after resignation, the RIO and any inquiry or investigation committee will use their best efforts to reach a conclusion concerning the allegations, noting in the report the respondent's failure to cooperate and its effect on the evidence.

B. Restoration of the Respondent's Reputation

Following a final finding of no research misconduct, including ORI concurrence where required by 42 CFR Part 93, the RIO must, at the request of the respondent, undertake all reasonable and practical efforts to restore the respondent's reputation.[39] Depending on the particular circumstances and the views of the respondent, the RIO should consider notifying those individuals aware of or involved in the investigation of the final outcome, publicizing the final outcome in any forum in which the allegation of research misconduct was previously publicized, and expunging all reference to the research misconduct allegation from the respondent's personnel file. Any institutional actions to restore the respondent's reputation should first be approved by the DO.

C. Protection of the Complainant, Witnesses, and Committee Members

During the research misconduct proceeding and upon its completion, regardless of whether the institution or ORI determines that research misconduct occurred, the RIO must undertake all reasonable and practical efforts to protect the position and reputation of, or to counter potential or actual retaliation against, any complainant who made allegations of research misconduct in good faith and of any witnesses and committee members who cooperate in good faith with the research misconduct proceeding.[40] The DO will determine, after consulting with the RIO, and with the complainant, witnesses, or committee members, respectively, what steps, if any, are needed to restore their respective positions or reputations or to counter potential or actual retaliation against them. The RIO is responsible for implementing any steps the DO approves.

D. Allegations Not Made in Good Faith

If relevant, the DO will determine whether the complainant's allegations of research misconduct were made in good faith, or whether a witness or committee member acted in good faith. If the DO determines that there was an absence of good faith he/she will determine whether any administrative action should be taken against the person who failed to act in good faith.

XII. APPENDIX A: RESEARCH INTEGRITY OFFICER RESPONSIBILITIES

I. General

The Research Integrity Officer has lead responsibility for ensuring that the institution:

- Takes all reasonable and practical steps to foster a research environment that promotes the responsible conduct of research, research training, and activities related to that research or research training, discourages research misconduct, and deals promptly with allegations or evidence of possible research misconduct.
- Has written policies and procedures for responding to allegations of research misconduct and reporting information about that response to ORI, as required by 42 CFR Part 93.
- Complies with its written policies and procedures and the requirements of 42 CFR Part 93.
- Informs its institutional members who are subject to 42 CFR Part 93 about its research misconduct policies and procedures and its commitment to compliance with those policies and procedures.
- Takes appropriate interim action during a research misconduct proceeding to protect public health, federal funds and equipment, and the integrity of the PHS supported research process.

II. Notice and Reporting to ORI and Cooperation with ORI

The RIO has lead responsibility for ensuring that the institution:

- Files an annual report with ORI containing the information prescribed by ORI.
- Sends to ORI with the annual report such other aggregated information as ORI may prescribe on the institution's research misconduct proceedings and the institution's compliance with 42 CFR Part 93.

- Notifies ORI immediately if, at any time during the research misconduct proceeding, it has reason to believe that health or safety of the public is at risk, HHS resources or interests are threatened, research activities should be suspended, there is reasonable indication of possible violations of civil or criminal law, federal action is required to protect the interests of those involved in the research misconduct proceeding, the institution believes that the research misconduct proceeding may be made public prematurely, or the research community or the public should be informed.
- Provides ORI with the written finding by the responsible institutional official that an investigation is warranted and a copy of the inquiry report, within 30 days of the date on which the finding is made.
- Notifies ORI of the decision to begin an investigation on or before the date the investigation begins.
- Within 120 days of beginning an investigation, or such additional days as may be granted by ORI, (or upon completion of any appeal made available by the institution) provides ORI with the investigation report, a statement of whether the institution accepts the investigation's findings, a statement of whether the institution found research misconduct and, if so, who committed it, and a description of any pending or completed administrative actions against the respondent.
- Seeks advance ORI approval if the institution plans to close a case at the inquiry, investigation, or appeal stage on the basis that the respondent has admitted guilt, a settlement with the respondent has been reached, or for any other reason, except the closing of a case at the inquiry stage on the basis that an investigation is not warranted or a finding of no misconduct at the investigation stage.
- Cooperates fully with ORI during its oversight review and any subsequent administrative hearings or appeals, including providing all research records and evidence under the institution's control, custody, or possession and access to all persons within its authority necessary to develop a complete record of relevant evidence.

III. Research Misconduct Proceeding

A. *General* The RIO is responsible for:

- Promptly taking all reasonable and practical steps to obtain custody of all research records and evidence needed to conduct the research misconduct proceeding, inventory the records and evidence, and sequester them in a secure manner.
- Taking all reasonable and practical steps to ensure the cooperation of respondents and other institutional members with research misconduct proceedings, including, but not limited to their providing information, research records and evidence.

- Providing confidentiality to those involved in the research misconduct proceeding as required by 42 CFR § 93.108, other applicable
  law, and institutional policy.
- Determining whether each person involved in handling an allegation
  of research misconduct has an unresolved personal, professional or
  financial conflict of interest and taking appropriate action, including
  recusal, to ensure that no person with such a conflict is involved in
  the research misconduct proceeding.
- Keeping the Deciding Official and others who need to know apprised of
  the progress of the review of the allegation of research misconduct.
- In cooperation with other institutional officials, taking all reasonable
  and practical steps to protect or restore the positions and reputations
  of good faith complainants, witnesses, and committee members and
  to counter potential or actual retaliation against them by respondents or other institutional members.
- Making all reasonable and practical efforts, if requested and as
  appropriate, to protect or restore the reputation of persons alleged to
  have engaged in research misconduct, but against whom no finding
  of research misconduct is made.
- Assisting the DO in implementing his/her decision to take administrative action against any complainant, witness, or committee member determined by the DO not to have acted in good faith.
- Maintaining records of the research misconduct proceeding, as
  defined in 42 CFR §93.317, in a secure manner for 7 years after completion of the proceeding, or the completion of any ORI proceeding
  involving the allegation of research misconduct, whichever is later,
  unless custody of the records has been transferred to ORI or ORI has
  advised that the records no longer need to be retained.
- Ensuring that administrative actions taken by the institution and
  ORI are enforced and taking appropriate action to notify other
  involved parties, such as sponsors, law enforcement agencies, professional societies, and licensing boards, of those actions.

B. *Allegation Receipt and Assessment*   The RIO is responsible for:

- Consulting confidentially with persons uncertain about whether to
  submit an allegation of research misconduct.
- Receiving allegations of research misconduct.
- Assessing each allegation of research misconduct to determine if an
  inquiry is warranted because the allegation falls within the definition of research misconduct, is within the jurisdictional criteria of
  42 CFR § 93.102(b), and is sufficiently credible and specific so that
  potential evidence of research misconduct may be identified.

C. *Inquiry*   The RIO is responsible for:

- Initiating the inquiry process if it is determined that an inquiry is
  warranted.

- At the time of, or before beginning the inquiry, making a good faith effort to notify the respondent in writing, if the respondent is known.
- On or before the date on which the respondent is notified, or the inquiry begins, whichever is earlier, taking all reasonable and practical steps to obtain custody of all research records and evidence needed to conduct the research misconduct proceeding, inventorying the records and evidence and sequestering them in a secure manner, except that where the research records or evidence encompass scientific instruments shared by a number of users, custody may be limited to copies of the data or evidence on the instruments, so long as those copies are substantially equivalent to the evidentiary value of the instruments.
- Appointing an inquiry committee and committee chair as soon after the initiation of the inquiry as is practical.
- Preparing a charge for the inquiry committee in accordance with the institution's policies and procedures.
- Convening the first meeting of the inquiry committee and at that meeting briefing the committee on the allegations, the charge to the committee, and the appropriate procedures for conducting the inquiry, including the need for confidentiality and for developing a plan for the inquiry, and assisting the committee with organizational and other issues that may arise.
- Providing the inquiry committee with needed logistical support, e.g., expert advice, including forensic analysis of evidence, and clerical support, including arranging witness interviews and recording or transcribing those interviews.
- Being available or present throughout the inquiry to advise the committee as needed and consulting with the committee prior to its decision on whether to recommend that an investigation is warranted on the basis of the criteria in the institution's policies and procedures and 42 CFR § 93.307(d).
- Determining whether circumstances clearly warrant a period longer than 60 days to complete the inquiry (including preparation of the final inquiry report and the decision of the DO on whether an investigation is warranted), approving an extension if warranted, and documenting the reasons for exceeding the 60-day period in the record of the research misconduct proceeding.
- Assisting the inquiry committee in preparing a draft inquiry report, sending the respondent a copy of the draft report for comment (and the complainant if the institution's policies provide that option) within a time period that permits the inquiry to be completed within the allotted time, taking appropriate action to protect the confidentiality of the draft report, receiving any comments from the respondent (and the complainant if the institution's policies provide that option), and ensuring that the comments are attached to the final inquiry report.

- Receiving the final inquiry report from the inquiry committee and forwarding it, together with any comments the RIO may wish to make, to the DO who will determine in writing whether an investigation is warranted.
- Within 30 days of a DO decision that an investigation is warranted, providing ORI with the written finding and a copy of the inquiry report and notifying those institutional officials who need to know of the decision.
- Notifying the respondent (and the complainant if the institution's policies provide that option) whether the inquiry found an investigation to be warranted and including in the notice copies of or a reference to 42 CFR Part 93 and the institution's research misconduct policies and procedures.
- Providing to ORI, upon request, the institutional policies and procedures under which the inquiry was conducted, the research records and evidence reviewed, transcripts or recordings of any interviews, copies of all relevant documents, and the allegations to be considered in the investigation.
- If the DO decides that an investigation is not warranted, securing and maintaining for 7 years after the termination of the inquiry sufficiently detailed documentation of the inquiry to permit a later assessment by ORI of the reasons why an investigation was not conducted.

D. *Investigation*   The RIO is responsible for:

- Initiating the investigation within 30 calendar days after the determination by the DO that an investigation is warranted.
- On or before the date on which the investigation begins: (1) notifying ORI of the decision to begin the investigation and providing ORI a copy of the inquiry report; and (2) notifying the respondent in writing of the allegations to be investigated.
- Prior to notifying respondent of the allegations, taking all reasonable and practical steps to obtain custody of and sequester in a secure manner all research records and evidence needed to conduct the research misconduct proceeding that were not previously sequestered during the inquiry.
- In consultation with other institutional officials as appropriate, appointing an investigation committee and committee chair as soon after the initiation of the investigation as is practical.
- Preparing a charge for the investigation committee in accordance with the institution's policies and procedures.
- Convening the first meeting of the investigation committee and at that meeting: (1) briefing the committee on the charge, the inquiry report and the procedures and standards for the conduct of the investigation, including the need for confidentiality and developing a specific plan for the investigation; and (2) providing committee

members a copy of the institution's policies and procedures and 42 CFR Part 93.

- Providing the investigation committee with needed logistical support, e.g., expert advice, including forensic analysis of evidence, and clerical support, including arranging interviews with witnesses and recording or transcribing those interviews.
- Being available or present throughout the investigation to advise the committee as needed.
- On behalf of the institution, the RIO is responsible for each of the following steps and for ensuring that the investigation committee: (1) uses diligent efforts to conduct an investigation that includes an examination of all research records and evidence relevant to reaching a decision on the merits of the allegations and that is otherwise thorough and sufficiently documented; (2) takes reasonable steps to ensure an impartial and unbiased investigation to the maximum extent practical; (3) interviews each respondent, complainant, and any other available person who has been reasonably identified as having information regarding any relevant aspects of the investigation, including witnesses identified by the respondent, and records or transcribes each interview, provides the recording or transcript to the interviewee for correction, and includes the recording or transcript in the record of the research misconduct proceeding; and (4) pursues diligently all significant issues and leads discovered that are determined relevant to the investigation, including any evidence of any additional instances of possible research misconduct, and continues the investigation to completion.
- Upon determining that the investigation cannot be completed within 120 days of its initiation (including providing the draft report for comment and sending the final report with any comments to ORI), submitting a request to ORI for an extension of the 120-day period that includes a statement of the reasons for the extension. If the extension is granted, the RIO will file periodic progress reports with ORI.
- Assisting the investigation committee in preparing a draft investigation report that meets the requirements of 42 CFR Part 93 and the institution's policies and procedures, sending the respondent (and complainant at the institution's option) a copy of the draft report for his/her comment within 30 days of receipt, taking appropriate action to protect the confidentiality of the draft report, receiving any comments from the respondent (and complainant at the institution's option) and ensuring that the comments are included and considered in the final investigation report.
- Transmitting the draft investigation report to institutional counsel for a review of its legal sufficiency.
- Assisting the investigation committee in finalizing the draft investigation report and receiving the final report from the committee.

- Transmitting the final investigation report to the DO and: (1) if the DO determines that further fact-finding or analysis is needed, receiving the report back from the DO for that purpose; (2) if the DO determines whether or not to accept the report, its findings and the recommended institutional actions, transmitting to ORI within the time period for completing the investigation, a copy of the final investigation report with all attachments, a statement of whether the institution accepts the findings of the report, a statement of whether the institution found research misconduct, and if so, who committed it, and a description of any pending or completed administrative actions against the respondent; or (3) if the institution provides for an appeal by the respondent that could result in a modification or reversal of the DO's finding of research misconduct, ensuring that the appeal is completed within 120 days of its filing, or seeking an extension from ORI in writing (with an explanation of the need for the extension) and, upon completion of the appeal, transmitting to ORI a copy of the investigation report with all attachments, a copy of the appeal proceedings, a statement of whether the institution accepts the findings of the appeal proceeding, a statement of whether the institution found research misconduct, and if so, who committed it, and a description of any pending or completed administrative actions against the respondent.
- When a final decision on the case is reached, the RIO will normally notify both the respondent and the complainant in writing and will determine whether law enforcement agencies, professional societies, professional licensing boards, editors of involved journals, collaborators of the respondent, or other relevant parties should be notified of the outcome of the case.
- Maintaining and providing to ORI upon request all relevant research records and records of the institution's research misconduct proceeding, including the results of all interviews and the transcripts or recordings of those interviews.

NOTES

Sections based on 42 CFR Part 93 have endnotes indicating the applicable section.

1. 42 CFR § 93.214.
2. 42 CFR § 93.102.
3. 42 CFR § 93.310(g).
4. 42 CFR §§ 93.304(c), 93.307(b).
5. 42 CFR §§ 93.304(e), 93.307(f).
6. 42 CFR § 308(a).
7. 42 CFR § 310(c).
8. 42 CFR § 310(g).

9. 42 CFR § 310(g).
10. 42 CFR §§ 93.304(f), 93.312(a).
11. 42 CFR § 93.316.
12. 42 CFR § 93.309(c).
13. 42 CFR § 93.304(k).
14. 42 CFR § 93.304(h).
15. 42 CFR § 93.318.
16. 42 CFR § 93.307(a).
17. 42 CFR § 93.307(c).
18. 42 CFR §§ 93.305, 93.307(b).
19. 42 CFR § 93.304(b).
20. 42 CFR § 93.307(g).
21. 42 CFR § 93.309(a).
22. 42 CFR § 93.308(a).
23. 42 CFR § 93.309(a) and (b).
24. 42 CFR § 93.310(a).
25. 42 CFR § 93.310(b) and (c).
26. 42 CFR § 93.310(d).
27. 42 CFR § 93.310(e).
28. 42 CFR § 93.310(f).
29. 42 CFR § 93.310(g).
30. 42 CFR § 93.310(h).
31. 42 CFR § 93.311.
32. 42 CFR § 93.313.
33. 42 CFR § 93.313(f).
34. 42 CFR §§ 93.312(a), 93.313(g).
35. 42 CFR § 93.315.
36. 42 CFR § 93.317(b).
37. 42 CFR §§ 93.300(g), 93.403(b) and (d).
38. 42 CFR § 93.316(a).
39. 42 CFR § 93.304(k).
40. 42 CFR § 93.304(1).

Appendix 2

Resources

CODES AND REGULATIONS

Code of Federal Regulations

21 CFR 50, 1997. Food and Drug Administration: Protection of Human Subjects.
21 CFR 54, 1998. Food and Drug Administration: Financial Disclosure by Clinical Investigators.
21 CFR 56, 1996. Food and Drug Administration: Institutional Review Boards.
42 CFR Parts 50 and 93, 2005. Department of Health and Human Services: Public Health Service Policies on Research Misconduct; Final Rule. http://ori.dhhs.gov/documents/42_cfr_parts_50_and_93_2005.pdf. Accessed on October 22, 2007.
45 CFR 46, part A, 1991. Department of Health and Human Services: Basic Policy for Protection of Human Research Subjects.
45 CFR 46, part B, 1991. Department of Health and Human Services: Additional Protections Relating to Research, Development, and Related Activities Involving Fetuses, Pregnant Women, and Human In Vitro Fertilization.
45 CFR 46, part C, 1991. Department of Health and Human Services: Additional Protections Pertaining to Biomedical and Behavioral Research Involving Prisoners as Subjects.
45 CFR 46, part D, 1991. Department of Health and Human Services: Additional Protections for Children Involved as Subjects in Research.

Federal Register

56 *Federal Register* 28012, 1991. The Common Rule.
59 *Federal Register* 14508, 1994. NIH Guidelines on the Inclusion of Women and Minorities in Research.
60 *Federal Register* 35810, 1995. Objectivity in Research.

Medical Associations

Council of the International Organization of Medical Sciences, 1993. International Ethical Guidelines for Biomedical Research Involving Human Subjects. CIOMS, Geneva.
World Medical Association, 1964, 1975, 1983, 1989, 1996, 2000. Declaration of Helsinki: Recommendations Guiding Physicians in Biomedical Research Involving Human Subjects. WMA, Ferney-Voltaire Cedex, France.

BOOKS AND ARTICLES

Association of Academic Health Centers, 1994. Conflicts of Interest in Institutional Decision-Making. AAHC, Washington, DC.
Beauchamp, T.L., and Childress, J.F., 2001. Principles of Biomedical Ethics, 5th ed. Oxford University Press, New York.
Bok, D., 2003. Universities in the Marketplace—the Commercialization of Higher Education. Princeton University Press, Princeton, NJ.

Cape, R., 1984. Academic and Corporate Values and Goals: Are They Really in Conflict? In Industrial-Academic Interfacing, D. Runser (ed.). American Chemical Society, Washington, DC, pp. 1–21.

Committee on Life Sciences and Health of the Federal Coordinating Council for Science, Engineering and Technology, 1993. Biotechnology for the 21st Century: Realizing the Promise. FCCSET, Washington, DC.

Davis, M., 1999. Ethics and the University. Routledge, New York.

Haber, E., 1996. Industry and the University. Nature Biotechnology 14: 1501–1502.

Huth, E., 1996. Conflicts of Interest in Industry-Funded Research. In Conflicts of Interest in Clinical Practice and Research, R. Spece, D. Shimm, and A. Buchanan (eds.). Oxford University Press, New York, pp. 407–417.

Kreeger, K.Y., 1997. Studies Call Attention to Ethics of Industry Support. The Scientist 11: 1–5.

Krimsky, S., 2003. Science in the Private Interest—Has the Lure of Profits Corrupted Bimedical Research? Rowman and Littlefield, Lanham, MD.

Lomasky, L., 1987. Public Money, Private Gain, Profit for All. Hastings Center Report 17(3): 5–7.

Macrina, F. (ed.), 2005. Scientific Integrity: Textbook and Cases in Responsible Conduct of Research, 3rd ed. American Society for Microbiology Press, Washington, DC.

McCain, K., 1996. Communication, Competition and Secrecy: The Production and Dissemination of Research Related Information in Genetics. Science, Technology and Human Values 16:492–510.

Snapper, J. (ed.), 1989. Owning Scientific and Technical Information. Rutgers University Press, Brunswick, NJ, pp. 29–39.

Williams, T., 1987. The History of Invention. Facts on File Publications, New York.

WEB SITES

American Association for the Advancement of Science Colloquium: Secrecy in Science. http://www.aaas.org/spp/secrecy/AAASMIT.htm.

Council of the International Organization of Medical Sciences (CIOMS): Guidelines. http://www.cioms.ch/frame_guidelines_nov_2002.htm.

Department of Health and Human Services: The Belmont Report. http://ohrp. osophs.dhhs.gov/humansubjects/guidance/belmont.htm.

Department of Health and Human Services: Institutional Review Board Guidebook. http://www.hhs.gov/ohrp/irb/irb_guidebook.htm.

Freedom of Information Act (5 U.S. Code, 552). 2007. http://www.usdoj.gov/oip/amended-foia-redlined.pdf.

International Committee of Medical Journal Editors: Uniform Requirements. 2007. http://www.icmje.org/.

National Institutes of Health: Other Data Sharing Documents and Resources. http://grants.nih.gov/grants/policy/data_sharing/data_sharing_resources. htm?print=yes&.

National Institutes of Health: A Guide to Training and Mentoring in the Intramural Research Program at NIH. http://www1.od.nih.gov/oir/sourcebook/ethic-con-duct/mentor-guide.htm.

National Academy of Sciences: Advisor, Teacher, Role Model, Friend: On Being a Mentor to Students in Science and Engineering. http://www.nap.edu/html/mentor/.

National Science Foundation: Women, Minorities, and Persons with Disabilities in Science and Engineering. http://www.nsf.gov/statistics/wmpd/.

Nuremberg Code. http://ohsr.od.nih.gov/guidelines/nuremberg.html.

Office for Human Research Protections. http://ohrp.osophs.dhhs.gov/index.htm.

Office of Research Integrity. http://ori.dhhs.gov/; see especially "About ORI." http://ori.dhhs.gov/about/index.shtml.

Online Ethics: Responsible Research. http://www.onlineethics.diamax.com/CMS/research.aspx.

Protection of Human Subjects (45 CFR 46). http://www.hhs.gov/ohrp/humansubjects/guidance/45cfr46.htm.

University of Maryland: COI Procedures. http://www.president.umd.edu/policies/docs/II-310B.pdf.

World Medical Association: Declaration of Helsinki. http://www.wma.net/e/policy/b3.htm.

References

References that we also suggest for further reading are marked with an asterisk (*).

Abby, M., Massey, M.D., Galandiuk, S., and Polk, H.C., Jr., 1994. Peer Review Is an Effective Screening Process to Evaluate Medical Manuscripts. Journal of the American Medical Association 272: 105–107.

Actuarial Foundation, 1998. Genetic Testing: Implications for Insurance. Actuarial Foundation, Schaumburg, IL.

Adler, R.G., 1993. Choosing the Form of Legal Protection. In Understanding Biotechnology Law, G.R. Peterson (ed.). Dekker, New York, pp. 63–86.

Advisory Committee on Human Radiation Experiments, 1995. Final Report. Stock No. 061-000-00-848-9. Superintendent of Documents, U.S. Government Printing Office, Washington, DC.

Agar, N., 2007. Embryonic Potential and Stem Cells. Bioethics 21: 198–207.

Agnew, B., 1999a. NIH Eyes Sweeping Reform of Peer Review. Science 286: 1074–1076.

Agnew, B., 1999b. NIH Invites Activists into the Inner Sanctum. Science 283: 1999–2001.

Agnew, B., 2000. Financial Conflicts Get More Scrutiny at Clinical Trials. Science 289: 1266–1267.

Aller, R., and Aller, C., 1997. An Institutional Response to Patient/Family Complaints. In Ethics in Neurobiological Research with Human Subjects, A.E. Shamoo (ed.). Gordon and Breach, Amsterdam, pp. 155–172.

Altman, L., 1995. Promises of Miracles: News Releases Go Where Journals Fear to Tread. New York Times, January 10, C2–C3.

Altman, L.K., 1997. Experts See Bias in Drug Data. New York Times, April 29, C1–C8.

American Association for the Advancement of Science, 1991. Misconduct in Science. AAAS, Washington, DC.

American Association for the Advancement of Science, 2000. Inheritable Genetic Modifications. AAAS, Washington, DC.

American Association for the Advancement of Science–American Bar Association, 1988. Project on Scientific Fraud and Misconduct. National Conference of Lawyers and Scientists, Report on Workshop No. 1. AAAS, Washington, DC.

American Association of University Professors, 2006. Research on Human Subjects: Academic Freedom and the Institutional Review Board. http://www.aaup.org/AAUP/About/committees/committee+repts/CommA/Research onHumanSubjects.htm. Accessed February 28, 2006.

American Chemical Society, 1994. Chemist's Code of Conduct. http://portal.acs.org/portal/acs/corg/content?_nfpb=true&_pageLabel=PP_ARTICLEMAIN&node_id=1095&content_id=CTP_004007&use_sec=true&sec_url_var=region1. Accessed August 5, 2008.

American Physical Society, 2002. Guidelines for Professional Conduct. http://www.aps.org/policy/statements/02_2.cfm. Accessed August 5, 2008.

American Psychological Association, 1992. Ethical Principles of Psychologists and Code of Conduct. American Psychologist 47: 1597–1611.

American Society for Biochemistry and Molecular Biology, 1998. Code of Ethics. http://www.asbmb.org/Page.aspx?id=70&terms=Code+of+Ethics. Accessed August 5, 2008.

*American Statistical Association, 1999. Ethical Guidelines for Statistical Practice. ASA, Alexandria, VA.

Ananworanich, J., Cheunyam, T., Teeratakulpisarn, S., Boyd, M., Ruxrungtham, K., Lange, P., et al., 2004. Creation of a Drug Fund for Post-clinical Trial Access to Antiretrovirals. Lancet 364: 101–102.

Anderson, M.S., Horn, A.S., Ronning, E.A., De Vries, R., and Martinson, B.C., 2007. What Do Mentoring and Training in the Responsible Conduct of Research Have to Do with Scientific Misbehavior? Findings from a National Survey of NIH-Funded Scientists. Academic Medicine 82(9): 853–860.

Anderson, W., 1989. Why Draw a Line? Journal of Medicine and Philosophy 14(4): 681–693.

Andrews, L., 2000. The Clone Age. Henry Holt, New York.

Andrews, L., 2002. Future Perfect. Columbia University Press, New York.

Andrews, L., and Nelkin, D., 2001. Body Bazaar. Crown, New York.

Angell, M., 1997a. The Ethics of Clinical Research in the Third World. New England Journal of Medicine 337: 847–849.

Angell, M., 1997b. Science on Trial. Norton, New York.

Angell, M., 2000. Is Academic Medicine for Sale? New England Journal of Medicine 342: 1516–1518.

Angell, M., 2001. Medicine in the Noise Age: What Can We Believe? Accountability in Research 8: 189–196.

Angell, M., 2004. The Truth about Drug Companies—How They Deceive Us and What to Do about It. Random House, New York.

Animal Welfare Act, 1966, 1996. Title 7 U.S. Code, 2131–2156. http://www.nal.usda.gov/awic/legislat/awa.htm. Accessed August 6, 2008.

Annas, G.J., 1994. Regulatory Model for Human Embryo Cloning: The Free Market Professional Guidelines and Government Restrictions. Kennedy Institute of Ethics Journal 4: 235–249.

Annas, G.J., and Elias, S., 1992. Gene Mapping. Oxford University Press, New York.

Anonymous, 1993. Toxicity Tests in Animals: Extrapolating to Human Risks. Environmental Health Perspectives 101: 396–401.

Anonymous, 1999. JAMA and Editorial Independence. Journal of the American Medical Association 281: 460.

Appelbaum, B.C., Appelbaum, P.S., and Grisso, T., 1998. Competence to Consent to Voluntary Psychiatric Hospitalization: A Test of a Standard Proposed by APA. Psychiatric Services 49(9): 1193–1196.

Appelbaum, P., Roth, L.H., Lidz, C.W., and Winslade, W., 1987. False Hopes and Best Data: Consent to Research and the Therapeutic Misconception. Hastings Center Report 17(2): 20–24.

Appelbaum, P.S., 2001. Competence to Consent to Research: A Critique of the Recommendations of the National Bioethics Advisory Committee. Accountability in Research 7: 265–276.

Aristotle, 330 B.C. [1984]. Nichomachean Ethics. In Complete Works of Aristotle, J. Barnes (ed.). Princeton University Press, Princeton, NJ.

Armstrong, D., 2005. How a Famed Hospital Invests in a Device It Uses and Promotes. Wall Street Journal, December 12. http://online.wsj.com/article_print/SB113435097142119825.html. Accessed January 12, 2006.

Armstrong, D., 2006. Cleveland Clinic to Tighten Its Disclosure Policies. Wall Street Journal, February 9, A3. http://online.wsj.com/article_email/SB113915792253969290-IMYQ. Accessed February 9, 2006.

Armstrong, J., 1997. Peer Review for Journals: Evidence of Quality Control, Fairness, and Innovation. Science and Engineering Ethics 3(1): 63–84.

Associated Press, 1997. Test of AIDS Vaccine Sought. Denver Post, September 22, A3.

Association for the Accreditation of Human Research Protection Programs, 2008. Homepage. http://www.aahrpp.org/www.aspx. Accessed May 9, 2008.

Association for the Assessment and Accreditation of Laboratory Animal Care, 2007. Homepage. http://www.aaalac.org/. Accessed October 21, 2007.

Association of Academic Health Centers, 1990. Conflict of Interest in Academic Health Centers. AAHC, Washington, DC.

Association of Academic Health Centers, 1994. Conflict of Interest in Institutional Decision-Making. AAHC, Washington, DC.

Association of Academic Health Centers, 2001. Task Force on Financial Conflict of Interest in Clinical Research, Protecting Subjects, Preserving Trust, Promoting Progress II, Principles and Recommendations for Oversight of Individual Financial Interests in Human Subjects Research. AAHC, Washington, DC.

Association of Academic Health Centers, 2002. Task Force on Financial Conflict of Interest in Clinical Research, Protecting Subjects, Principles and Recommendations for Oversight of an Institution's Financial Interests in Human Subjects Research, Preserving Trust, Promoting Progress II. AAHC, Washington, DC.

Association of American Medical Colleges, 2002. Protecting Subjects, Preserving Trust, Promoting Progress II: Principles and Recommendations for Oversight of an Institution's Financial Interests in Human Subjects Research. http://www.aamc.org/research/coi/2002coireport.pdf. Accessed July 18, 2008.

Austin, C., Battey, J., Bradley, A., Bucan, M., Capecchi, M., Collins, F., et al., 2004. The Knockout Mouse Project. Nature Genetics 36: 921–924.

Babbage, C., 1830 [1970]. Reflections on the Decline of Science in England. Augustus Kelley, New York.

Bailar, J., 1986. Science, Statistics, and Deception. Annals of Internal Medicine 105: 259–260.

Baldessarini, R.J., and Viguera, A.C., 1995. Neuroleptic Withdrawal in Schizophrenic Patients. Archives of General Psychology 52: 189–191.

Ballantyne, A., 2005. HIV International Clinical Research: Exploitation and Risk. Bioethics 19: 476–491.

Baltimore, D., 1991. Baltimore Declares O'Toole Mistaken. Nature 351: 341–343.

Banoub-Baddour, S., and Gien, L.T., 1991. Student-Faculty Joint-Authorship: Mentorship in Publication. Canadian Journal of Nursing Research 23: 5–14.

Barbash, F., 1996. Piltdown Meltdown: A Hoaxer Revealed. Washington Post, May 24, A1, A34.

Barber, B., 1961. Resistance by Scientists to Scientific Discovery. Science 134: 596–602.

Barinaga, M., 2000. Soft Money's Hard Realities. Science 289: 2024–2028.

*Barnard, N., and Kaufman, S., 1997. Animal Research Is Wasteful and Misleading. Scientific American 276(2): 80–82.

Basic Books, Inc. v. Kinko's Graphics Corp., 1991. 758 F. Supp. 1522.

Baughman, F., Jr., 2006. The ADHD Fraud. Traffordi, Victoria, BC.

Bayh-Dole Act, 1980. 35 U.S. Code, 200.

Bayles, M., 1988. Professional Ethics, 2nd ed. Wadsworth, Belmont, CA.

Baylis, F., and Robert, J., 2004. The Inevitability of Genetic Enhancement Technologies. Bioethics 18: 1–26.

Bean, W.B., 1977. Walter Reed and the Ordeal of Human Experiments. Bulletin of the History of Medicine 51: 75–92.

Beardsley, T., 1994. Big-Time Biology. Scientific American 271(5): 90–97.

Beauchamp, T.L., 1996. Looking Back and Judging Our Predecessors. Kennedy Institute of Ethics Journal 6: 251–270.

Beauchamp, T.L., and Childress, J.F., 2001. Principles of Biomedical Ethics, 5th ed. Oxford University Press, New York.

*Beaudet A.L., Scriver, C.R., Sly, W.S., and Valle, D., 2001. Genetics, Biochemistry and Molecular Basis of Variant Human Phenotypes. In The Metabolic and Molecular Bases of Inherited Disease, 8th ed., C.R. Scriver, A.R. Beaudet, W. Sly, and D. Valle (eds.). McGraw-Hill, New York, pp. 3–45.

Beaudette, C.G., 2000. Excess Heat—Why Cold Fusion Research Prevailed. Oak Grove Press, South Bristol, ME.

Beecher, H., 1966. Ethics and Clinical Research. New England Journal of Medicine 274: 1354–1360.

Bekelman, J.E., Li, Y., and Gross, G.P., 2003. Scope and Impact of Financial Conflicts of Interest in Biomedical Research. Journal of the American Medical Association 289: 454–465.

Benatar, S.R., 2000. Avoiding Exploitation in Clinical Research. Cambridge Quarterly of Healthcare Ethics 9: 562–565.

Bennett, B.T., 1994. Regulations and Requirements. In Essentials for Animals Research, T. Bennett, M.J. Brown, and J.C. Schofield (eds.). National Agricultural Library, Beltsville, MD.

Bentham, J., 1789 [1988]. Introduction to Principles of Morals and Legislation. New York, Penguin.

Berg, J., Applebaum, P., Parker, L., and Lidz, C., 2001. Informed Consent: Legal Theory and Clinical Practice. Oxford University Press, New York.

Berger, E., and Gert, B., 1997. Institutional Responsibility. In Research Ethics: A Reader, D. Elliot and J. Stern (eds.). University Press of New England, Hanover, NH, pp. 197–212.

Bernard, C., 1865 [1957]. An Introduction to the Study of Experimental Medicine, H. Green (trans). Dover, New York.

Bero, L.A., Glantz, S.A., and Rennie, D., 1994. Publication Bias and Public Health on Environmental Tobacco Smoke. Journal of the American Medical Association 272: 133–136.

Bhutta, Z.A., 2004. Beyond Informed Consent. Bulletin of the World Health Organization 82: 771–777.

Bingham, C., 2000. Peer Review and the Ethics of Internet Publishing. In Ethical Issues in Biomedical Publication, A. Jones and F. McLellan (eds.). Johns Hopkins University Press, Baltimore, MD, pp. 85–112.

Bird, S., 1993. Teaching Ethics in Science: Why, How, and What. In Ethics, Values, and the Promise of Science. Sigma Xi, The Scientific Research Society, Research Triangle Park, NC, pp. 228–232.

Blank, R., 1991. The Effects of Double Blind versus Single Blind Reviewing: Experimental Evidence from American Economic Review. American Economic Review 81: 1041–1067.

Blinderman, C., 1986. The Piltdown Inquest. Prometheus Books, Buffalo, NY.

Blumenthal, D., 1995. Academic-Industry Relationships in the 1990s: Continuity and Change. Paper presented at the symposium Ethical Issues in Research Relationships between Universities and Industry, Baltimore, MD, November 3–5.

*Blumenthal, D., 1997. Withholding Research Results in Academic Life Science: Evidence from a National Survey of Faculty. Journal of the American Medical Association 277: 1224–1228.

Blumenthal, D., Campbell, E.G., Causino, N., and Louis, K.S., 1996a. Participation of Life-Science Faculty in Relationships with Industry. New England Journal of Medicine 335: 1734–1739.

Blumenthal, D., Campbell, E., Gokhale, M., Yucel, R., Clarridge, B., Hilgartner, S., et al., 2006. Data Withholding in Genetics and the Other Life Sciences: Prevalences and Predictors. Academic Medicine 81: 137–145.

Blumenthal, D., Causino, N., Campbell, E.G., and Louis, K.S., 1996b. Relationship between Academic Institutions and Industry in the Life Sciences—an Industry Survey. New England Journal of Medicine 334: 368–373.

Blumenthal, D., Gluck, M., Louis, K.S., Stoto, M.A., and Wise, D., 1986. University-Industry Research Relationships in Biotechnology: Implications for the University. Science 232: 1361–1366.

Bodenheimer, T., 2000. Uneasy Alliance: Clinical Investigators and the Pharmaceutical Industry. New England Journal of Medicine 342: 1539–1544.

Bok, D., 1994. The Commercialized University. In N. Bowie, University-Business Partnerships: An Assessment. Rowman and Littlefield, Lanham, MD, pp. 116–122.

Bok, D., 2003. Universities in the Marketplace—the Commercialization of Higher Education. Princeton University Press, Princeton, NJ.

Bonner, J., 1980. The Evolution of Culture in Animals. Princeton University Press, Princeton, NJ.

Borry, P., Schotsmans, P., and Dierickx, K., 2006. Author, Contributor or Just a Signer? A Quantitative Analysis of Authorship Trends in the Field of Bioethics. Bioethics 20: 213–220.

Botkin, J., 2001. Protecting the Privacy of Family Members in Survey and Pedigree Research. Journal of the American Medical Association 285: 207–211.

Botting, J., and Morrison, A., 1997. Animal Research Is Vital to Medicine. Scientific American 276(2): 83–85.

Bower, B., 1991. Peer Review under Fire. Science News 139: 394–395.

Bowie, N., 1994. University-Business Partnerships: An Assessment. Rowman and Littlefield, Lanham, MD.

Boyd, R., 1988. How to Be a Moral Realist. In Essay on Moral Realism, J. Sayre-McCord (ed.). Cornell University Press, Ithaca, NY, pp. 181–228.

Bradley, G., 2000. Managing Conflicting Interests. In Scientific Integrity, F. Macrina (ed.). American Society for Microbiology Press, Washington, DC, pp. 131–156.

Bradley, S.G., 1995. Conflict of Interest. In Scientific Integrity, F.L. Macrina (ed.). American Society for Microbiology Press, Washington, DC.

Brennan, T.A., et al., 2007. Health Industry Practices That Create Conflicts of Interest—a Policy Proposal for Academic Medical Centers. Journal of the American Medical Association 295: 429–433.

*Broad, W., and Wade, N., 1982 [1993]. Betrayers of the Truth: Fraud and Deceit in the Halls of Science. Simon and Schuster, New York.

Broad, W.J., 1981. The Publishing Game: Getting More for Less. Science 211: 1137–1139.

Bronowski, J., 1956. Science and Human Values. Harper and Row, New York.

Browing, T., 1995. Reaching for the "Low Hanging Fruit": The Pressure for Results in Scientific Research—a Graduate Student's Perspective. Science and Engineering Ethics 1: 417–426.

Brown, J., 2000. Privatizing the University—the New Tragedy of the Commons. Science 290: 1701–1702.

Buchanan, A., and Brock, A., 1990. Deciding for Others: The Ethics of Surrogate Decision Making. Cambridge University Press, Cambridge.

Buchanan, A., Brock, D., Daniels, N., and Wikler, D., 2001. From Chance to Choice: Genetics and Justice. Cambridge University Press, Cambridge.

Buck v. Bell, 1927. 274 U.S. 200.

Bulger, R., 1987. Use of Animals in Experimental Research: A Scientist's Perspective. Anatomical Record 219: 215–220.

Burke, J., 1995. The Day the Universe Changed. Little, Brown, Boston.

Burnham, J.C., 1990. The Evolution of Editorial Peer Review. Journal of the American Medical Association 263: 1323–1329.

Butler, D., 1999a. NIH Plan Brings Global Electronic Journal a Step Nearer Reality. Nature 398: 735.

Butler, D., 1999b. The Writing Is on the Web for Science Journals in Print. Nature 397: 195–199.

Butler, D., 2007. Copycat Trap. Nature 448: 633.

Buyse, M., and Evans, S., 2001. Fraud in Clinical Trials. In Biostatistics in Clinical Trials, T. Colton and C. Redmond (eds.). Wiley, New York, pp. 432–469.

Buzzelli, D., 1993. NSF's Approach to Misconduct in Science. Accountability in Research 3: 215–222.

Callahan, M., Baxt, W.G., Waeckerle, J.F., and Wears, R.L., 1998. Reliability of Editors' Subjective Quality Rating of Peer Review of Manuscripts. Journal of the American Medical Association 280: 229–231.

Callahan, S., 1998. The Ethical Challenge of the New Reproductive Technology. In Health Care Ethics—Critical Issues for the 21st Century, J. Monagle and D. Thomasma (eds.). Aspen Publishers, Boulder, CO, pp. 45–55.

Campbell, E., Clarridge, B., Gokhale, M., Birenbaum, L., Hilgartner, S., Holtzman, N., et al., 2002. Data Withholding in Academic Genetics: Evidence from a National Survey. Journal of the American Medical Association 287: 473–480.

Campbell, E., Moy, B., Feibelmann, S., Weissman, J., and Blumenthal, D., 2004. Institutional Academic Industry Relationship: Results of Interviews with University Leaders. Accountability in Research 11: 103–118.

Campbell, N., and Reece, J., 2004. Biology, 7th ed. Benjamin Cummings, San Francisco.

Caplan, A., 1983. Beastly Conduct: Ethical Issues in Animal Experimentation. Annals of the New York Academy of Science 406: 159–169.

Caplan, A.L. (ed.), 1992. When Medicine Went Mad—Bioethics and the Holocaust. Humana Press, Totowa, NJ.

*Capron, A.M., 1989. Human Experimentation. In Medical Ethics, R.M. Veatch (ed.). Jones and Bartlett, Boston, pp. 125–172.

Carey, J., Freudlich, N., Flynn, J., and Gross, N., 1997. The Biotech Century. Business Week, March 10, 79–88.

Carruthers, P., 1992. The Animals Issue. Cambridge University Press, Cambridge.

Carson, R., 1961. Silent Spring. Houghton Mifflin, Boston.

Cassell, E., 1991. The Nature of Suffering. Oxford University Press, New York.

Cech, T., and Leonard, J., 2001. Conflicts of Interest: Moving beyond Disclosure. Science 291: 989.

Celera Genomics, 2008. Corporate Profile. http://phx.corporate-ir.net/phoenix. zhtml?c=129200&p=irol-irhome. Accessed July 21, 2008.

Chalmers, T.C., Frank, C.S., and Reitman, D., 1990. Minimizing the Three Stages of Publication Bias. Journal of the American Medical Association 263: 1392–1395.

Chapman, A. (ed.), 1999. Perspectives on Genetic Patenting. AAAS, Washington, DC.

Check, E., and Cyranoski, D., 2005. Korean Scandal Will Have Global Fallout. Nature 438: 1056–1057.

*Cheny, D. (ed.), 1993. Ethical Issues in Research. University Publishing Group, Frederick, MD.

Chernow, R., 1998. Titan—the Life of John D. Rockefeller, Jr. Vantage Books, New York.

Chickering, R.B., and Hartman, S., 1980. How to Register a Copyright and Protect Your Creative Work. Charles Scribner's Sons, New York.

Chiong, W., 2006. The Real Problem with Equipoise. American Journal Bioethics 6(4): 37–47.

Cho, M.K., 1997. Letters to the Editor, Disclosing Conflicts of Interest. Lancet 350: 72–73.

Cho, M.K., 1998. Fundamental Conflict of Interest. HMS Beagle: The BioMedNet Magazine, issue 24, http://biomednet.com/hmsbeagle/1998/24/people/op-ed.htm. Accessed January 15, 2007.

Cho, M.K., and Bero, L.A., 1996. The Quality of Drug Studies Published in Symposium Proceedings. Annals of Internal Medicine 124: 485–489.

Cho, M.K., and Billings, P., 1997. Conflict of Interest and Institutional Review Boards. Journal of Investigative Medicine 45: 154–159.

Cho, M., Shohara, R., Schissel, A., and Rennie, D., 2000. Policies on Faculty Conflicts of Interest at US Universities. Journal of the American Medical Association 284: 2203–2208.

Chou, R., and Helfand, M., 2005. Challenges Reviews That Assess Treatment Harm. Annals of Internal Medicine 142: 1090–1099.

Chubb, S., 2000. Introduction to the Series of Papers in Accountability in Research Dealing with "Cold Fusion." Accountability in Research 8: 1–18.

Chubin, D., and Hackett, E., 1990. Peerless Science. State University of New York Press, Albany.

Citizens for Responsible Care and Research, 2007. Homepage. http://www.circare. org. Accessed February 10, 2008.

Clayton, E., 2001. Bioethics of Genetic Testing. Encyclopedia of Life Sciences. Nature Publishing Group, Macmillan, London. http://www.els.net. Accessed August 7, 2008.

Clayton, E., et al., 1995. Informed Consent for Genetic Research on Stored Tissue Samples. Journal of the American Medical Association 274: 1786–1792.

Clinical Laboratory Improvement Amendments, 1988. Public Law 100–578.

Cohen, C., 1986. The Case for the Use of Animals in Biomedical Research. New England Journal of Medicine 315: 865–870.

Cohen, J., 1991. What Next in the Gallo Case? Science 254: 944–949.

Cohen, J., 1994. US-French Patent Dispute Heads for Showdown. Science 265: 23–25.

Cohen, L., and Hahn, R., 1999. A Solution to Concerns over Public Access to Scientific Data. Science 285: 535–536.

Cohen, S., 1995: Human Relevance of Animal Carcinogenicity Studies. Regulatory Toxicology and Pharmacology 21: 75–80.

Cole, S., and Cole, J.R., 1981. Peer Review in the NSF: Phase Two. National Academy of Sciences, Washington, DC.

Cole, S., Rubin, L., and Cole, J.R., 1978. Peer Review in the NSF: Phase One. National Academy of Sciences, Washington, DC.

Collins, F., and McKusick, V., 2001. Implications of the Human Genome Project for Medical Science. Journal of the American Medical Association 285: 540–544.

Collins, F., Morgan, M., and Patrinos, A., 2003. The Human Genome Project: Lessons from Large-Scale Biology. Science 300: 286–290.

Commission on Research Integrity, 1995. Integrity and Misconduct in Research. U.S. Department of Health and Human Services, Washington, DC.

Committee on Publication Ethics, 2007. Guidelines on Good Publication and the Code of Conduct. http://www.publicationethics.org.uk/guidelines. Accessed November 25, 2007.

Conference on Ethical Aspects of Research in Developing Countries, 2004. Moral Standards for Research in Developing Countries: From "Reasonable Availability" to "Fair Benefits." Hastings Center Report 34(3): 17–27.

Conrad, P., 2007. The Medicalization of Society. Johns Hopkins University Press, Baltimore, MD.

Copernicus, N., 1542 [1995]. On the Revolutions of the Heavenly Spheres, C. Wallis (trans.). Prometheus Books, Amherst, NY.

Council for International Organizations of Medical Sciences (CIOMS), 2002. International Ethical Guidelines for Biomedical Research Involving Human Subjects. Geneva, CIOMS.

Couzin, J., 2002. A Call for Restraint on Biological Data. Science 297: 749–748.

Couzin, J., 2006a. Don't Pretty Up That Picture Just Yet. Science 314: 1866–1868.

Couzin, J., 2006b. Truth and Consequences. Science 313: 1222–1226.

Crewdson, J., 2003. Science Fictions: A Scientific Mystery, a Massive Cover-up and the Dark Legacy of Robert Gallo. Back Bay Books, Boston, MA.

Crossen, C., 1994. Tainted Truth. Simon and Schuster, New York.

Crow, T.J., MacMillian, J.F., Johnson, A.L., and Johnstone, B.C., 1986. A Randomised Controlled Trial of Prophylatic Neuroleptic Treatment. British Journal of Psychiatry 148: 120–127.

Culliton, B.J., 1990. Gallo Inquiry Takes Puzzling Turn. Science 250: 202–203.

Culliton, J., 1977. Harvard and Monsanto: The $23 Million Alliance. Science 195: 759–763.

Cummings, M., 2002. Informed Consent and Investigational New Drug Abuses in the U.S. Military. Accountability in Research 9: 93–103.

Dalton, R., 1999. Kansas Kicks Evolution out the Classroom. Nature 400: 701.

Dalton, R., 2001. Peers under Pressure. Nature 413: 102–104.

Davidoff, F., 1998. Masking, Blinding, and Peer Review: The Blind Leading the Blinded. Annals of Internal Medicine 128: 66–68.

Davidson, R., 1986. Source of Funding and Outcome of Clinical Trials. Journal of General Internal Medicine 1: 155–158.

Davis, J., 2003. Self-experimentation. Accountability in Research 10: 175–187.

Davis, M., 1982. Conflict of Interest. Business and Professional Ethics Journal 1(4): 17–27.

*Davis, M., 1990. The Discipline of Science: Law or Profession. Accountability in Research 1: 137–145.

Davis, M., 1991. University Research and the Wages of Commerce. Journal of College and University Law 18: 29–39.

Davis, M., 1995a. Panel Discussion. Held at the symposium Ethical Issues in Research Relationships between Universities and Industry. Baltimore, MD. [A summary of the discussion is provided in Rule and Shamoo (1997), listed later in this section.]

Davis, M., 1995b. A Preface to Accountability in the Professions. Accountability in Research 4: 81–90.

DeBakey, L., 1990. Journal Peer Reviewing. Archives of Ophthalmology 108: 345–349.

De Cock, K., Fowler, M., Mercier, E., de Vincenzi, I., Saba, J., Hoff, E., et al., 2000. Prevention of Mother-to-Child HIV Transmission in Resource-Poor Countries: Translating Research into Policy and Practice. Journal of the American Medical Association 283: 1175–1182.

De George, R., 1995. Business Ethics, 4th ed. Prentice-Hall, Englewood Cliffs, NJ.

DeGrazia, D., 1991. The Moral Status of Animals and Their Use in Research. Kennedy Institute of Ethics Journal 1: 48–70.

De Melo-Martín, I., Palmer, L., and Fins, J., 2007. Viewpoint: Developing a Research Ethics Consultation Service to Foster Responsive and Responsible Clinical Research. Academic Medicine 82: 900–904.

DeMets, D., 1999. Statistics and Ethics in Medical Research. Science and Engineering Ethics 5: 97–117.

DeMets, D., Fost, N., and Powers, M., 2005. An Institutional Review Board Dilemma: Responsible for Safety Monitoring but Not in Control. Clinical Trials 3: 142–148.

Deming, N., et al., 2007. Incorporating Principles and Practical Wisdom in Research Ethics Education: A Preliminary Study. Academic Medicine 82: 18–23.

Dennett, D., 1995. Darwin's Dangerous Idea. Simon and Schuster, New York.

Descartes, R., [1970]. Descartes: Philosophical Letters, A. Kenny (trans. and ed.). Oxford University Press, Oxford.

de Waal, F., 1996. Good Natured: The Origins of Right and Wrong in Humans and Other Animals. Harvard University Press, Cambridge, MA.

Deyo, R.A., Psaty, B.M., Simon, G., Wagner, E.H., and Omenn, G.S., 1997. The Messenger under Attack—Intimidation of Research by Special-Interest Groups. New England Journal of Medicine 336: 1176–1180.

Diamond v. Chakrabarty, 1980. U.S. Supreme Court Cases and Opinions 447: 303–310.

Dickson, D., 1988. The New Politics of Science. University of Chicago Press, Chicago.

Dickson, D., 1995. Between Politics and Science. Cambridge University Press, Cambridge.

Diguisto, E., 1994. Equity in Authorship: A Strategy for Assigning Credit When Publishing. Social Science and Medicine 38: 55–58.

Djerassi, C., 1999. Who Will Mentor the Mentors? Nature 397: 291.

Doucet, M.S., Bridge, A., Grimlund, R.A., and Shamoo, A.E., 1994. An Application of Stratified Sampling Techniques for Research Data. Accountability in Research 3: 237–247.

Doukas, D., 2003. Genetics Providers and the Family Covenant: Connecting Individuals with Their Families. Genetic Testing 7: 315–321.

Doumbo, O., 2005. It Takes a Village: Medical Research and Ethics in Mali. Science 307: 679–681.

Drenth, J., 1996. Proliferation of Authors on Research Reports in Medicine. Science and Engineering Ethics 2: 469–480.

Dresser, R., 1992. Wanted: Single, White Male for Medical Research. Hastings Center Report 22(1): 24–29.

Dresser, R., 2001. When Science Offers Salvation. Oxford University Press, New York.

Dreyfuss, R., 2000. Collaborative Research: Conflicts on Authorship, Ownership, and Accountability. Vanderbilt Law Review 53: 1161–1232.

DuBois, J.M., 2008. Ethics in Mental Health Research. Oxford University Press, New York.

Ducor, P., 2000. Coauthorship and Coinventorship. Science 289: 873–875.

Duft, B.J., 1993. Preparing Patent Application. In Understanding Biotechnology Law, Gale R. Peterson (ed.). Dekker, New York, pp. 87–186.

Duke University, 2007. Duke Policies, Regulations, and Committees. http://www.gradschool.duke.edu/policies_and_forms/responsible_conduct_of_research/duke_policies_regulations_and_committees.html. Accessed November 25, 2007.

Dustira, A.K., 1992. The Funding of Basic and Clinical Biomedical Research. In Biomedical Research: Collaboration and Conflict of Interest, R.J. Porter and T.E. Malone (eds.). Johns Hopkins University Press, Baltimore, MD, pp. 33–56.

Edwards, S., Kirchin, S., and Huxtable, R., 2004. Research ethics committees and paternalism. Journal of Medical Ethics 30: 88–91.

Egilman, D., Wallace, W., Stubbs, C., and Mora-Corrasco, F., 1998a. Ethical Aerobics: ACHREs Fleight from Responsibility. Accountability in Research 6: 15–62.

Egilman, D., Wallace, W., Stubbs, C., and Mora-Corrasco, F., 1998b. A Little Too Much of the Buchenwald Touch? Military Radiation Research at the University of Cincinnati, 1960–1972. Accountability in Research 6: 63–102.

Ehringhause, S.H., Weissman, J.S., Sears, J.L., Goold, S.D., Feibelmann, S., and Campbell, E., G., 2008. Responses of Medical Schools to Institutional Conflicts of Interest. Journal of the American Medical Association 299: 665–671.

Eisen, H.N., 1991. Origins of MIT Inquiry. Nature 351: 343–344.

Eisenberg, R., 1995. Patenting Organisms. In Encyclopedia of Bioethics, rev. ed. Simon and Schuster, New York, pp. 1911–1914.

Elliott, K.C., 2008. Scientific Judgment and the Limits of Conflict-of-Interest Policies. Accountability in Research 15: 1–29.

Emanuel, E., and Miller, F., 2003. The Ethics of Placebo-Controlled Trials—a Middle Ground. New England Journal of Medicine 345: 915–919.

Emanuel, E., Wendler, D., and Grady, C., 2000. What Makes Clinical Research Ethical? Journal of the American Medical Association 283: 2701–2711.

Engler, R.L., Covell, J.W., Friedman, P.J., Kitcher, P.S., and Peters, R.M., 1987. Misrepresentation and Responsibility in Medical Research. New England Journal of Medicine 317: 1383–1389.

Enriquez, J., 1998. Genomics and the World's Economy. Science 281: 925–926.

Etzioni, A., 2004. The Common Good. Polity Press, New York.

Etzkowitz, H., Kemelgor, C., Neuschatz, M., Uzzi, B., and Alonzo, J., 1994. The Paradox of Critical Mass for Women in Science. Science 266: 51–54.

Evans, D., 2005. A series of articles at Bloomberg.com: http://bloomberg.com/apps/news?pid=specialreport&sid=aF3ZWezawt5k&refer=news#; http://bloomberg.com/apps/news?pid=specialreport&sid=amgWGeDrNJtg&refer=news#. Accessed December 2, 2005.

Faden, R.R., and Beauchamp, T.L., 1986. A History of Informed Consent. Oxford University Press, New York.

Feinberg, J., 1973. Social Philosophy. Prentice-Hall, Englewood Cliffs, NJ.

Ferber, D., 2002. HHS Intervenes in Choice of Study Section Members. Science 298: 1323–1323.

Fiedler, F., 1967. A Theory of Leadership Effectiveness. McGraw-Hill, New York.

Fields, K.L., and Price, A.R., 1993. Problems in Research Integrity Arising from Misconceptions about the Ownership of Research. Academic Medicine Supplement 3: S60–S64.

Fitzgerald, D., Marotte, C., Verdier, R., Johnson, W., Jr., and Pape, J., 2002. Comprehension during Informed Consent in a Less-Developed Country. Lancet 360: 1301–1302.

Flanagin, A., et al., 1998. Prevalence of Articles with Honorary and Ghost Authors in Peer-Reviewed Medical Journals. Journal of the American Medical Association 280: 222–224.

Fleischmann, M., 2000. Reflections on the Sociology of Science and Social Responsibility in Science, in Relationship to Cold Fusion. Accountability in Research 8: 19–54.

Fletcher, R., and Fletcher, S., 1997. Evidence for the Effectiveness of Peer Review. Science and Engineering Ethics 3(1): 35–50.

Food and Drug Administration, 2002. Best Pharmaceuticals for Children Act. http://www.fda.gov/opacom/laws/pharmkids/pharmkids.html. Accessed December 19, 2007.

Food and Drug Administration, 2003. Drug Research and Children. http://www.fda.gov/fdac/features/2003/103_drugs.html. Accessed November 9, 2007.

Food and Drug Administration, 2007a. The Clinical Impact of Adverse Event Reporting—Postmarketing Reporting of Adverse Events. http://www.fda.gov/medwatch/articles/medcont/postrep.htm. Accessed August 15, 2007.

Food and Drug Administration, 2007b. Draft Guidance for the Public, FDA Advisory Committee Members, and FDA Staff on Procedures for Determining Conflict of Interest and Eligibility for Participation in FDA Advisory Committees. http://www.fda.gov/oc/advisory/waiver/COIguidedft.html. Accessed October 29, 2007.

Forkenflick, D., 1996. Academic Career Unravels. Baltimore Sun, July 24, 1A, 7A.

Foster, F., and Shook, R., 1993. Patents, Copyrights, and Trademarks, 2nd ed. Wiley, New York.

Fox, J.L., 1981. Theory Explaining Cancer Partly Retracted. Chemical and Engineering News 59: 35–36.

Fox, R., and DeMarco, J., 1990. Moral Reasoning. Holt, Rinehart, and Winston, Chicago.

Fraedrich, F., 1991. Business Ethics. Houghton Mifflin, Boston.

Frankena, W., 1973. Ethics, 2nd ed. Prentice-Hall, Englewood Cliffs, NJ.

Freedman, B., 1987. Equipoise and the Ethics of Clinical Research. New England Journal of Medicine 317: 141–145.

Freedom of Information Act, 1966, 2002. 5 U.S. Code, 552.

Frey, R., 1980. Interests and Rights: The Case against Animals. Oxford University Press, New York.

Frey, R., 1994. The Ethics of the Search for Benefits: Experimentation in Medicine. In Principles of Health Care Ethics, R. Gillon (ed.). Wiley, Chichester, UK, pp. 1067–1075.

Friedberg, M., Saffran, B., Stinson, T., Nelson, W., and Bennett, C., 1999. Evaluation of Conflict of Interest in New Drugs Used in Oncology. Journal of the American Medical Association 282: 1453–1457.

Friedly, J., 1996a. After 9 Years, a Tangled Case Lurches toward a Close. Science 272: 947–948.

Friedly, J., 1996b. How Congressional Pressure Shaped the Baltimore Case. Science 273: 873–875.

Friedman, P.J., Dunwoody, S., and Rogers, C.L., 1990. Correcting the Literature following Fraudulent Publication. Journal of the American Medical Association 263: 1416–1419.

Friedman, S., et al. (eds.), 1999. Communicating Uncertainty: Media Coverage of New and Controversial Science. Lawrence Erlbaum, Mahwah, NJ.

Friedman Ross, L., 2006. Children in Research. Oxford University Press, New York.

Fukuyama, F., 2003. Our Posthuman Future. New York: Picador.

Funk, C.L., Barrett, K.A., and Macrina, F.L., 2007. Authorship and Publication: Evaluation of the Effect of Responsible Conduct of Research Instruction to Postdoctoral Trainees. Accountability in Research 14(4): 269–305.

Funk Brothers Seed Co. v. Kalo Inoculant Co., 1948. U.S. Supreme Court Cases and Opinions 333: 127–132.

Garfield, E., 1987. The Anomie-Deviant Behavior Connection: The Theories of Durkheim, Merton, and Srole. Current Comments, September 28, no. 39: 3–12.

*Garfield, E., 1990. The Impact of Fraudulent Research on the Scientific Literature— the Stephen Breuning Case. Journal of the American Medical Association 263: 1424–1426.

Garfunkel, J.M., Ulshen, M.H., Hamrick, H.J., and Lawson, E.E., 1994. Effect of Institutional Prestige on Reviewer's Recommendations and Editorial Decisions. Journal of the American Medical Association 272: 137–138.

Garner, B.A. (ed.), 1999. Black's Law Dictionary, 7th ed. West Group, St. Paul, MN.

Geison, G.L., 1978. Pasteur's Work on Rabies: Reexamining the Ethical Issues. Hastings Center Report 8: 26–33.

Geison, G.L., 1995. The Private Science of Louis Pasteur. Princeton University Press, Princeton, NJ.

Gelsinger, P., and Shamoo, A.E., 2008. Eight Years after Jesse's Death, Are Human Research Subjects Any Safer? Hastings Center Report 38(2): 25–27.

Gert, B., 2007. Common Morality: Deciding What to Do. New York: Oxford University Press, New York.

Gibbard, A., 1990. Wise Choices, Apt Feelings. Harvard University Press, Cambridge, MA.

Gibbs, W., 1996. The Price of Silence: Does Profit-Minded Secrecy Retard Scientific Progress? Scientific American 275(5): 15–16.

Giere, R., 1991. Understanding Scientific Reasoning, 3rd ed. Holt, Rinehart and Winston, Chicago.

Gilbert, P.L., Harris, M.J., McAdams, L.A., and Jeste, D.V., 1995. Neuroleptic Withdrawal in Schizophrenic Patients: A Review of the Literature. Archives of General Psychiatry 52: 173–188.

Glantz, L., 1998. Research with Children. American Journal of Law and Medicine 24(2/3): 213–244.

Glantz, S.A., and Bero, L.A., 1994. Inappropriate and Appropriate Selection of "Peers" in Grant Review. Journal of the American Medical Association 272: 114–116.

Glick, J.L., 1992. Scientific Data Audit—a Key Management Tool. Accountability in Research 2: 153–168.

Glick, J.L., 1993. Perceptions Concerning Research Integrity and the Practice of Data Audit in the Biotechnology Industry. Accountability in Research 3: 187–195.

Glick, J.L., and Shamoo, A.E., 1991. Auditing Biochemical Research Data: A Case Study. Accountability in Research 1: 223–243.

Glick, J.L., and Shamoo, A.E., 1993. A Call For the Development of "Good Research Practices" (GRP) Guidelines. Accountability in Research 2: 231–235.

Glick, J.C.L., and Shamoo, A.E., 1994. Results of a Survey on Research Practices, Completed by Attendees at the Third Conference on Research Policies and Quality Assurance. Accountability in Research 3: 275–280.

Godlee, F., 2000. The Ethics of Peer Review. In Ethical Issues in Biomedical Publication, A. Jones and F. McLellan (eds.). Johns Hopkins University Press, Baltimore, MD, pp. 59–84.

Godlee, F., 2002. Making Reviewers Visible—Openness, Accountability, and Credit. Journal of the American Medical Association 287: 2762–2765.

Goldstein, T., 1980. Dawn of Modern Science. Houghton Mifflin, Boston.

*Goode, S., 1993. Trying to Declaw the Campus Copycats. Insight Magazine, April 18, 10–29.

Goodman, S., 1994. Manuscript Quality before and after Peer Review and Editing at the Annals of Internal Medicine. Annals of Internal Medicine 121: 11–21.

Goodstein, D.L., 1994. Whatever Happened to Cold Fusion? American Scholar 63: 527.

Gøtzsche, P., Hróbjartsson, A., Johansen, H., Haahr, M., Altman, D., and Chan, A., 2007. Ghost Authorship in Industry-Initiated Randomized Trials. PLoS Medicine 4(1): e19.

Grady, C., 2005. Payment of Clinical Research Subjects. Journal of Clinical Investigation 115: 1681–1687.

Graham, G., 1999. The Internet: A Philosophical, Inquiry. Routledge, London.

Gray, S., Hlubocky, F., Ratain, M., and Daugherty, C., 2007. Attitudes toward Research Participation and Investigator Conflicts of Interest among Advanced Cancer Patients Participating in Early Phase Clinical Trials. Journal of Clinical Oncology 25: 3488–3494.

Green, R., 2001. The Human Embryo Research Debates. Oxford University Press, New York.

Green, R., 2007. Can We Develop Ethically Universal Embryonic Stem-Cell Lines? Nature Reviews Genetics 8: 480–485.

Greenberg, D., 2001. Science, Money, and Politics. University of Chicago Press, Chicago.

Griffin, P., 1992. Animal Minds. University of Chicago Press, Chicago.

Grimes v. Kennedy Krieger Institute, Inc., 2001. Atlantic Reporter, 2nd Series 782: 807 (Maryland).

Grimlund, R.A., and Doucet, M.S., 1992. Statistical Auditing Techniques for Research Data: Financial Auditing Parallels and New Requirements. Accountability in Research 2: 25–53.

Grinnell, F., 1992. The Scientific Attitude, 2nd ed. Guilford Press, New York.

Grisso, T., and Appelbaum, P.S., 1998. Assessing Competence to Consent to Treatment: A Guide for Physicians and Other Health Professionals. Oxford University Press, New York.

Gross, C., Anderson, G.F., and Powe, N.R., 1999. The Relation between Funding at the National Institutes of Health and the Burden of Disease. New England Journal of Medicine 340: 1881–1887.

Gruen, L., and Grabel, L., 2006. Scientific and Ethical Roadblocks to Human Embryonic Stem Cell Therapy. Stem Cells 24: 2162–2169.

Guston, D., 2000. Between Politics and Science. Cambridge University Press, Cambridge.

Guttman, D., 1998. Disclosure and Consent: Through the Cold War Prism. Accountability in Research 6: 1–14.

Haack, S., 2003. Defending Science within Reason. Prometheus Books, New York.

Haidt, J., 2007. The New Synthesis in Moral Psychology. Science 316: 998–1002.

Hall, M., and Rich, S., 2000a. Genetic Privacy Laws and Patients' Fear of Discrimination by Health Insurers: The View from Genetic Counselors. Journal of Law, Medicine, and Ethics 28: 245–257.

Hall, M., and Rich, S., 2000b. Patients' Fear of Genetic Discrimination by Health Insurers: The Impact of Legal Protections. Genetics in Medicine 2: 214–221.

Hambruger, P., 2005. The New Censorship: Institutional Review Boards. Law School, University of Chicago. http://www.law.uchicago.edu/academics/publiclaw/95-ph-censorship.pdf. Accessed March 15, 2007.

Hamilton, D.P., 1991. NIH Finds Fraud in Cell Paper. Science 25: 1552–1554.

Hammurabi, Code of, 1795–1750 b.c. http://www.wsu.edu/~dee/MESO/CODE. HTM. Accessed April 7, 2006.

Harman, G., 1977. The Nature of Morality. Oxford University Press, New York.

Harris, J., 2007. Enhancing Evolution: The Ethical Case for Making Better People. Princeton University Press, Princeton, NJ.

Harvard University, 2004. Policy on Conflict of Interest and Commitment. http://www.hms.harvard.edu/integrity/conf.html. Accessed May 8, 2008.

Harvey, A., Zhang, H., Nixon, J., and Brown, C.J., 2007. Comparison of Data Extraction from Standardized versus Traditional Narrative Operative Reports for Database-Related Research and Quality Control. Surgery 141: 708–714.

Harvey, W., 1628 [1993]. On the Motion of the Heart and Blood in Animals. Prometheus Books, Buffalo, NY.

Hauser, M., 2006. Moral Minds: How Nature Designed Our Universal Sense of Right and Wrong. Harper Collins, New York.

Hellemans, A., 2003. Drug Trials without Drugs? Science 300: 1212–1213.

Hempel, C., 1965. Philosophy of Natural Science. Prentice-Hall, Englewood Cliffs, NJ.

Herrnstein, R., and Murray, C., 1994. The Bell Curve: Intelligence and Class Structure in American Life. Free Press, New York.

Hilts, P., 1991. The Doctor's World; U.S. and France Finally Agree in Long Feud on AIDS Virus. New York Times, May 7, A1.

Hilts, P., 1992. American Co-discover of HIV Is Investigated Anew. New York Times, March 2, A1.

Hilts, P., 1994. Philip Morris Blocked '83 Paper Showing Tobacco Is Addictive, Panel Finds. New York Times, April 1, A1.

Hinman, L., 2002. Ethics: A Pluralistic Approach to Moral Theory. Wadsworth, Belmont, CA.

Hixson, J., 1976. The Patchwork Mouse. Doubleday, Garden City, NJ.

Hobbes, T., 1651 [1962]. Leviathan. Macmillan, New York.

Holden, C., 2000. NSF Searches for Right Way to Help Women. Science 289: 379–381.

Holden, C., 2007. Prominent Researchers Join the Attack on Stem Cell Patents. Science 317: 187–187.

Hollander, R., Johnson, D., Beckwith, J., and Fader, B., 1995. Why Teach Ethics in Science and Engineering? Science and Engineering Ethics 1: 83–87.

Holton, G., 1978. Subelectrons, Presuppositions, and the Millikan-Ehrenhaft Dispute. Historical Studies in the Physical Sciences 9: 166–224.

Holtug, N., 1997. Altering Humans—the Case for and against Human Gene Therapy. Cambridge Quarterly of Healthcare Ethics 6: 157–174.

Hoppe, S.K., 1996. Institutional Review Boards and Research on Individuals with Mental Disorders. Accountability in Research 4: 187–196.

*Hornblum, A.M., 1998. Acres of Skin. Routledge, New York.

Hudson, K., Holohan, M., and Collins, F., 2008. Keeping Pace with the Times—the Genetic Non-Discrimination Act of 2008. New England Journal of Medicine 358: 2661–2663.

Hull, D., 1988. Science as a Process. University of Chicago Press, Chicago.

Hulley, S.B., Cummings, S.R., Browner, W.S., Grady, D., Hearst, N., and Newman, T.B., 2001. Designing Clinical Research, 2nd ed. Lippincott Williams and Wilkins, Philadelphia, PA.

Human Genome Project, 2003. History of the Human Genome Project. http://www.ornl.gov/TechResources/Human_Genome/project/hgp.html. Accessed November 7, 2007.

Human Genome Project Information, 2007. Gene Testing. http://www.ornl.gov/sci/techresources/Human_Genome/medicine/genetest.shtml#whatis. Accessed November 3, 2007.

Hurt, R., and Robertson, C., 1998. Prying Open the Door to the Tobacco Industry's Secrets about Nicotine. Journal of the American Medical Association 280: 1173–1181.

Huth, E., 2000. Repetitive and Divided Publication. In Ethical Issues in Biomedical Publication, A. Jones and F. McLellan (eds.). Johns Hopkins University Press, Baltimore, MD, pp. 112–136.

Hwang, W., Roh, S., Lee, B., Kang, S., Kwon, D., Kim, S., et al., 2005. Patient-Specific Embryonic Stem Cells Derived from Human SCNT Blastocysts [Retracted]. Science 308: 1777–1783.

Hwang, W., Ryu, Y., Park, J., Park, E., Lee, E., Koo, J., et al., 2004. Evidence of a Pluripotent Human Embryonic Stem Cell Line Derived from a Cloned Blastocyst [Retracted]. Science 303: 1669–1674.

Hyder, A.A., and Wali, S.A., 2006. Informed Consent and Collaborative Research: Perspectives from the Developing World. Developing World Bioethics 6: 33–40.

Iltis, A., 2000. Bioethics as Methodological Case Resolution: Specification, Specified Principlism and Casuistry. Journal of Medicine and Philosophy 25: 271–284.

Imanishi-Kari, T., 1991. OSI's Conclusions Wrong. Nature 351: 344–345.

Inadomi, J., 2006. Implementing a Research Project: The Nitty Gritty. Gastrointestinal Endoscopy 64:S7–S10.

*Inglefinger, F., 1972. Informed (but Uneducated) Consent. New England Journal of Medicine 287: 465–466.

Institute of Medicine, 2002a. Integrity in Scientific Research: Creating Environment That Promote Responsible Conduct. National Academy Press, Washington, DC.

Institute of Medicine, 2002b. Responsible Research: A System Approach to Protecting Research Participants. National Academy Press, Washington, DC.

Institute of Medicine, 2004. Ethical Conduct of Clinical Research Involving Children. National Academy Press, Washington, DC.

International Committee of Medical Journal Editors, 2007. Uniform Requirements for Manuscripts Submitted to Biomedical Journals. http://www.icmje.org/ Accessed November 24, 2007.

International Conference on Harmonization, 1996. Guidance for Industry: Good Clinical Practice. http://www.fda.gov/cder/guidance/959fnl.pdf. Accessed November 12, 2006.

International Society for Stem Cell Research, 2007. The ISSCR Guidelines for Human Embryonic Stem Cell Research. Science 315: 603–604.

Irving, D.N., and Shamoo, A.E., 1993. Which Ethics for Science and Public Policy. Accountability in Research 3: 77–100.

Isaacson, W., 2007. Einstein: His Life and Universe. Simon and Schuster, New York.

ISLAT Working Group, 1998. ART into Science: Regulation of Fertility Techniques. Science 281: 651–652.

Jaffe, A., 1996. Trends and Patterns in Research and Development Expenditures the United States. Proceedings of the National Academy of Sciences of the USA 93: 12658–12663.

Jasanoff, S., 1990. The Fifth Branch: Science Advisors as Policy Makers. Harvard University Press, Cambridge, MA.

Jasanoff, S., 1995. Science at the Bar: Law, Science, and Technology in America. Harvard University Press, Cambridge, MA.

Johnsen, A., and Toulmin, S., 1988. The Abuse of Casuistry: A History of Moral Reasoning. University of California Press, Berkeley.

Johnson, H., 1993. The Life of a Black Scientist. Scientific American 268(1): 160.

Jonas, H., 1980. Philosophical Essays. University of Chicago Press, Chicago.

*Jonas, H., 1992. Philosophical Reflections on Experimenting with Human Subjects. Reprinted in R. Munson, Intervention and Reflection, 4th ed. Wadsworth, Belmont, CA, pp. 362–371.

Jones, A., 2000. Changing Traditions of Authorship. In Ethical Issues in Biomedical Publication, A. Jones and F. McLellan (eds.). Johns Hopkins University Press, Baltimore, MD, pp. 3–29.

Jones, J.H., 1981. Bad Blood. Free Press, London.

Journal of Cell Biology, 2007. Image Manipulation. http://www.jcb.org/misc/ifora. shtml#image_aquisition. Accessed November 26, 2007.

Journal of the American Medical Association, 2007. Instructions for Authors. http:// jama.ama-assn.org/misc/ifora.dtl#AuthorshipCriteriaandContributionsandAuth orshipForm. Accessed October 8, 2007.

Justice, A., Cho, M.K., Winker, M.A., Berlin, J.A., and Rennie, D., 1998. Does Masking the Author Identity Improve Peer Review Quality? Journal of the American Medical Association 280: 240–242.

Kahn, J., Mastroianni, A., and Sugarman, J., 1998. Beyond Consent: Seeking Justice in Research. Oxford University Press, New York.

Kaiser, J., 2005a. NIH Chief Clamps Down on Consulting and Stock Ownership. Science 307: 824–825.

Kaiser, J., 2005b. NIH Wants Public Access to Papers "as Soon as Possible." Science 307: 825.

Kaiser, J., 2006. Court Decides Tissue Samples Belong to University, Not Patients. Science 312: 346–346.

Kant, I., 1753 [1981]. Grounding for the Metaphysics of Morals, J. Ellington (trans.). Hackett, Indianapolis, IN.

*Kass, L., 1985. Toward a More Natural Science. Free Press, New York.

Kass, L., 2004. Life, Liberty and the Defense of Dignity. American Enterprise Institute, Washington, DC.

Katz, D., Caplan, A., and Merz, J., 2003. All Gifts Large and Small: Toward an Understanding of the Ethics of Pharmaceutical Industry Gift-Giving. American Journal of Bioethics 3(3): 39–46.

Katz, J., 1972. Experimentation with Human Beings. Russell Sage Foundation, New York.

*Katz, J., 1993. Human Experimentation and Human Rights. Saint Louis University Law Journal 38: 7–54.

Katz, J., 1996. Ethics in Neurobiological Research with Human Subjects—Final Reflections. Accountability in Research 4: 277–283.

Kayton, I., 1995. Patent Practice, Vol. 1. Patent Resources Institute, Charlottesville, VA.

Kennedy, D., 2001. Accepted Community Standards. Science 291: 789.

Kennedy, D., 2006. Responding to Fraud. Science 316: 1353.

*Kevles, D., 1995. In the Name of Eugenics. Harvard University Press, Cambridge, MA.

Kevles, D.J., 1996. The Assault on David Baltimore. New Yorker, May 27, 94–109.

Kevles, D., 1998. The Baltimore Case: A Trial of Politics, Science, and Character. Norton, New York.

King, N., 1995. Experimental Treatment: Oxymoron or Aspiration? Hastings Center Report 25(4): 6–15.

Kinitisch, E., 2005. Researcher Faces Prison for Fraud in NIH Grant Applications and Papers. Science 307: 1851.

Kirk, R., 1995. Experimental Design, 3rd ed. Brooks/Cole, New York.

Kitcher, P., 1993. The Advancement of Science. Oxford University Press, New York.

Kitcher, P., 1997. The Lives to Come: The Genetic Revolution and Human Possibilities. Simon and Schuster, New York.

Knoll, E., 1990. The Communities of Scientists and Journal Peer Review. Journal of the American Medical Association 263: 1330–1332.

Kodish, E., 2005. Editor, Ethics and Research with Children—a Case-Based Approach. Oxford University Press, New York.

Koenig, R., 1997. Panel Calls Falsification in German Case "Unprecedented." Science 277: 894.

Koenig, R., 1999. European Researchers Grapple with Animal Rights. Science 284: 1604–1606.

Koenig, R., 2001. Scientific Misconduct: Wellcome Rules Widen the Net. Science 293: 1411–1412.

Koestenbaum, P., 1991. Leadership—the Inner Side of Greatness. Jossey-Bass, Oxford, UK.

*Kong, D., and Whitaker, R., 1998. Doing Harm: Research on the Mentally Ill. Boston Globe, November 15, A1.

Koocher, G.P., 2005. Behavioral Research with Children: The Fenfluramine Challenge. In Ethics and Research with Children—a Case-Based Approach, E. Kodish (ed.). Oxford University Press, New York, chap. 10, pp. 179–193.

Kopelman, L., 1986. Consent and Randomized Clinical Trials: Are There Moral or Design Problems? Journal of Medicine and Philosophy 11(4): 317–345.

Kopelman, L., 1999. Values and Virtues: How Should They Be Taught? Academic Medicine 74(12): 1307–1310.

Kopelman, L., 2000a. Children as Research Subjects: A Dilemma. Journal of Medicine and Philosophy 25: 745–764.

Kopelman, L., 2000b. Moral Problems in Assessing Research Risk. IRB: A Review of Human Subjects Research 22(5): 7–10.

Kopelman, L.M., 2002. Group Benefit and Protection of Pediatric Research: Grimes v. Kennedy Krieger and the Lead Abatement Study. Accountability in Research 9: 177–192.

Koppelman-White, E., 2006. Research Misconduct and the Scientific Process: Continuing Quality Improvement. Accountability in Research 13: 325–346.

Korsgaard, C., 1996. The Sources of Normativity. Cambridge University Press, Cambridge.

Krimsky, S., 1996. Financial Interest of Authors in Scientific Journals: A Pilot Study of 14 Publications. Science and Engineering Ethics 2: 1–13.

Krimsky, S., 2003. Science in the Private Interest—Has the Lure of Profits Corrupted Biomedical Research? Rowman and Littlefield, Lanham, MD.

Krimsky, S., 2007. Defining Scientific Misconduct: When Conflict-of-Interest Is a Factor in Scientific Misconduct. Medicine and Law 26: 447–463.

Krimsky, S., and Rothenberg, L., 2001. Conflict of Interest Policies in Science and Medical Journals: Editorial Practices and Author Disclosures. Science and Engineering Ethics 7: 205–218.

Krimsky, S., Rothenberg, L.S., Stott, P., and Kyle, G., 1996. Financial Interests of Authors in Scientific Publications. Science and Engineering Ethics 2(4): 396–410.

Krimsky, S., and Simocelli, T., 2007. Testing Pesticides in Humans—of Mice and Men Divided by Ten. Journal of the American Medical Association 297: 2405–2407.

Kronic, D.A., 1990. Peer Review in 18th-Century Scientific Journalism. Journal of the American Medical Association 263: 1321–1322.

Kuflik, A., 1989. Moral Foundations of Intellectual Property Rights. In Owning Scientific and Technical Information, V. Weil and J. Snapper (eds.). Rutgers University Press, Brunswick, NJ, pp. 29–39.

Kuhn, T., 1970. The Structure of Scientific Revolutions, 2nd ed. University of Chicago Press, Chicago.

Kuznik, F., 1991. Fraud Buster. Washington Post Magazine, April 14, 22–26, 31–33.

Laband, D.N., and Piette, M.J., 1994. A Citation Analysis of the Impact of Blinded Peer Review. Journal of the American Medical Association 272: 147–149.

*LaFollette, H., and Shanks, S., 1996. Brute Science. Routledge, New York.

*LaFollette, M.C., 1992. Stealing into Print—Fraud, Plagiarism, and Misconduct in Scientific Publishing. University of California Press, Berkeley.

LaFollette, M., 1994a. Measuring Equity—the U.S. General Accounting Office Study of Peer Review. Science Communication 6: 211–220.

LaFollette, M.C., 1994b. The Pathology of Research Fraud: The History and Politics of the US Experience. Journal of Internal Medicine 235: 129–135.

LaFollette, M.C., 1994c. Research Misconduct. Society 31(3): 6–10.

LaFollette, M.C., 2000. The Evaluation of the "Scientific Misconduct" Issues: An Historical Overview. Proceedings for the Society for Experimental Biology and Medicine 224: 211–215.

LaFuze, W.L., and Mims, P.E., 1993. Ownership of Laboratory Discoveries and Work Product. In Understanding Biotechnology Law, G.R. Peterson (ed.). Dekker, New York, pp. 203–238.

Lasagna, L., 1992. Some Ethical Problems in Clinical Investigation. In Intervention and Reflection, 4th ed., R. Munson (ed.). Wadsworth, Belmont, CA, pp. 356–362.

Lawler, A., 2000. Silent No Longer: A "Model Minority" Mobilizes. Science 290: 1072–1077.

Lawrence, P., 2003. The Politics of Publication. Nature 422: 259–261.

Lebacqz, K., 1999. Geron Ethics Advisory Board. Hastings Center Report 29: 31–36.

Lederer, S., 1995. Subjected to Science: Human Experimentation in America before the Second World War. Johns Hopkins University Press, Baltimore, MD.

Lentz, E.T., 1993. Inventorship in the Research Laboratory. In Understanding Biotechnology Law, G.R. Peterson (ed.). Dekker, New York, pp. 187–202.

Levine, C., Faden, R., Grady, C., Hammerschmidt, D., Eckemweiler, L., and Sugarman, J., 2004. The Limitations of "Vulnerability" as a Protection for Human Research Participants. American Journal of Bioethics 4(3): 44–49.

Levine, R.J., 1988. Ethics and Regulation of Clinical Research, 2nd ed. Yale University Press, New Haven, CT.

Liddell, M., Lovestone, S., and Owen, M., 2001. Genetic Risk of Alzheimer's Disease: Advising Relatives. British Journal of Psychiatry 178: 7–11.

Lo, B., Dornbrand, L., and Dubler, N., 2005. HIPAA and Patient Care: The Role for Professional Judgment. Journal of the American Medical Association 293: 1766–1771.

Lock, S., 1991. A Difficult Balance: Editorial Peer Review in Medicine. BMJ, London.

Lock, S., 1993. Research Misconduct: A Resume of Recent Events. In Fraud and Misconduct in Medical Research, S. Lock and F. Wells (eds.). BMJ, London, pp. 5–24.

Locke, J., 1764 [1980]. Second Treatise of Government, C. Macpherson (ed.). Hackett, Indianapolis, IN.

Loeb, S.E., and Shamoo, A.E., 1989. Data Audit: Its Place in Auditing. Accountability in Research 1: 23–32.

London, A., 2001. Equipoise and International Human-Subjects Research. Bioethics 15: 312–332.

London, A., 2005. Justice and the Human Development Approach to International Research. Hastings Center Report 35(1): 24–37.

Longino, H., 1990. Science as Social Knowledge. Princeton University Press, Princeton, NJ.

Lowrance, W.W., and Collins, F.S., 2007. Identifiability in Genomic Research. Science 317: 600–602.

Lucky, R., 2000. The Quickening Pace of Science Communication. Science 289: 259–264.

Lurie, P., Almeida, C.M., Stine, N., Stine, A.R., and Wolfe, S.M., 2007. Financial Conflict of Interest Disclosure and Voting Patterns at Food and Drug

Administration Drug Advisory Committee Meetings. Journal of the American Medical Association 295: 1921–1928.

Lurie, P., and Wolfe, S., 1997. Unethical Trials of Interventions to Reduce Perinatal Transmission of the Human Immunodeficiency Virus in Developing Countries. New England Journal of Medicine 337: 853–856.

Macilwain, C., 1999. Scientists Fight for the Right to Withhold Data. Nature 397: 459.

Macklin, R., 1994. Splitting Embryos on the Slippery Slope. Ethics and Public Policy. Kennedy Institute of Ethics Journal 4: 209–225.

Macklin, R., 1999a. Against Relativism. Oxford University Press, New York.

Macklin, R., 1999b. International Research: Ethical Imperialism or Ethical Pluralism? Accountability in Research 7: 59–83.

Macklin, R., 2003. Bioethics, Vulnerability, and Protection. Bioethics 17: 472–486.

MacQueen, K., and Buehler, J., 2004. Ethics, Practice, and Research in Public Health. American Journal of Public Health 94: 928–931.

Macrina, F. (ed.), 2005. Scientific Integrity: Textbook and Cases in Responsible Conduct of Research, 3rd ed. American Society for Microbiology Press, Washington, DC.

Madey v. Duke University, 2002. Federal Reporter, Third Series 307: 1351 (Federal Circuit).

Malakoff, D., 2000a. Does Science Drive the Productivity Train? Science 289: 1274–1276.

Malakoff, D., 2000b. Researchers Fight Plan to Regulate Mice, Birds. Science 290: 23.

Malakoff, D., and Marshall, E., 1999. NIH Wins Big as Congress Lumps Together Eight Bills. Science 282: 598.

Mangan, K.S., 2000. Harvard Weighs a Change in Conflict-of-Interest Rules. Chronicle of Higher Education, May 9, A47–A48.

Mann, H., and Shamoo, A.E., 2006. Introduction to Special Issue of Accountability in Research on the Review and Approval of Biomedical Research Proposals: A Call for a Centralized National Human Research Protections System. Accountability in Research 13: 1–10.

Manning, K., 1998. Science and Opportunity. Science 282: 1037–1038.

Marshall, E., 1983. The Murky World of Toxicity Testing. Science 220: 1130–1132.

Marshall, E., 1997. NIH Plans Peer Review Overhaul. Science 276: 888–889.

Marshall, E., 1998. Embargoes: Good, Bad, or "Necessary Evil"? Science 282: 860–865.

Marshall, E., 1999a. E-Biomed Morphs to E-Biosci, Focus Shift to Reviewed Papers. Science 285: 810–811.

Marshall, E., 1999b. A High-Stakes Gamble on Genome Sequencing. Science 284: 1906–1909.

Marshall, E., 1999c. NIH Weighs Bold Plan for Online Preprint Publishing. Science 283: 1610–1611.

Marshall, E., 1999d. Two Former Grad Students Sue over Alleged Misuse of Ideas. Science 284: 562–563.

Marshall, E., 2000a. Gene Therapy on Trial. Science 288: 951–957.

Marshall, E., 2000b. Patent Suit Pits Postdoc against Former Mentor. Science 287: 2399–2400.

Marshall, E., 2000c. The Rise of the Mouse, Biomedicine's Model Animal. Science 288: 248–257.

Marshall, E., 2001. Sharing the Glory, Not the Credit. Science 291: 1189–1193.

Martinson, B.C., Anderson, M.S., and de Vries, R., 2005. Scientists Behaving Badly. Nature 435: 737–738.

May, R., 1998. The Scientific Investments of Nations. Science 281: 49–51.

Mayr, E., 1982. The Growth of Biological Thought. Harvard University Press, Cambridge, MA.

Mbidde, E., 1998. Bioethics and Local Circumstances. Science 279: 155.

McCallin, A.M., 2006. Interdisciplinary Researching: Exploring the Opportunities and Risks of Working Together. Nursing and Health Sciences 8: 88–94.

McCrary, S.V., Anderson, C.B., Jakovljevic, J., Khan, T., McCullough, L.B., Wray, N.P., and Brody, B.A., 2000. A National Survey of Policies on Disclosure of Conflicts of Interest in Biomedical Research. New England Journal of Medicine 343: 1621–1626.

McGee, G., 1997. The Perfect Baby. Rowman and Littlefield, Lanham, MD.

McGuire, A., and Gibbs, R., 2006. No Longer De-identified. Science 312: 370–371.

McKeon, R. (ed.), 1947. Introduction to Aristotle. Modern Library, New York.

McLean, B., and Elkind, P., 2004. The Smartest Guys in the Room: The Amazing Rise and Scandalous Fall of Enron. Portfolio Trade, New York.

McLellin, F., 1995. Authorship in Biomedical Publications: How Many People Can Wield One Pen? American Medical Writers Association 10: 11.

McNutt, R., Evans, A.T., Fletcher, R.H., and Fletcher, S.W., 1990. The Effects of Blinding on the Quality of Peer Review: A Randomized Trial. Journal of the American Medical Association 263: 1371–1376.

Meadows, J., 1992. The Great Scientists. Oxford University Press, New York.

Menikoff, J., 2006. What the Doctor Didn't Say: The Hidden Truth behind Medical Experimentation. Oxford University Press, New York.

Merritt, M., and Grady, C., 2006. Reciprocity and Post-trial Access for Participants in Antiretroviral Therapy Trials. AIDS 20: 1791–1794.

Merton, R., 1973. The Sociology of Science. University of Chicago Press, Chicago.

Mervis, J., 1999. Efforts to Boost Diversity Face Persistent Problems. Science 284: 1757–1758.

Mervis, J., 2002. Science with an Agenda: NSF Expands Centers Program. Science 297: 506–507.

Milgram, S., 1974. Obedience to Authority. Harper and Row, New York.

Mill, J., 1861 [1979]. Utilitarianism, G. Sher (ed.). Hackett, Indianapolis, IN.

Miller, A., and Davis, M., 2000. Intellectual Property. West, St. Paul, MN.

Miller, F., and Brody, H., 2002. What Makes Placebo-Controlled Trials Unethical? American Journal of Bioethics 2(2): 3–9.

Miller, K., and Levine, J., 2005. Biology. Prentice-Hall, Upper Saddle River, NJ.

Miller, L., and Bloom, F., 1998. Publishing Controversial Research. Science 282: 1045.

Miller, S., and Selgelid, M., 2007. Ethical and Philosophical Consideration of the Dual-Use Dilemma in the Biological Sciences. Science and Engineering Ethics 13: 523–580.

Monson, N., 1991. How the Scientific Community Protects Its Black Sheep at the Expense of the Whistleblowers and the Public. Health Watch, July/August, 25–33.

Mooney, C., 2005. The Republican War on Science. Basic Books, New York.

Moore v. Regents of the University of California, 1990. 51 California Reports, Third Series 51: 120, 134–147.

Moreno, J., Caplan, A.L., and Wolpe, P.R., 1998. Updating Protections for Human Subjects Involved in Research. Journal of the American Medical Association 280: 1951–1958.

Moreno, J., and Hynes, R., 2005. Guidelines for Human Embryonic Stem Cell Research. Nature Biotechnology 23(7): 793–794.

Moreno, J.D., 1998. In J. Kahn, A. Mastroianni, and J. Sugarman, Beyond Consent: Seeking Justice in Research. Oxford University Press, New York, chap. 7.

Moreno, J.D., 2000. Undue Risk—Secret State Experiments on Humans. W.H. Freeman, New York.

Morreim, E., 2005. The Clinical Investigator as Fiduciary: Discarding a Misguided Idea. Journal of Law, Medicine, and Ethics 33: 586–598.

Moses, H., and Martin, B., 2001. Academic Relationships with Industry: A New Model for Biomedical Research. Journal of the American Medical Association 285: 933–935.

Mosimann, J.E., Wieseman, C.V., and Edelman, R.E., 1995. Data Fabrication: Can People Generate Random Digits? Accountability in Research 4: 31–56.

Moskop, J., 1998. A Moral Analysis of Military Medicine. Military Medicine 163: 76–79.

Müller-Hill, B., 1992. Eugenics: The Science and Religion of the Nazis. In When Medicine Went Mad, A. Caplan (ed.). Humana Press, Totowa, NJ, pp. 43–53.

Munson, R., 1992. Intervention and Reflection, 4th ed. Wadsworth, Belmont, CA.

Murphy, P., 1998. 80 Exemplary Ethics Statements. University of Notre Dame Press, Notre Dame, IN.

National Academy of Sciences, 1992. Responsible Science—Ensuring the Integrity of the Research Process, Vol. 1, Panel on Scientific Responsibility and the Conduct of Research. NAS, Washington, DC.

National Academy of Sciences, 1994. On Being a Scientist. NAS, Washington, DC.

*National Academy of Sciences, 1997. Advisor, Teacher, Role Model, Friend: On Being a Mentor to Students in Science and Engineering. NAS, Washington, DC.

National Academy of Sciences, 2002. Integrity in Scientific Research: Creating an Environment That Promotes Responsible Conduct. NAS, Washington, DC.

*National Bioethics Advisory Commission, 1997. Report on Cloning Human Beings. NBAC, Rockville, MD.

*National Bioethics Advisory Commission, 1998. Research Involving Persons with Mental Disorders That May Affect Decisionmaking Capacity, Vol. 1, Report and Recommendations. NBAC, Rockville, MD.

*National Bioethics Advisory Commission, 1999. Report on Ethical Issues in Human Stem Cell Research, Vol. 1. NBAC, Rockville, MD.

*National Bioethics Advisory Commission, 2001. Ethical and Policy Issues in Human in Research Involving Human Participants, Vol. 1. NBAC, Bethesda, MD.

National Commission for the Protection of Human Subjects of Biomedical and Behavioral Research, 1979. The Belmont Report. U.S. Department of Health, Education, and Welfare, Washington, DC.

National Human Genome Research Institute, 2007. Genetic Discrimination. http://www.genome.gov/10002077 Accessed November 3, 2007.

National Institutes of Health, 1998a. Consensus Report of the Working Group on "Molecular and Biochemical Markers of Alzheimer's Disease." The Ronald and Nancy Reagan Research Institute of the Alzheimer's Association and the National Institute on Aging Working Group. Neurobiology of Aging 19: 109–116.

National Institutes of Health, 1998b. Report of the National Institute of Health Working Group on Research Tools. http://www.nih.gov/news/researchtools/index.htm.

National Institutes of Health, 1999. Principles and Guidelines for Recipients of NIH Grants and Contracts on Obtaining and Disseminating Biomedical Research Resources. http://grants.nih.gov/grants/intell-property_64FR72090.pdf. Accessed October 15, 2007.

National Institutes of Health, 2000. Guide to Mentoring and Training. NIH, Bethesda, MD.

National Institutes of Health, 2003. NIH Data Sharing and Policy Implementation Guidance. http://grants.nih.gov/grants/policy/data_sharing/data_sharing_guidance.htm. Accessed October 15, 2007.

National Institutes of Health, 2005. NIH Roadmap on Translational Research. http://nihroadmap.nih.gov/overview.asp. Accessed October 12, 2007.

National Institutes of Health, 2007a. Award Data. http://grants.nih.gov/grants/award/success/Success_ByIC.cfm. Accessed December 26, 2007.

National Institutes of Health, 2007b. Enhancing Peer Review. http://enhancing-peer-review.nih.gov/meetings/073007-summary.html. Accessed October 15, 2007.

National Institutes of Health, 2007c. Genome-Wide Association Studies: Frequently Asked Questions. http://grants.nih.gov/grants/gwas/GWAS_faq.htm. Accessed April 30, 2008.

National Institutes of Health, 2007d. Guidelines for the Conduct of Research in the Intramural Research Program at NIH, 4th ed. http://www1.od.nih.gov/oir/sourcebook/ethic-conduct/Conduct%20Research%206-11-07.pdf. Accessed March 17, 2008.

National Institutes of Health, 2008. NIH Roadmap for Medical Research. http://nihroadmap.nih.gov/clinicalresearch/overview-translational.asp. Accessed July 21, 2008.

National Research Act, 1974. Title II, Public Law 93–348.

National Research Council, 1996. Guide for the Care and Use of Laboratory Animals. National Academy Press, Washington, DC.

National Research Council, 2003. Biotechnology Research in an Age of Terrorism: Confronting the Dual Use Dilemma. National Academy of Sciences, Washington, DC.

National Science Foundation, 1997. Scientists and Engineers Statistical Data System. Science Resources Studies Division, SESTAT, NSF, Washington, DC.

National Science Foundation, 2007. America Competes Act, Public Law 110–69, August 9. http://www.itpnm.com/whats-new-archives/p-1-110-69-america-competes-act.pdf. Accessed November 7, 2007.

Nature, 1999. Policy on Papers' Contributors. Nature 399: 393.

Nature, 2007. Authorship in Nature Journals. http://www.nature.com/authors/editorial_policies/authorship.html. Accessed October 8, 2007.

Newton, I., 1687 [1995]. The Principia, A. Motte (trans.). Prometheus Books, Amherst, NY.

New York Times, 2007. A Cleaner Food and Drug Agency [Editorial]. March 23.

New York Times News Service, 1997. University Settles Case over Theft of Data. Baltimore Sun, August 10, 9A.

New York Times News Service, 2004. Cold Fusion Is Offered a Second Look. Baltimore Sun, March 25, 7A.

Normile, D., Vogel, G., and Couzin, J., 2006. South Korean Team's Remaining Human Stem Cell Claim Demolished. Science 311: 156–157.

Nuffield Council, 2002. The Ethics of Patenting DNA. Nuffield Council, London.

Nundy, S., and Gulhati, C., 2005. A New Colonialism? Conducting Clinical Trials in India. New England Journal of Medicine 352: 1633–1635.

Nuremberg Code, 1949. http://ohsr.od.nih.gov/guidelines/nuremberg.html. Accessed November 12, 2007.

Office of Human Research Protections, 2007. International Compilation of Human Research Protections. http://www.hhs.gov/ohrp/international/HSPCompilation. pdf. Accessed November 12, 2007.

Office of Research Integrity, 1996. Annual Report, 1995. Department of Health and Human Services, Rockville, MD.

Office of Research Integrity, 1999. Scientific Misconduct Investigations, 1993–1997. ORI, Rockville, MD.

Office of Research Integrity, 2001. Guidelines for Assessing Possible Research Misconduct in Clinical Research and Clinical Trials. http://www.ehcca.com/presentations/ressummit2/1_03_1.pdf. Accessed July 24, 2008.

Office of Research Integrity, 2003. Survey of Research Measures Utilized in Biomedical Research Laboratories: Final Report—Executive Summary. http:// ori.dhhs.gov/documents/research/intergity_measures_final_report_11_07_03. pdf. Accessed October 23, 2007.

Office of Research Integrity, 2005a Case Summary—Eric Poehlman. http://ori.hhs. gov/misconduct/cases/poehlman_notice.shtml. Accessed October 23, 2007.

Office of Research Integrity, 2005b. Press Release—Dr. Eric T. Poehlman. http:// ori.hhs.gov/misconduct/cases/press_release_poehlman.shtml. Accessed October 23, 2007.

Office of Research Integrity, 2007a. About ORI. http://ori.dhhs.gov/about/index. shtml. Accessed October 23, 2007.

Office of Research Integrity, 2007b. Forensic Tools. http://ori.hhs.gov/tools/index. shtml. Accessed August 13, 2008.

Office of Research Integrity, 2008. About ORI—History, the Beginning. http://ori. hhs.gov/about/history.shtml. Accessed August 15, 2008.

Office of Science and Technology Policy, 1991. Common Rule. Federal Register 56: 28012–28018. http://www.hhs.gov/ohrp/references/frcomrul.pdf. Accessed August 13, 2008.

Office of Technology Assessment, 1986. Alternatives to Animal Use in Research, Testing, and Education. OTA, Washington, DC.

Office of Technology Assessment, 1990. New Developments in Biotechnology— Patenting Life. Dekker, New York.

O'Harrow, R., Jr., 2000. Harvard Won't Ease Funding Restrictions. Washington Post, May 26, A16.

Oliver, D.T., 1999. Animal Rights—the Inhumane Crusade. Capital Research Center, Merrill Press, Washington, DC.

Ollove, M., 2001. The Lessons of Lynchburg. Baltimore Sun, May 6, 7F.

Orlans, F., Beauchamp, T., Dresser, R., Morton, D., and Gluck, J., 1998. The Human Use of Animals. Oxford University Press, New York.

Osler, W., 1898. The Principles and Practice of Medicine, 3rd ed. Appleton, New York.

Oxman, A., et al., 1991. Agreement among Reviewers of Review Articles. Journal of Clinical Epidemiology 44: 91–98.

Parens, E. (ed.), 1999. Enhancing Human Traits: Ethical and Social Implications. Georgetown University Press, Washington, DC.

Parens, E., and Asch, A. (eds.), 2000. Prenatal Testing and Disability Rights. Georgetown University Press, Washington, DC.

Pascal, C., 1999. The History and Future of the Office of Research Integrity: Scientific Misconduct and Beyond. Science and Engineering Ethics 5: 183–198.

*Pascal, C., 2000. Scientific Misconduct and Research Integrity for the Bench Scientist. Proceedings of the Society for Experimental Biology and Medicine 224: 220–230.

Pasquerella, L., 2002. Confining Choices: Should Inmates' Participation in Research Be Limited? Theoretical Medicine and Bioethics 23: 519–536.

Pellegrino, E., 1992. Character and Ethical Conduct of Research. Accountability in Research 2: 1–2.

Pence, G., 1995. Animal Subjects. In Classic Cases in Medical Ethics. McGraw-Hill, New York, pp. 203–224.

Pence, G., 1996. Classic Cases in Medical Ethics, 2nd ed. McGraw-Hill, New York.

Pence, G., 1998. Who's Afraid of Human Cloning? Rowman and Littlefield, Lanham, MD.

Penslar, R., 1995. Research Ethics: Cases and Materials. Indiana University Press, Bloomington, IN.

Peters, D., and Ceci, S., 1982. Peer-Review Practices of Psychological Journals: The Fate of Published Articles, Submitted Again. Behavioral and Brain Sciences 5: 187–195.

Pharmaceutical Research and Manufacturers Association (PHRMA), 2008. About PHRMA. http://www.phrma.org/. Accessed May 1, 2008.

*Pimple, K., 1995. General Issues in Teaching Research Ethics. In Research Ethics: Case and Materials, R. Penslar (ed.). Indiana University Press, Indianapolis, pp. 3–16.

Plant Variety Protection Act, 1930. 9 U.S. Code, 2321.

Plemmons, D., Brody, S., and Kalichman, M., 2006. Student Perceptions of the Effectiveness of Education in the Responsible Conduct of Research. Science and Engineering Ethics 12: 571–582.

Plous, S., and Herzog, H., 2000. Poll Shows Researchers Favor Animal Lab Protection. Science 290: 711.

Pojman, J., 1995. Ethics. Wadsworth, Belmont, CA.

Popper, K., 1959. The Logic of Scientific Discovery. Routledge, London.

*Porter, D., 1993. Science, Scientific Motivation, and Conflict of Interest in Research. In Ethical Issues in Research, D. Cheny (ed.). University Publishing Group, Frederick, MD, pp. 114–125.

Porter, R., 1997. The Greatest Benefit to Mankind. Norton, New York.

Porter, R.J., 1992. Conflicts of Interest in Research: The Fundamentals. In Biomedical Research: Collaboration and Conflict of Interest, R.J. Porter and T.E. Malone (eds.). Johns Hopkins University Press, Baltimore, MD, pp. 122–161.

Powell, S., Allison, M., and Kalichman, M., 2007. Effectiveness of a Responsible Conduct of Research Course: A Preliminary Study. Science and Engineering Ethics 13: 249–264.

Pramik, M.J., 1990. NIH Redrafts Its Guidelines to Cover Conflicts of Interest. Genetic and Engineering News 10: 1, 24.

President's Commission for the Study of Ethical Problems in Medicine and Biomedical and Behavioral Research, 1983a. Splicing Life: A Report on the Social and Ethical

Issues of Genetic Engineering with Human Beings. U.S. Government Printing Office, Washington, DC.

President's Commission for the Study of Ethical Problems in Medicine and Biomedical and Behavioral Research, 1983b. Summing Up. U.S. Government Printing Office, Washington, DC.

President's Council on Bioethics, 2002. Human Cloning and Human Dignity. President's Council on Bioethics, Washington, DC.

President's Council on Bioethics, 2003. Beyond Therapy: Technology and the Pursuit of Happiness. President's Council on Bioethics, Washington, DC.

President's Council on Bioethics, 2004. Reproduction and Responsibility: The Regulation of New Biotechnologies. Washington, DC: President's Council on Bioethics, Washington, DC.

Press, E., and Washburn, J., 2000. The Kept University. Atlantic Monthly 285(3): 39–54.

Préziosi, M., Yam, A., Ndiaye, M., Simaga, A., Simondon, F., and Wassilak, S., 1997. Practical Experiences in Obtaining Informed Consent for a Vaccine Trial in Rural Africa. New England Journal of Medicine 336: 370–373.

Price, A.R., and Hallum, J.V., 1992. The Office of Scientific Integrity Investigations: The Importance of Data Analysis. Accountability in Research 2: 133–137.

Proctor, R., 1988. Radical Hygiene—Medicine under the Nazis. Harvard University Press, Cambridge, MA.

Proctor, R., 1999. The Nazi War on Cancer. Princeton University Press, Princeton, NJ.

Psaty, B.M., et al., 1995. The Risk of Myocardial Infarction Associated with Antihypertensive Drug Therapies. Journal of the American Medical Association 274: 620–625.

Psaty, B.M., Furberg, C.D., Ray, W.A., and Weiss, N.S., 2004. Potential for Conflict of Interest in the Evaluation of Suspected Adverse Drug Reactions—Use of Cervastatin and Risk of Rhabdomyolysis. Journal of the American Medical Association 292: 2622–2631.

Public Health Service, 2000a. Draft Interim Guidance: Financial Relationships in Clinical Research. PHS, Washington, DC.

Public Health Service, 2000b. Policy on the Humane Care and Use of Laboratory Animals. PHS, Bethesda, MD.

Pucéat, M., and Ballis, A., 2007. Embryonic Stem Cells: From Bench to Bedside. Clinical Pharmacology and Therapeutics 82: 337–339.

Putnam, H., 1981. Reason, Truth, and History. Cambridge University Press, Cambridge.

Quine, W., 1969. Ontological Relativity. Columbia University Press, New York.

Rawls, J., 1971. A Theory of Justice. Harvard University Press, Cambridge, MA.

Ready, T., 1999. Science for Sale. Boston Phoenix, April 29, 60–62.

Redman, B., and Merz, J., 2006. Research Misconduct Policies of Highest Impact Biomedical Journals. Accountability in Research 13: 247–258.

Regaldo, A., 1995. Multiauthor Papers on the Rise. Science 268: 25–27.

Regan, T., 1983. The Case for Animal Rights. University of California Press, Berkeley.

Regan, T., and Singer, P. (eds.), 1989. Animal Rights and Human Obligations, 2nd ed. Prentice-Hall, Englewood Cliffs, NJ.

Reiss, M., and Straughan, R., 1996. Improving Nature? The Science and Ethics of Genetic Engineering. Cambridge University Press, Cambridge.

Relman, A., 1999. The NIH "E-Biomed" Proposal—a Potential Threat to the Evaluation and Orderly Dissemination of New Clinical Studies. New England Journal of Medicine 340: 1828–1829.

Renegar, G., Webster, C.J., Stuerzebecher, S., Harty, L., Ide, S.E., Balkite, B., et al., 2006. Returning Genetic Research Results to Individuals: Points-to-Consider. Bioethics 20: 24–36.

Rennie, D., 1989a. Editors and Auditors [See Comments]. Journal of the American Medical Association 261: 2543–2545.

Rennie, D., 1989b. How Much Fraud? Let's Do an Experimental Audit. AAAS Observer, January, no. 3: 4.

Rennie, D., Yank, V., and Emanuel, L., 1997. When Authorship Fails: A Proposal to Make Contributors Accountable. Journal of the American Medical Association 278: 579–585.

Rescher, N., 1965. The Ethical Dimension of Scientific Research. In Beyond the Edge of Certainty: Essays on Contemporary Science and Philosophy, N. Rescher (ed.). Prentice-Hall, Englewood Cliffs, NJ, pp. 261–276.

Resnik, D., 1994. Methodological Conservatism and Social Epistemology. International Studies in the Philosophy of Science 8: 247–264.

Resnik, D., 1996a. Data Falsification in Clinical Trials. Science Communication 18(1): 49–58.

Resnik, D., 1996b. Ethical Problems and Dilemmas in the Interaction between Science and the Media. In Ethical Issues in Physics: Workshop II Proceedings, M. Thomsen and B. Wylo (eds.). Eastern Michigan University, Ypsilanti, pp. 80–112.

Resnik, D., 1996c. Social Epistemology and the Ethics of Research. Studies in the History and Philosophy of Science 27: 565–586.

Resnik, D., 1997. A Proposal for a New System of Credit Allocation in Science. Science and Engineering Ethics 3: 237–243.

*Resnik, D., 1998a. Conflicts of Interest in Science. Perspectives on Science 6(4): 381–408.

Resnik, D., 1998b. The Ethics of HIV Research in Developing Nations. Bioethics 12(4): 285–306.

Resnik, D., 1998c. The Ethics of Science. Routledge, New York.

Resnik, D., 1999a. Industry-Sponsored Research: Secrecy versus Corporate Responsibility. Business and Society Review 99: 31–34.

Resnik, D., 1999b. Privatized Biomedical Research, Public Fears, and the Hazards of Government Regulation: Lesson from Stem Cell Research. Health Care Analysis 23: 273–287.

Resnik, D., 2000. Statistics, Ethics, and Research: An Agenda for Education and Reform. Accountability in Research 8: 163–188.

*Resnik, D., 2001a. Bioethics of Gene Therapy. In Encyclopedia of Life Sciences. Nature Publishing Group, Macmillan, London.

Resnik, D., 2001b. Developing Drugs for the Developing World: An Economic, Legal, Moral, and Political Dilemma. Developing World Bioethics 1: 11–32.

*Resnik, D., 2001c. DNA Patents and Scientific Discovery and Innovation: Assessing Benefits and Risks. Science and Engineering Ethics 7(1): 29–62.

Resnik, D., 2001d. Ethical Dilemmas in Communicating Medical Information to the Public. Health Policy 55: 129–149.

*Resnik, D., 2001e. Financial Interests and Research Bias. Perspectives on Science 8(3): 255–285.

Resnik, D., 2003a. Exploitation in biomedical research. Theoretical Medicine and Bioethics 24: 233–259.

Resnik, D., 2003b. From Baltimore to Bell Labs: Reflections on Two Decades of Debate about Scientific Misconduct. Accountability in Research 10: 123–135.

Resnik, D., 2003c. A Pluralistic Account of Intellectual Property. Journal of Business Ethics 46: 319–335.

Resnik, D., 2003d. Strengthening the United States' Database Protection Laws: Balancing Public Access and Private Control. Science and Engineering Ethics 9: 301–318.

Resnik, D., 2004a. Disclosing Conflicts of Interest to Research Subjects. Accountability in Research 11: 141–159.

Resnik, D., 2004b. Disclosing Conflicts of Interest to Research Subjects: An Ethical and Legal Analysis. Accountability in Research 11: 141–159.

Resnik, D., 2004c. The Distribution of Biomedical Research Resources and International Justice. Developing World Bioethics 4: 42–57.

Resnik, D.B., 2004d. Liability and Institutional Review Boards. Journal of Legal Medicine 25: 131–184.

Resnik, D., 2004e. Owning the Genome: A Moral Analysis of DNA Patenting. SUNY Press, Albany, NY.

Resnik, D., 2005. Affirmative Action in Science and Engineering. Science and Education 14: 75–93.

Resnik, D., 2006. A Survey of Research Practices at the National Institutes of Environmental Health Sciences. Proceedings from 2006 ORI Conference on Research on Research Integrity. Office of Research Integrity, Potomac, MD.

Resnik, D., 2007a. The New EPA Regulations for Protecting Human Subjects: Haste Makes Waste. Hastings Center Report 37(1): 17–21.

Resnik, D., 2007b. The Price of Truth: How Money Affects the Norms of Science. Oxford University Press, New York.

Resnik, D., and DeVille, K., 2002. Bioterrorism and Patent Rights: "Compulsory Licensure" and the Case of Cipro. American Journal of Bioethics 2(3): 29–39.

Resnik, D., and Langer, P., 2001. Human Germline Gene Therapy Reconsidered. Human Gene Therapy 12: 1449–1458.

Resnik, D., and Portier, C., 2005. Pesticide Testing on Human Subjects: Weighing Benefits and Risks. Environmental Health Perspectives 113: 813–817.

Resnik, D., and Shamoo, A.E., 2002. Conflict of Interest and the University. Accountability in Research 9: 45–64.

Resnik, D., and Shamoo, A.E., 2005. Bioterrorism and the Responsible Conduct of Biomedical Research. Drug Development Research 63: 121–133.

Resnik, D., Shamoo, A., and Krimsky, S., 2006. Fraudulent Human Embryonic Stem Cell Research in South Korea: Lesson Learned. Accountability in Research 13: 101–109.

Resnik, D., Steinkraus, H., and Langer, P., 1999. Human Germline Gene Therapy: Scientific, Moral, and Political Issues. RG Landes, Austin, TX.

Resnik, D., and Vorhaus, D., 2006. Genetic Modification and Genetic Determinism. Philosophy, Ethics, and Humanities in Medicine 1: 9.

Rhoades, L.J., 2004. ORI Closed Investigation into Misconduct Allegations Involving Research Supported by the Public Health Service: 1994–2003. http://ori.dhhs.gov/publications/documents/Investigations1994-2003-2.pdf. Accessed October 23, 2007.

Rhodes, R., 2006. Why Test Children for Adult-Onset Genetic Diseases? Mount Sinai Journal of Medicine 73: 609–616.

Richardson, H., 2000. Specifying, Balancing, and Interpreting Bioethical Principles. Journal of Medicine and Philosophy 25: 285–307.

Ridker, P.M., and Torres, J., 2006. Reported Outcomes in Major Cardiovascular Clinical Trials Funded by For-Profit and Not-for-Profit Organizations: 2000–2005. Journal of the American Medical Association 295: 2270–2274.

Rifkin, J., 1983. Algeny. Viking Press, New York.

Rifkin, J., 1998. The Biotech Century. Tarcher/Putnam, New York.

Robert, J., 2006. The Science and Ethics of Making Part-Human Animals in Stem Cell Biology. FASEB J 20: 838–845.

Robert, J., and Baylis, F., 2003. Crossing Species Boundaries. American Journal of Bioethics 3(3): 1–13.

Robert, L., 2001. Controversial from the Start. Science 291: 1182–1188.

Robertson, H., and Gorovitz, S., 2000. Pesticide Toxicity, Human Subjects, and the Environmental Protection Agency's Dilemma. Journal of Contemporary Health Law and Policy 16: 427–458.

Robertson, J., 1994. Children of Choice. Princeton University Press, Princeton, NJ.

Robson, M., and Offit, K., 2007. Management of an Inherited Predisposition to Breast Cancer. New England Journal of Medicine 357: 154–162.

Rochon, P.A., et al., 1994. A Study of Manufacturer-Supported Trials of Nonsteroidal Anti-inflammatory Drugs in the Treatment of Arthritis. Archives of Internal Medicine 154: 157–163.

Roe v. Wade, 1973. 410 U.S. 113.

Rollin, B., 1989. The Unheeded Cry: Animal Consciousness, Animal Pain, and Science. Oxford University Press, New York.

Rollin, B., 1992. Animal Rights and Human Morality, 2nd ed. Oxford University Press, Oxford.

Rollin, B., 1995. The Frankenstein Syndrome. Cambridge University Press, Cambridge.

Ronan, C., 1982. Science: Its History and Development among the World Cultures. Facts on File, New York.

Rose, M., and Fischer, K., 1995. Policies and Perspectives on Authorship. Science and Engineering Ethics 1: 361–370.

Rosenberg, S.A., 1996. Sounding Board—Secrecy in Medical Research. New England Journal of Medicine 334: 392–394.

Ross, W., 1930. The Right and the Good. Clarendon Press, Oxford, UK.

Rothman, S., and Rothman, D., 2004. The Pursuit of Perfection. Vintage Books, New York.

Roy, R., 1993. Science Publishing Is Urgently in Need of Reform. The Scientist, September 6, 11, 22.

Rule, J.T., and Shamoo, A.E., 1997. Ethical Issues in Research Relationship between Universities, and Industry. Accountability in Research 5: 239–250.

Russell, W., and Birch, R., 1959. Principles of Humane Animal Experimentation. Charles C. Thomas, Springfield, IL.

Sandel, M., 2007. The Case against Perfection. Harvard University Press, Cambridge, MA.

Savulescu, J., and Spriggs, M., 2002. The Hexamethonium Asthma Study and the Death of a Normal Volunteer in Research. Journal of Medical Ethics 28: 3–4.

Schaffner, K., 1986. Ethical Problems in Clinical Trials. Journal of Medicine and Philosophy 11(4): 297–315.

Scheetz, M., 1999. Office of Research Integrity: A Reflection of Disputes and Misunderstandings. Croatian Medical Journal 40: 321–325.

Schreier, A., Wilson, K., and Resnik, D., 2006. Academic Research Record-Keeping: Best Practices for Individuals, Group Leaders, and Institutions. Academic Medicine 81: 42–47.

Science, 2007. Information for Authors. http://www.sciencemag.org/about/authors/ prep/gen_info.dtl#datadep. Accessed November 26, 2007.

Seife, C., 2002. Heavy-Element Fizzle Laid to Falsified Data. Science 297: 313–315.

Service, R., 2005. Pharmacogenomics: Going from Genome to Pill. Science 308: 158–160.

Service, R., 2006. The Race for the $1000 Genome. Science 311: 1544–1546.

Service, R.F., 2002. Bell Labs Fires Star Physicist Found Guilty of Forging Data. Science 298: 30–31.

Service, R.F., 2007. Sonofusion Back on the Firing Line as Misconduct Probe Reopens. Science 316: 964–964.

Service, R.F., 2008. The Bubble Bursts. ScienceNOW Daily News, July 18. http:// sciencenow.sciencemag.org/cgi/content/full/2008/718/1. Accessed August 15, 2008.

Shah, S., 2006. Body Hunters: How the Drug Industry Tests Its Products on the World's Poorest Patients. New Press, New York.

Shah, S., Whittle, A., Wilfond, B., Gensler, G., and Wendler, D., 2004. How Do Institutional Review Boards Apply the Federal Risk and Benefit Standards for Pediatric Research? Journal of the American Medical Association 291: 476–482.

Shamoo, A.E., 1988. We Need Data Audit. AAAS Observer, November 4, 4.

Shamoo, A.E., 1989. Principles of Research Data Audit. Gordon and Breach, New York.

Shamoo, A.E., 1991a. Policies and Quality Assurance in the Pharmaceutical Industry. Accountability in Research 1: 273–284.

Shamoo, A.E., 1991b. Quality Assurance. Quality Assurance: Good Practice, Regulation, and Law 1: 4–9.

*Shamoo, A.E., 1992. Role of Conflict of Interest in Scientific Objectivity—a Case of a Nobel Prize. Accountability in Research 2: 55–75.

Shamoo, A.E., 1993. Role of Conflict of Interest in Public Advisory Councils. In Ethical Issues in Research, D. Cheney (ed.). University Publishing Group, Frederick, MD, pp. 159–174.

Shamoo, A.E., 1994a. Editors, Peer Reviews, and Ethics. AAAS Perspectives 14: 4–5.

Shamoo, A.E., 1994b. Our Responsibilities toward Persons with Mental Illness as Human Subjects in Research. Journal of the California Alliance for the Mentally Ill 5: 14–16.

Shamoo, A.E., 1995. Scientific Evidence and the Judicial System. Accountability in Research 4: 21–30.

Shamoo, A.E., 1997a. Brain Disorders—Scientific Facts, Media, and Public Perception. Accountability in Research 5: 161–174.

Shamoo, A.E., 1997b. The Ethical Import of Creation. Baltimore Sun, March 2, F1, F7.

Shamoo, A.E. (ed.), 1997c. Ethics in Neurobiological Research with Human Subjects. Gordon and Breach, Amsterdam.

Shamoo, A.E., 1997d. The Unethical Use of Persons with Mental Illness in High Risk Research Experiments. Testimony to National Bioethics Advisory Commission, January 9, 1997. National Bioethics Advisory Commission, Washington, DC.

Shamoo, A.E., 1999. Institutional Review Boards (IRBs) and Conflict of Interest. Accountability in Research 7: 201–212.

Shamoo, A.E., 2000. Future Challenges to Human Subject Protection. The Scientist, June 26, 35.

Shamoo, A.E., 2001. Adverse Events Reporting—the Tip of an Iceberg. Accountability in Research 8: 197–218.

Shamoo, A.E., 2002. Ethically Questionable Research with Children: The Kennedy Krieger Lead Abatement Study. Accountability in Research 9: 167–175.

Shamoo, A.E., 2005. Debating Moral Issues in Developing Countries. Applied Clinical Trials (actmagazine.com) 14(6): 86–96.

Shamoo, A.E., 2006. Ethical Issues in Laboratory Management. In Laboratory Management—Prinicples and Processes, 2nd ed., D. Harmening (ed.). D.H. Publishing, St. Petersburg, FL, pp. 325–345.

Shamoo, A.E., 2007. Deregulating Low-Risk Research. Chronicle of Higher Education, August 3, 53: B16.

Shamoo, A.E., and Annau, Z., 1987. Ensuring Scientific Integrity [Letter]. Nature 327: 550.

Shamoo, A.E., and Annau, Z., 1989. Data Audit—Historical Perspective. In Principles of Research Data Audit, A.E. Shamoo (ed.). Gordon and Breach, New York, pp. 1–12.

Shamoo, A.E., and Cole, J., 2004. Ethics of Genetic Modification of Behavior. Accountability in Research 11: 201–214.

Shamoo, A.E., and Davis, S., 1989. The Need for Integration of Data Audit Research and Development Operations. In Principles of Research Data Audit, A. Shamoo (ed.). Gordon and Breach, Amsterdam, pp. 119–128.

Shamoo, A.E., and Davis, S., 1990. The Need for Integration of Data Audit into Research and Development Operation. Accountability in Research 1: 119–128.

*Shamoo, A.E., and Dunigan, C., 2000. Ethics in Research. Proceedings of the Society for Experimental Biology and Medicine 224: 205–210.

Shamoo, A.E., and Irving, D.N., 1993. Accountability in Research Using Persons with Mental Illness. Accountability in Research 3: 1–17.

Shamoo, A.E., Irving, D.N., and Langenberg, P., 1997. A Review of Patient Outcomes in Pharmacological Studies from the Psychiatric Literature, 1996–1993. Science and Engineering Ethics 3: 395–406.

Shamoo, A.E., and Katzel, L., 2007. Urgent Ethical Challenges in Human Subjects Protections. Journal of Clinical Research Best Practices 3(3). http://www.first-clinical.com/journal/2007/0703_Urgent.pdf. Accessed February 11, 2008.

Shamoo, A.E., and Katzel, L., 2008. How Should Adverse Events Be Reported in US Clinical Trials?: Ethical Consideration. Clinical Pharmacology and Therapeutics, February 27. DOI: 10.1038/clpt.2008.14, Ethics.

*Shamoo, A.E., and Keay, T., 1996. Ethical Concerns about Relapse Studies. Cambridge Quarterly of Health Care Ethics 5: 373–386.

Shamoo, A.E., and Moreno, J., 2004. Ethics of Research Involving Mandatory Drug Testing of High School Athletes in Oregon. American Journal of Bioethics 4: 25–31.

Shamoo, A.E., and O'Sullivan, J.L., 1998. The Ethics of Research on the Mentally Disabled. In Health Care Ethics—Critical Issues for the 21st Century,

J.F. Monagle and D.C. Thomasma (eds.). Aspen Publishers, Gaithersburg, MD, pp. 239–250.

Shamoo, A.E., and Resnik, D.B., 2006a. Ethical Issues for Clinical Research Managers. Drug Information Journal 40: 371–383.

Shamoo, A.E., and Resnik, D.B., 2006b. Strategies to Minimize Risks and Exploitation in Phase One Trials on Healthy Subjects. American Journal of Bioethics 6(3): W1–W13.

Shamoo, A.E., and Schwartz, J., 2007. Universal and Uniform Protections of Human Subjects in Research. American Journal of Bioethics 7(12): 7–9.

Shamoo, A.E., and Strause, L., 2004. Closing the Gap in Regulations for the Protection of Human Subjects in Research. Monitor, Spring, 15–18.

Shamoo, A.E., and Tauer, C.A., 2002. Ethically Questionable Research with Children: The Fenfluramine Study. Accountability in Research 9: 143–166.

Shamoo, A.E., and Teaf, R., 1990. Data Ownership. CBE Views 13: 112–114.

Shamoo, A.E., and Woeckner, E., 2007. Ethical Flaws in the TeGenero Trial. American Journal of Bioethics 7: 2, 90–92.

Sharav, V.H., and Shamoo, A.E., 2000. Are Experiments That Chemically Induce Psychosis in Patients Ethical? BioLaw 2(1): S1–S36.

Shenk, D., 1999. Money and Science = Ethics Problems on Campus. The Nation, May 22, 11–18.

Sheremeta, L., and Knoppers, B., 2003. Beyond the Rhetoric: Population Genetics and Benefit-Sharing. Health Law Journal 11: 89–117.

Shrader-Frechette, K., 1994. Ethics of Scientific Research. Rowman and Littlefield, Boston.

Shulman, S., 2007. Undermining Science: Suppression and Distortion in the Bush Administration. University of California Press, Berkeley.

Sigma Xi, 1986. Honor in Science. Sigma Xi, The Scientific Research Society, Research Triangle Park, NC.

Sigma Xi, 1999. The Responsible Research: Paths and Pitfalls. Sigma Xi, The Scientific Research Society, Research Triangle Park, NC.

Silver, L., 1998. Remaking Eden. Harper Perennial, New York.

*Singer, P., 1975 [1990]. Animal Liberation, 2nd ed. Random House, New York.

Singer, P., 1985. In Defense of Animals. Harper and Row, New York.

Smith, R., 2004. Open Access Publishing Takes Off. British Medical Journal 328: 1–3.

Smith, R., 2006. Research Misconduct: The Poisoning of the Well. Journal of the Royal Society of Medicine 99: 232–237.

Snow, C., 1964. The Two Cultures and the Scientific Revolution. Cambridge University Press, Cambridge.

Sobel, A., 1978. Deception in Social Science Research: Is Informed Consent Possible? Hastings Center Report 8(5): 40–45.

Solter, D., and Gearhart, J., 1999. Putting Stem Cells to Work. Science 283: 1468–1470.

Spece, R., Shimm, D.S., and Buchanan, A.E. (eds.), 1996. Conflicts of Interest in Clinical Practice and Research. Oxford University Press, New York.

Spier, R., 1995. Ethical Aspects of the University-Industry Interface. Science and Engineering Ethics 1: 151–162.

Sprague, R., 1991. One Man's Saga with the University Federal System. Paper presented at the Second Conference on Research Policies and Quality Assurance, Rome, May 6–7.

Sprague, R.L., 1993. Whistleblowing: A Very Unpleasant Avocation. Ethics and Behavior 3: 103–133.

Steinbrook, R., 2002. Improving Protection for Human Research Subjects. New England Journal of Medicine 346: 1425–1430.

Steiner, J.F., Lanphear, B.P., Vu, K.O., and Main, D.S., 2004. Assessing the Role of Influential Mentors in the Research Development of Primary Care Fellows. Academic Medicine 79: 865–872.

Stelfox, H.T., Chua, G., O'Rourke, K., and Detsky, A.S., 1998. Conflict of Interest in the Debate over Calcium-Channel Antagonists. New England Journal of Medicine 338: 101–106.

Steneck, N., 1999. Confronting Misconduct in Science in the 1980s and 1990s: What Has and Has Not Been Accomplished. Science and Engineering Ethics 5: 161–176.

*Steneck, N., 2000. Assessing the Integrity of Publicly Funded Research. In Proceedings from the ORI Conference on Research Integrity. Office of Research Integrity, Washington, DC, pp. 1–16.

Steneck, N., 2006. ORI Introduction to Responsible Conduct of Research. Office of Research Integrity, Washington, DC.

Steneck, N., and Bulger, R., 2007. The History, Purpose, and Future of Instruction in the Responsible Conduct of Research. Academic Medicine 82: 829–834.

Stephens, T., and Brynner, R., 2001. Dark Remedy: The Impact of Thalidomide and Its Revival as a Vital Medicine. Perseus, New York.

Sterckx, S., 2004. Patents and Access to Drugs in Developing Countries: An Ethical Analysis. Developing World Bioethics 4: 58–75.

Stewart, W.W., and Feder, N., 1987. The Integrity of the Scientific Literature. Nature 325: 207–214.

Stewart, W.W., and Feder, N., 1991. Analysis of a Whistle-Blowing. Nature 351: 687–691.

Stock, G., 2003. Redesigning Humans. Mariner Books, New York.

Stokstad, E., 1999. Humane Science Finds Sharper and Kinder Tools. Science 286: 1068–1071.

Stout, D., 2006. In First Veto, Bush Blocks Stem Cell Bill. New York Times, July 19, A1.

Strong, C., 2000. Specified Principlism: What Is It, and Does It Really Resolve Cases Better Than Casuistry? Journal of Medicine and Philosophy 25: 323–341.

Swazey, J., 1993. Teaching Ethics: Needs, Opportunities, and Obstacles. In Ethics, Values, and the Promise of Science. Sigma Xi, The Scientific Research Society, Research Triangle Park, NC, pp. 233–242.

Swazey, J., and Bird, S., 1997. Teaching and Learning Research Ethics. In Research Ethics: A Reader, D. Elliot and J. Stern (eds.). University Press of New England, Hanover, NH, pp. 1–19.

Swazey, J.P., Anderson, M.S., and Louis, K.S., 1993. Ethical Problems in Academic Research. American Scientist 81: 542–553.

Swisher, K., 1995. What Is Sexual Harassment? Greenhaven Press, San Diego.

Tagney, J.P., 1987. Fraud Will Out—or Will It? New Scientist, August 6, 62–63.

Taleyarkhan, R.P., West, C.D., Cho, J.S., Lahey, R.T., Jr., Nigmatulin, R.I., and Block, R.C., 2002. Evidence for Nuclear Emissions during Acoustic Cavitation. Science 295: 1868–1873.

Tanne, J., 2005. US National Institutes of Health Issue New Ethics Guidelines. BMJ 331: 472.

*Tauer, C.A., 1999. Testing Drugs in Pediatric Populations: The FDA Mandate. Accountability in Research 7: 37–58.

Taylor, P., 1986. Respect for Nature. Princeton University Press, Princeton, NJ.

Technology Transfer Act, 1986. Public Law 99–502.

Teitelman, R., 1994. The Profits of Science. Basic Books, New York.

*Thompson, D., 1993. Understanding Financial Conflict of Interest. New England Journal of Medicine 329: 573–576.

Thomson, J., Itskovitz-Eldor, J., Shapiro, S., Waknitz, M., Swiergiel, J., Marshall, V., et al., 1998. Embryonic Stem Cell Lines Derived from Human Blastocysts. Science 282: 1145–1147.

Tsuchiya, T., 2008. Imperial Japanese Medical Atrocities and Their Enduring Legacy in Japanese Research Ethics. In Contemporary Issues in Bioethics, T. Beauchamp, L. Walters, J. Kahn, and A. Mastroianni (eds.). Wadsworth, Belmont, CA, pp. 56–64.

U.K. Patent Office, 2001. http://www.ipo.gov.uk/about-history-patent.htm. Accessed August 13, 2008.

University of Maryland, Baltimore, 1998. Policies and Procedures concerning Misconduct in Scholarly Work. UMB, Baltimore. http://cits-cf.umaryland.edu/hrpolicies/section3/misconduct.htm. Accessed October 22, 2007.

U.S. Congress, Committee on Government Operations, 1990. Are Scientific Misconduct and Conflict of Interests Hazardous to Our Health? Report 101–688. U.S. Government Printing Office, Washington, DC.

U.S. Copyright Office, 2001. http://lcweb.loc.gov/copyright/. Accessed March 16, 2001.

U.S. Department of Health and Human Services, 1989a. 42 CFR 50, Policies of General Applicability. Subpart A, Responsibility of PHS Awardee and Applicant Institutions for Dealing with and Reporting Possible Misconduct in Science. Sec. 493, Public Health Service Act, as amended, 99 Stat. 874–875 (42 U.S.C. 289b); Sec. 501(f), Public Health Service Act, as amended, 102 Stat., 4213 [42 U.S.C. 290aa(f)]. Federal Register 54: 32449–32451. http://ori.dhhs.gov/misconduct/reg_subpart_a.shtml. Accessed August 13, 2008.

U.S. Department of Health and Human Services, 1989b. PHS Misconduct Regulation. 42 CFR 50.101. Federal Register 54: 32446–32451.

U.S. Department of Health and Human Services, 1995. Objectivity in Research; Investigatory Financial Disclosure Policy; Final Rule and Notice. 45 CFR parts 50 and 64. Federal Register 60: 35809–35819.

U.S. Department of Health and Human Services, 2001. Report on Expert Panel Review under Subpart D of 45 CFR 46. Precursors to Diabetes in Japanese American Youth. http://www.hhs.gov/ohrp/pdjay/expert.htm. Accessed December 18, 2007.

U.S. Department of Health and Human Services, 2002. Meeting of the National Human Research Protections Committee (NHRPAC), January 29 and July 31 transcripts. http://www.hhs.gov/ohrp/nhrpac/mtg01–02/0129NHR.txt and http://www.hhs.gov/ohrp/nhrpac/mtg07–02/0731NHR2.txt. Accessed June 3, 2007.

U.S. Department of Health and Human Services, 2005a. Public Health Service Policies on Research Misconduct; Final Rule. 42 CFR parts 50 and 93. Federal Register 70: 28369–28400. http://ori.dhhs.gov/documents/42_cfr_parts_50_and_93_2005.pdf. Accessed October 22, 2007.

U.S. Department of Health and Human Services, 2005b. Supplemental Standards of Ethical Conduct and Financial Disclosure Requirements for Employees of the Department of Health and Human Services 5 CFR Parts 5501 and 5502. Federal

Register 70: 51559–51574. http://www.nih.gov/about/ethics/08252005supplement alfinancialdisclosure.pdf. Accessed October 29, 2007.

U.S. Department of Health and Human Services, Inspector General, 1998. Institutional Review Boards: A System in Jeopardy? DHHS, Washington, DC.

U.S. Department of Health, Education, and Welfare, 1973. Final Report of the Tuskegee Syphilis Study Ad Hoc Advisory Panel. DHEW, Washington, DC.

U.S. Department of Justice, 2007. Freedom of Information Act. http://www.usdoj. gov/oip/index.html. November 27, 2007.

U.S. General Accounting Office, 1994. Peer Review—Reforms Needed to Ensure Fairness in Federal Agency Grant Selection. GAO/PEMD-94-1. GAO, Washington, DC.

U.S. Patent and Trademark Office, 2001. http://www.uspto.gov/. Accessed March 16, 2001.

van Rooyen, S., Godlee, F., Evans, S., Smith, R., and Black, N., 1998. Effect of Blinding and Unmasking on the Quality of Peer Review: A Randomized Trial. Journal of the American Medical Association 280: 234–237.

Varmus, H., and Satcher, D., 1997. Ethical Complexities of Conducting Research in Developing Countries. New England Journal of Medicine 337: 1000–1005.

Vasgird, D., 2007. Prevention over Cure: The Administrative Rationale for Education in the Responsible Conduct of Research. Academic Medicine 82: 835–837.

Veatch, R., 1987. The Patient as Partner: A Theory of Human Experimentation Ethics. Indiana University Press, Bloomington.

Veatch, R., 2007. The Irrelevance of Equipoise. Journal of Medicine and Philosophy 32: 167–183.

Venter, C. et al., 2001. The Sequence of the Human Genome. Science 291: 1304–1351.

Vroom, V.H., and Jago, A.G., 2007. The Role of the Situation in Leadership. American Psychologist 62: 17–24.

Wade, N., 1981. The Rise and Fall of a Scientific Superstar. New Scientist 91: 781–782.

Wade, N., 2001. Genome Feud Heats Up as Academic Team Accuses Commercial Rival of Faulty Work. New York Times, May 2, A15.

Wadman, M., 1996. Drug Company Suppressed Publication of Research. Nature 381: 4.

Wadman, M., 1999. NIH Strives to Keep Resources Sharing Alive. Nature 399: 291.

Wager, E., Parkin, E.C., and Tamber, P.S., 2006. Are Reviewers Suggested by Authors as Good as Those Suggested by Editors? Results of a Rater-Blinded, Retrospective Study. BMC Medicine 4: 13. http://www.biomedcentral.com/1741-7015/4/13. Accessed August 13, 2008.

Wake Forest University School of Medicine, 2007. Data Ownership Guidelines. http://www1.wfubmc.edu/OR/Policies+and+Procedures/Manual/Section+4+-+Data+Ownership.htm. Accessed July 18, 2008.

Walsh, M.E., Graber, G.C., and Wolfe, A.K., 1997. University-Industry Relationships in Genetic Research: Potential Opportunities and Pitfalls. Accountability in Research 5: 265–282.

*Walters, L., 1986. The Ethics of Human Gene Therapy. Nature 320: 225–227.

Walters, L., 1989. Genetics and Reproductive Technologies. In Medical Ethics, Robert M. Veatch (ed.). Jones and Bartlett, Boston, pp. 201–231.

Walters, L., and Palmer, J., 1997. The Ethics of Human Gene Therapy. Oxford University Press, New York.

Wampold, B., Minami, T., Tierney, S., Baskin, T., and Bhati, K., 2005. The Placebo Is Powerful: Estimating Placebo Effects in Medicine and Psychotherapy from Randomized Clinical Trials. Journal of Clinical Psychology 61: 835–854.

Warsh, D., 1989. Conflict-of-Interest Guidelines on Research Projects Stir Up a Hornets Nest at NIH. Washington Post, December 27, F3.

Washburn, J., 2006. University, Inc.: The Corporate Corruption of American Higher Education. Basic Books, New York.

Washington, H.A., 2006. Medical Apartheid—the Dark History of Medical Experimentation on Black Americans from Colonial Times to the Present. Doubleday, New York.

Waugaman, P.C., and Porter, R.J., 1992. Mechanism of Interactions between Industry and the Academic Medical Center. In Biomedical Research: Collaboration and Conflict of Interest, R.J. Porter and T.E. Malone (eds.). Johns Hopkins University Press, Baltimore, MD, pp. 93–118.

Weaver, D., Reis, M.H., Albanese, C., Costantini, F., Baltimore, D., and Imanishi-Kari, T., 1986. Altered Repertoire of Endogenous Immunoglobulin Gene Expression in Transgenic Mice Containing a Rearranged Mu Heavy Chain Gene. Cell 45: 247–259.

Webster v. Reproductive Health Services, 1989. Supreme Court Cases and Opinions 492: 490.

Weijer, C., and Emanuel, E., 2000. Protecting Communities in Biomedical Research. Science 289: 1142–1144.

Weil, V., and Arzbaecher, R., 1997. Relationships in Laboratories and Research Communities. In Research Ethics: A Reader, D. Elliott and J. Stern (eds.). University of New England Press, Hanover, NH, pp. 69–90.

Wein, L., and Liu, Y., 2005. Analyzing a Bioterror Attack on the Food Supply: The Case of Botulinum Toxin in Milk. Proceedings of the National Academy of Sciences of the USA 102: 9984–9989.

Weir, R., Olick, R., and Murray, J., 2004. The Stored Tissue Issue. Oxford University Press, New York.

Weiss, R., 1997. Thyroid Drug Study Reveals Tug of War over Privately Financed Research. Washington Post, April 16, A3.

Weiss, R., 2007. Role of Gene Therapy in Death Called Unclear. Washington Post, September 18, A1.

Weissman, P., 2006. Politic Stem Cells. Nature 439: 145–147.

Welsome, E., 1999. The Plutonium File. Dial Press, New York.

Wendler, D., 1996. Deception in Medical and Behavioral Research: Is It Ever Acceptable? Milbank Quarterly 74: 87–114.

Wendler, D., 2006. Three Steps to Protecting Pediatric Research Participants from Excessive Risks. PLoS Clinical Trials 1(5): e25. DOI: 10.1371/journal.pctr.

Wendler, D., Belsky, L., Thompson, K.M., and Emanuel, E.J., 2005. Quantifying the Federal Minimal Risk Standard—Implications for Pediatric Research without a Prospect of Direct Benefit. Journal of the American Medical Association 29: 826–832.

Wenger, N., Korenman, S.G., Berk, R., and Liu, H., 1999. Reporting Unethical Research Behavior. Evaluation Reviews 23: 553–570.

Wertheimer, A., 1999. Exploitation. Princeton University Press, Princeton, NJ.

Whitbeck, C., 1996. Ethics as Design. Doing Justice to Moral Problems. Hastings Center Reports 26(3): 9–16.

Whitbeck, C., 1998. Ethics in Engineering Practice and Research. Cambridge University Press, Cambridge.

White, M., 2007. A Right to Benefit from International Research: A New Approach to Capacity Building in Less-Developed Countries. Accountability in Research 14: 73–92.

Wilcox, B.L., 1992. Fraud in Scientific Research: The Prosecutors Approach. Accountability in Research 2: 139–151.

Wilcox, L., 1998. Authorship: The Coin of the Realm, the Source of Complaints. Journal of the American Medical Association 280: 216–217.

Williams, R., Flaherty, L., and Threadgill, D., 2003. The Math of Making Mutant Mice. Genes, Brain, and Behavior 2: 191–200.

Williams, T., 1987. The History of Invention. Facts on File, New York.

Willman, D., 2005. NIH Inquiry Shows Widespread Ethical Lapses, Lawmaker Says. Los Angeles Times, July 14, A-1.

Wilmut, I., 1997. Cloning for Medicine. Scientific American 279(6): 58–63.

Wilson, K., Schreier, A., Griffin, A., and Resnik, D., 2007. Research Records and the Resolution of Misconduct Allegations at Research Universities. Accountability in Research 14: 57–71.

Whon, D., and Normile, D., 2006. Hwang Indicted for Fraud, Embezzlement. ScienceNOW, May 12. http://sciencenow.sciencemag.org/cgi/content/full/2006/512/1. Accessed August 13, 2008.

Wong, G.H.W., 2006. Five Attributes of a Successful Manager in a Research Organization. Nature Biotechnology 24: 1171–1174.

Wood, A., and Darbyshire, J., 2006. Injury to Research Volunteers—the Clinical-Research Nightmare. New England Journal of Medicine 354: 1869–1871.

Woolf, P., 1986. Pressure to Publish and Fraud in Science. Annals of Internal Medicine 104: 254–263.

World Medical Association, 2004. Declaration of Helsinki: Ethical Principles for Medical Research Involving Human Subjects. http://www.wma.net/e/policy/b3.htm. Accessed November 12, 2007.

World Trade Organization, 2007. TRIPS: Agreement on Trade-Related Aspects of Intellectual Property Rights. Part I—General Provisions and Basic Principles. http://www.wto.org/english/tratop_e/trips_e/t_agm2_e.htm. Accessed November 17, 2007.

Wyatt, R.J., 1986. Risks of Withdrawing Antipsychotic Medication. Archives of General Psychiatry 52: 205–208.

Wyatt, R.J., Henter, I.D., and Bartko, J.J., 1999. The Long-Term Effects of Placebo in Patients with Chronic Schizophrenia. Biological Psychiatry 46: 1092–1105.

Ziman, J., 1984. Introduction to Science Studies. Cambridge University Press, Cambridge.

Zolla-Parker, S., 1994. The Professor, the University, and Industry. Scientific American 270(3): 120.

Zuckerman, H., 1977a. Deviance Behavior and Social Control in Science. In Deviance and Social Change, E. Sagrin (ed.). Sage, Beverly Hills, CA, pp. 87–138.

Zuckerman, H., 1977b. Scientific Elite: Nobel Laureates in the United States. Free Press, New York.

Zuckerman, S., 1966. Scientists and War. Harper and Row, New York.

Zurer, P., 1990. Conflict of Interest: NIH Rules Go Back to Drawing Board. Chemical and Engineering News, January 8, 4–5.

Zurer, P., 1993. NIH Peer Reviewers to Watch for High-Risk, High-Payoff Proposals. Chemical and Engineering News, June 7, 25–26.

Index